Worcester Exc… Fargo D.T. (Da… …ed gentleman clos… repeater. *Cour… …on, Montana Historical Society, Montana.*

BULLARD ARMS

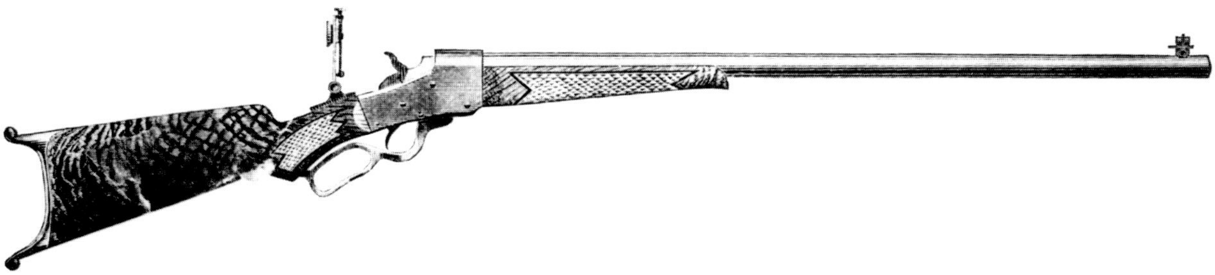

BULLARD

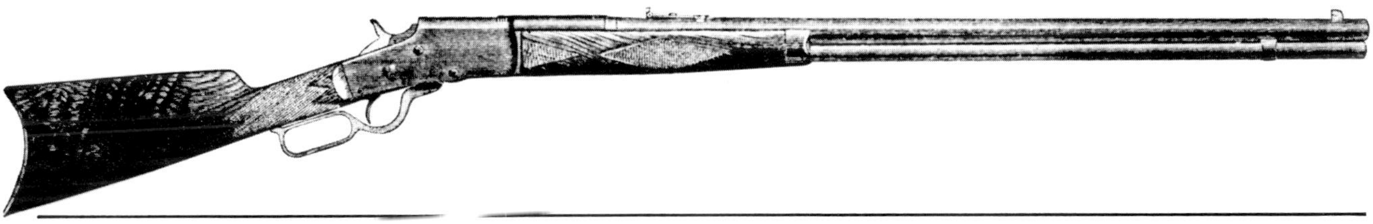

ARMS

G. Scott Jamieson

THE BOSTON MILLS PRESS

To my wife,
Frances Evelyn,
who rendered valuable advice, assistance,
and who was frequently a writer's widow
without complaint

CANADIAN CATALOGUING IN PUBLICATION DATA

Jamieson, G. Scott (Garth Scott), 1947-
Bullard Arms

Bibliography: p.
ISBN 0-919783-65-1

1. Bullard Arms Company - History. 2. Bullard, James Herbert, 1842-1914. 3. Firearms - Massachusetts - Springfield - History. I. Title.

HD9743.U64B85 1988 338.7'6834'00974426
C88-093953-2

Published by:
THE BOSTON MILLS PRESS
132 Main Street
Erin, Ontario N0B 1T0
(519) 833-2407
FAX (519) 833-2195

American Association for State and Local History Award of Merit

Winners of the Heritage Canada Communications Award

Design by John Denison
Cover design by Gill Stead
Typeset by Speed River Graphics
Printed by Ampersand, Guelph

The publisher wishes to acknowledge the financial assistance and encouragement of The Canada Council, the Ontario Arts Council and the Office of the Secretary of State.

CONTENTS

FOREWORD

The book you now hold in your hands is the culmination of 14 years of work. I do not pretend that the research actually consumed the full 14 years past, but I can say truthfully that Bullards and everything about them were never far from my thoughts. Writing this book was truly a rewarding experience.

Most rewarding were the many friendships I developed in countries around the world based on a mutual thirst for knowledge about some of the finest rifles ever made. The knowledge I slowly gained and amassed is something I am very proud to pass on to all, in the hope that the Bullard Repeating Arms Company will now take its rightful place in the annals of firearm history.

Without the support and contributions of hundreds of people, some now departed, this book would have been impossible to write. I mean that sincerely. Many people contributed photographs and artwork. The results of this combined effort will, I hope, please both my contributors and readers alike. I have endeavored to be as accurate as possible with all my information and to that end have included many footnotes in the text.

Of course this book is not the final word on Bullards and some will say that it leaves too many questions unanswered. My reply to that is: "If we know the questions, can the answers be far behind?"

This edition will, I am sure, bring forth much more information on Bullard rifles which will find its way into a later edition. I will always be glad to hear from anyone who can contribute to our knowledge of Bullard rifles. Thank you all.

April 1988
G. Scott Jamieson

Plate 1-1. Bullard family Coat of Arms.

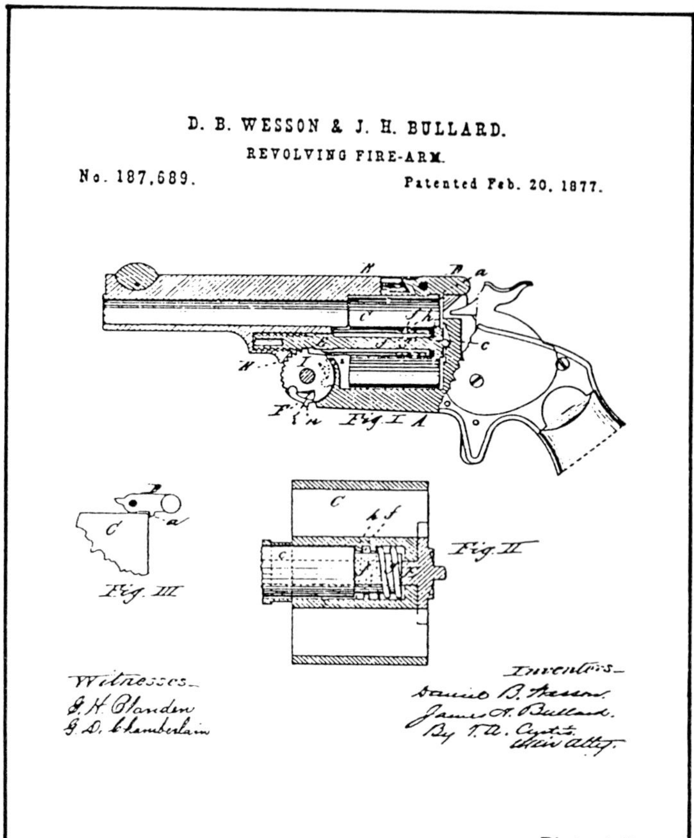
D. B. WESSON & J. H. BULLARD.
REVOLVING FIRE-ARM.
No. 187,689. Patented Feb. 20, 1877.

Fig. I

Fig. II

Fig. III

Witnesses_
G. H. Blanden
G. D. Chamberlain

Inventors_
Daniel B. Wesson
James H. Bullard.
By T. A. Curtis
their Atty.

Plate 1-2.

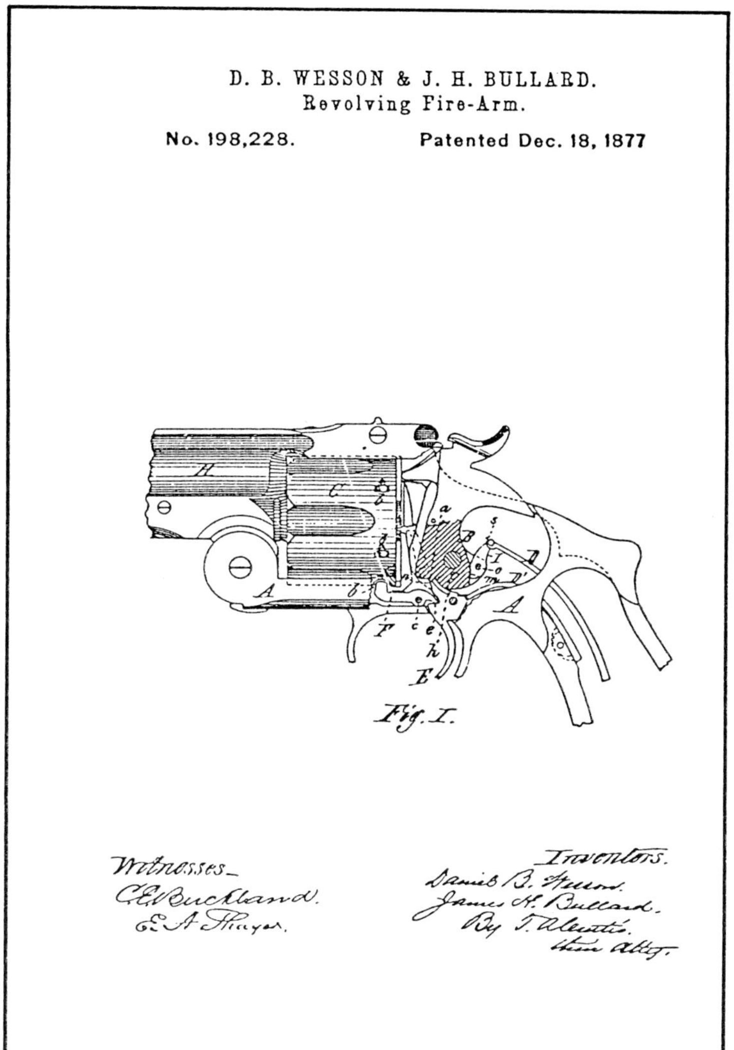
D. B. WESSON & J. H. BULLARD.
Revolving Fire-Arm.
No. 198,228. Patented Dec. 18, 1877

Fig. I.

Witnesses_
C. E. Buckland.
E. A. Thayer.

Inventors.
Daniel B. Wesson.
James H. Bullard.
By T. A. Curtis.
their Atty.

Plate 1-3.

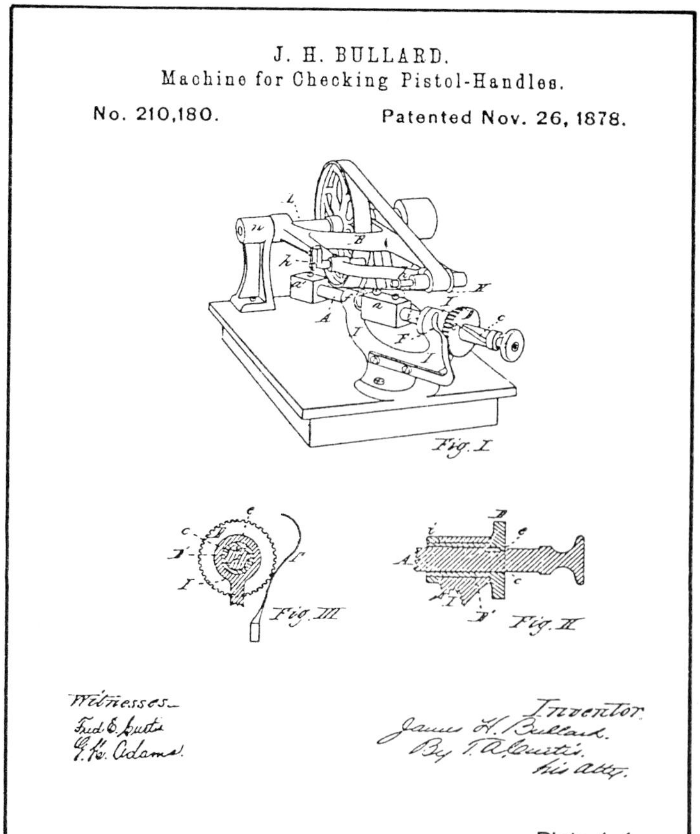
J. H. BULLARD.
Machine for Checking Pistol-Handles.
No. 210,180. Patented Nov. 26, 1878.

Fig. I

Fig. II

Fig. III

Witnesses_
Fred. C. Curtis
G. K. Adams.

Inventor.
James H. Bullard.
By T. A. Curtis.
his Atty.

Plate 1-4.

1

JAMES HERBERT BULLARD

May 14, 1842 - March 26, 1914

James Bullard was a man of prolific genius; his inventions ranged from automobiles to sewing needles. Had he been endowed with a better managerial sense and the ability to cling tenaciously to a sound concept and see it through to its successful conclusion, the companies he founded would still be flourishing. If the firms of Knox, Duryea Bros., and Bullard had survived, Springfield, Massachusetts, would not only be remembered as the historic birthplace of the American automobile, but also would have become as important to the industrial might of America as Detroit is today.

Bullard's firearms inventions, though, are the prime concern of this book. His repeating rifle patent claim was filed on May 5, 1879, and granted on August 16, 1881. This repeater was to gain a reputation for the smoothest lever action ever made. His rifle was also one of the first to positively extract the government .45-70 copper cartridges then in use.

James Bullard's inventive life began in Poultney, Vermont, on May 14, 1842. He was the son of Ezekiel W. and Miriam (Foster) Bullard. His father was also an inventor, mainly of agricultural implements, such as the Bullard Hay Tedder,[1] horse rake and hay cutter.

Until he was 15, James attended school in Barre, Massachusetts. Upon leaving school, he went to Smithville to work for the firm of C.W. and J.E. Smith, a mercantile house. He remained in their employ approximately three years. Just before the outbreak of the Civil War, he was engaged in making Shaker Hoods[2] for southern use, but hostilities ended that venture and he joined the ranks of the unemployed.

In 1861 James tried to enlist in the 15th Massachusetts Regiment but was refused enlistment. This was to happen to him three more times during the course of the war. The cause was apparently a birth defect or crippled foot.[3]

Frustrated in his attempts to join the Union Army, he went to Chicopee Falls (opposite Springfield, Massachusetts) as the master mechanic of the Belcher and Taylor Manufacturing Company, manufacturer of the Bullard Hay Tedder. He worked there for a number of years before moving to the Lamb Knitting Machine Company of Chicopee Falls, contracting to manufacture that company's knitting machine needles. Later he accepted a similar position with the Wheeler and Wilson Service Machine Company, for whom he made a number of machine models. He then returned to the Lamb Company, where he made drawings for an automated machine to make metallic seals for Canadian Postal Department mail pouches.

After leaving the Lamb Company for the second time, James Bullard went to work for Smith & Wesson as master mechanic of the works.

Mr. Bullard's employment record contained both the best and worst traits of his inventive character. The best traits embodied his consummate skill as a master mechanic and draftsman who was always in demand in an industrial area filled with

J. H. BULLARD.
Revolving Fire-Arm.

No. 227,481. Patented May 11, 1880.

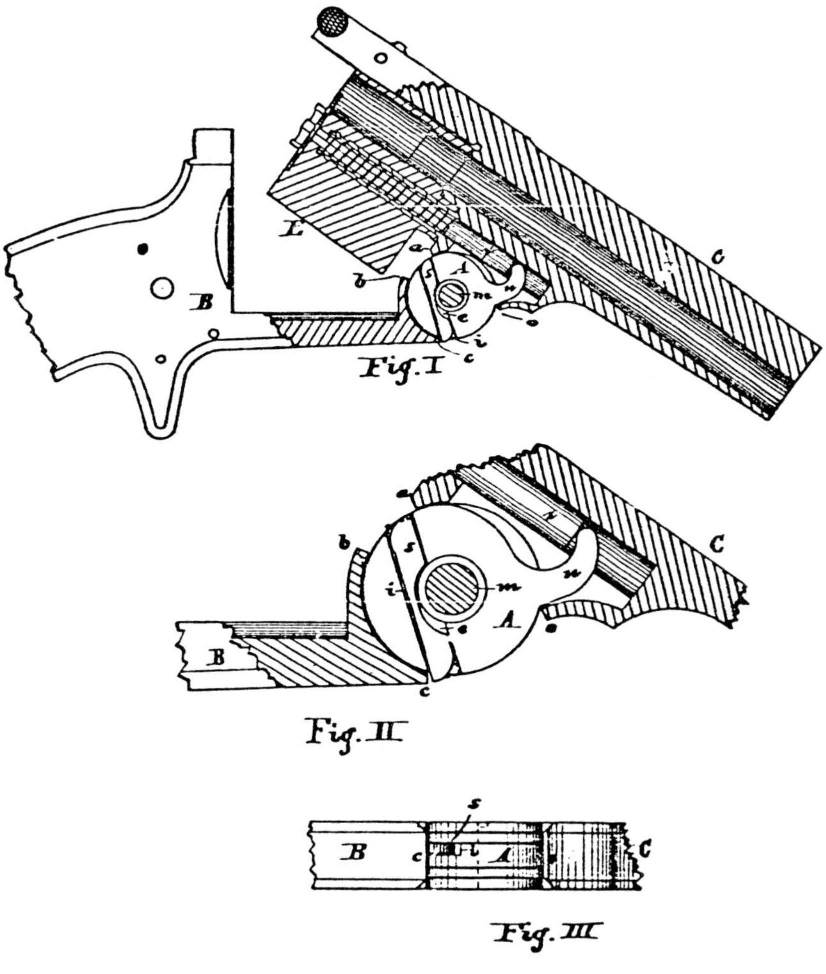

Witnesses.
F. E. Hall
E. E. Curtis

Inventor.
James H. Bullard
By T. A. Curtis.
his atty.

Plate 1-5.

UNITED STATES PATENT OFFICE.

JAMES H. BULLARD, OF SPRINGFIELD, MASSACHUSETTS, ASSIGNOR TO DANIEL B. WESSON.

REVOLVING FIRE-ARM.

SPECIFICATION forming part of Letters Patent No. 227,481, dated May 11, 1880.

Application filed December 29, 1879.

To all whom it may concern:

Be it known that I, JAMES H. BULLARD, of Springfield, in the county of Hampden and State of Massachusetts, have invented a new and useful Improvement in Revolving Fire-Arms; and I do hereby declare that the following is a full, clear, and exact description thereof, reference being had to the accompanying drawings, and to the letters of reference marked thereon.

The object of my invention is to provide an economical and effective means of operating the extractor-lifter which operates the extractor of a revolving fire-arm; and to this end my invention consists of a slide arranged to move to and fro in the lifter and operated by shoulders made upon the arm in the ordinary movements of tilting the barrel upon the frame in extracting the shells from the cylinder, as will be more fully hereinafter set forth.

Figure I is a vertical longitudinal section of a revolving fire-arm through the barrel and that portion of the frame in which the lifter is secured, showing my invention applied thereto. Fig. II is an enlarged longitudinal sectional view of the arm at the joint, showing the lifter, its slide, and that portion of the arm immediately adjacent thereto. Fig. III is a bottom view of the joint, showing the lifter, and the end view of the slide operating therein.

In the drawings, C represents the barrel of a revolving fire-arm; E, the cylinder secured thereto, and B the frame, in the forward part of which, in the barrel-hinge, is placed the lifter A, secured in place by the hinge-pivot *m*, upon which it turns and operates the extractor by its finger *n*.

A recess, *i*, is made in the side of the lifter, in which is fitted, to move to and fro, a slide, *s*, having a portion of its edge at *c* cut away, so that the hinge-pivot *m*, which partially fills the cavity *c*, may limit the movement of the slide.

The upper end of the slide may be rounded or beveled, so that as the rear end of the barrel is tilted down to close the breech the upper end of the slide, as it is moved down by contact with the shoulder *a* of the barrel, may readily pass under and the slide be easily forced in by the shoulder *b* on the frame, and the lower end of the slide may be beveled or rounded for the same purpose of easy movement by contact with the shoulder *o* on the barrel.

The lifter, otherwise than being arranged to contain the slide *s*, is made in the ordinary manner, being arranged to turn on the pivot *m*, and provided with a finger, *n*, to bear against the end of the extractor.

The operation of my invention is as follows: When the barrel is in a position with its rear part secured or latched to the upper part of the frame, and it is desired to extract the empty shells from the chambers of the cylinder, the barrel is unlatched from the frame and its forward end tilted down, which causes the forward end of the extractor-stem to bear against the finger *n* of the lifter and turn the latter on the pivot *m*. This causes the lower side of the slide, at its forward end, to strike against the shoulder *c* on the frame, which prevents the lifter from turning farther until the extractor has been forced rearward by the finger *n* sufficiently to extract the shells from the cylinder, and when the shoulder *o* on the barrel passes down it strikes the beveled forward end of the slide and forces it upward sufficiently to pass inside the shoulder *c*, when the lifter is then released, and the extractor-spring forces the extractor quickly forward, and the lifter flies around on the pivot *m* until the upper part of the rear end of the slide strikes against the shoulder *a* on the barrel, where it is stopped. The rear end of the barrel being then tilted down, the shoulder *a* carries down the rear end of the slide and turns the lifter back to its first position again, ready to extract the shells when the forward end of the barrel is again tilted down.

In order that the slide may remain in any position into which it may be moved, a small portion of its side may be cut away and a friction-spring be inserted; or the slide itself may be made slightly curved and made elastic, and then be forced into its straight recess in the lifter; or the slide may be made straight and fit its recess in the lifter sufficiently close for that purpose.

Having thus described my invention, what I claim as new is—

The combination, with the extractor of a fire-arm and its actuating-spring, of the lifter A, located at the hinge which secures the barrel to the frame, and a slide, *s*, arranged to be moved longitudinally therein by contact of the shoulders of the frame and barrel with the ends of the slide when the barrel is tilted upon its hinge, substantially as described.

JAMES H. BULLARD.

Witnesses:

T. A. CURTIS,

E. E. CURTIS.

skilled workmen. His worst traits included his inability to stay with any one company or to apply himself in any specific endeavor for long. Possibly this was the direct result of his lack of managerial skill at the large factory level.

While with Smith & Wesson he collaborated with D.B. Wesson on several patents involving handguns, from approximately 1877 to 1880. Three patents involved handguns directly and one involved a machine to facilitate the checking of pistol grips (Plates 1-2 through 1-5).

It was during this time of feverish activity on Bullard's part that he applied for his patent on what was to become the Bullard repeating rifle. He applied for this patent on May 5, 1879. As significant as these patents were, two other Bullard patents were applied for at this time that were to have a more profound and lasting effect upon the munitions industry. Bullard was the father of the modern solid-head cartridge as we know it today. This cartridge, in .50-115, was also semi-rimless (rimless), certainly a first in America and probably a first in the world.

Sometime between mid-1880 (the time of his last patent with Smith & Wesson) and early 1881, Bullard was offered the superintendency of the Springfield Sewing Machine Company works, and he assigned to them his patent for a shuttle-actuating mechanism for sewing machines on February 1, 1881. He followed this with four more patents assigned to the company in 1881 and 1882. When the company moved to Cleveland, Ohio, it did so without James Bullard, who was by now a stalwart citizen of Springfield.

His patent (#245700, filed May 5, 1879) for a magazine rifle was secured on August 16, 1881, so that when the Springfield Sewing Machine Company left Springfield, Bullard was able to devote his full energies to the task of setting up what would become the Bullard Repeating Arms Association. He surely used his time while still the superintendent of the Springfield Sewing Machine Company to make contacts with the people who would eventually finance the factory and its machinery, something well beyond Bullard's financial means to do so.[4] The company was set up in 1882 under the name Bullard Repeating Arms Association, but here the story becomes muddled. There are several different versions, all of which will be examined in the next chapter.

From 1875 until almost the end of his life, Bullard lived at 777 State Street in Springfield, not far from the site of the factory.

The Springfield City Clerk Directory of 1881-82 listed J.H. Bullard as superintendent of the Springfield Sewing Machine Company, and in the 1882-83 directory he was still listed in that capacity. The 1883-84 directory showed no change in J.H. Bullard's status, but his invalid brother, William A., had become an employee of the Bullard Repeating Arms Association. In the 1884-85 directory J.H. was listed as employed by the Bullard Repeating Arms Association. Wm. A. Bullard became plant superintendent in 1885-86, and his brother E.W. Bullard also became an employee during that time.

In 1886-87 J.H. Bullard was again listed by the directory, but only in the capacity of mechanic.[5] He apparently spent the summer of 1885 in England studying all sorts of burner and mechanical devices. The primary object of his trip to England, however, was the negotiation of a license agreement with an English house to manufacture Bullard rifles. (For legal reasons the author cannot publish information directly from this source but will provide the source so that any readers wishing to study the subject further may write for the pertinent documents.)

Upon returning to Springfield, James Bullard severed his connections with the company that bore his name.[6] He now began to turn his attention to the design of appliances for supplying steam road vehicles with heat generated by petroleum rather than solid fuels.

UNITED STATES PATENT OFFICE.

JAMES H. BULLARD, OF SPRINGFIELD, MASSACHUSETTS.

MAGAZINE-GUN.

SPECIFICATION forming part of Letters Patent No. 287,229, dated October 23, 1883.

Application filed June 14, 1883. (No model.)

To all whom it may concern:

Be it known that I, JAMES H. BULLARD, of Springfield, in the county of Hampden and State of Massachusetts, have invented a new and useful Improvement in Fire-Arms, of which the following is a description and specification.

My invention is an improvement upon the device shown in the patent to B. F. Henry, of October 16, 1860; and its object is to remove much of the friction in the operation of cocking the hammer, in moving the operating-lever forward, and thereby moving back the breech-pin and piston, and to render the movement of the operating-lever more uniform as to the power required to operate it and the mechanism connected therewith; and I accomplish this by the mechanism substantially as hereinafter described, and illustrated in the accompanying drawings, in which—

Figure I is a longitudinal vertical section of so much of the arm as is required to show my invention as applied thereto at line A. Fig. II is a horizontal section of the same at line B, both these illustrations showing the operating mechanism in position with the hammer thrown forward. Fig. III is a longitudinal section with the operating mechanism in position, with the hammer thrown back or in position of full-cock; and Fig. IV is a modification of the invention.

In the drawings, 2 represents the frame of the arm; 3, the operating-lever; 4, the carrier-block; 6, the arm operating the carrier; 5, the hollow breech-pin; 7, the piston with the firing-pin forward; 10, the link, pivoted at its forward end to the breech-pin; 12, the brace, pivoted at its forward end to the rear end of the link, and at its rear end to the frame, and which is also provided with a slot which engages with a pin, 16, in the upper end of the operating-lever 3, this link, slotted brace, and pin in the upper end of the operating-lever, combined, operating to move back the hollow breech-pin and piston, and, as heretofore constructed, to cock the hammer by forcing back the piston against its nose. All these parts before mentioned, as ordinarily constructed, are well known, and are shown in the patent to B. F. Henry, before mentioned; in that to O. F. Winchester, of September 4, 1866, and others, and require no further description, as the accompanying illustration, with the exception of my invention applied thereto, as hereinafter described, shows the operating mechanism of the arm as ordinarily and heretofore constructed.

It will be evident upon inspection that as the link 10 and brace 12 are drawn downward by the upper end of the operating-lever 3, and as the upper end of the latter moves down into the position shown in Fig. III, a very great increase of force or power is required to be exerted against the operating-lever to bring the hammer into the position shown in Fig. III. This will be obvious upon examination of the parts shown in Fig. III, in which the upper end of the operating-lever 3 is in a position, during the latter portion of the movement of the parts in cocking the hammer by moving forward the lower end of the operating-lever and drawing down the link 10 and brace 12, to force the piston backward against the nose of the hammer. In this position it will be seen that the upper end of the operating-lever extends substantially at right angles to the length of the link 10—the medium through which the breech-pin is drawn back by the upper end of the operating-lever, and consequently the medium through which the piston 7 is forced back against the nose of the hammer to cock it, as ordinarily constructed. Of course, under that construction, the power is applied to the operating-lever to cock the hammer under the greatest disadvantage, and more especially is this the case when it is considered that as the lower part of the lever is being moved forward and the hammer commences its backward movement, the mainspring becomes more and more rigid and stiff as the backward movement of the hammer straightens it. The consequence is that the operating-lever moves forward with a very great increase of friction and difficulty of movement as the hammer begins and continues its backward movement.

My invention obviates all this objection and difficulty, and renders the whole forward movement of the operating-lever even and uniform in the amount of power required to be exerted, and with my invention applied the friction of the working parts, together with the force required to move the hammer backward, seems to have very little effect upon the forward movement of the lever.

In the upper end of the operating-lever 3, I pivot at 15 an arm, 14, which curves around the pin 16, which engages in the slot of the brace 12, to operate the latter with the link 10, and the rear end of this arm 14, I pivot to a link, 18, which is pivoted to the hammer in the recess 17, in the front side thereof. This link 18, however, is not essential to the successful operation of the invention, and is only for the purpose of holding up the rear end of the arm 14 in certain positions of the operating-lever 3, as when the hammer is at full-cock and the operating-lever is thrown only partially forward. In such case, if no link 18 were used, the rear end of the arm 14 would drop a little, although it would be held partially up to its place by resting on the pin 16, and if the hammer were let forward the rear end of the arm 14 would strike the front side of the hammer and would slide up along the front side of the hammer as the latter was moved forward until it reached and abutted against the top of the recess 17 in the hammer.

When constructed in this manner, as shown in Fig. IV, and without the link, if the parts should be operated very fast, this falling of the rear end of the arm 14 and striking against the front side of the hammer and sliding up against the shoulder 17, as shown in Fig. IV, into position to again force back the hammer to the position of full-cock, would be accompanied with more or less noise, and I prefer to pivot the rear end of the arm 14 to the link 18, which serves the purpose of a guide to hold up the rear end of said arm and guide it directly to its position against the hammer without first striking the latter lower down and sliding up to position, and preventing the noise resulting from the action of those parts. As the mainspring is weakest when the hammer is thrown forward, a certain amount of power is required to be exerted against the lever 3 in moving it forward, and as the hammer continues to move back the mainspring becomes more and more rigid and stiff; but as the upper end of the operating-lever continues to move back, the pivot 15, which secures the arm 14 to the upper end of the operating-lever, is brought nearer and nearer into line with the pivot upon which the operating-lever moves, and the point of pressure of the rear end of the arm 14 against the hammer until the latter is brought back into a position of full-cock, where the position of the lever 3 and arm 14 and hammer is shown in Fig. III. Inasmuch as the power is applied directly from the lever 3 to and against the hammer, through the medium of the arm 14, in the operation of cocking the hammer, no pressure of the piston against the nose of the hammer is required or can occur, because with my invention applied the hammer commences its backward movement in advance of the backward movement of the piston 7, and owing to the direct backward movement of the arm 14 the hammer moves backward, even after it starts, much faster than the piston moves, and arrives at its position of full-cock slightly in advance of the piston, the rear end of the latter being cut away at 8 on the under side, so that it may not strike against the nose of the hammer as the piston arrives at its backward position. In this construction the nose of the hammer and the rear end of the piston are not worn or injured, as no contact of the two occurs except when the arm is discharged.

Having thus described my invention, what I claim as new is—

1. The combination, with the frame of a fire-arm, of the sliding breech-pin 5, containing the firing-pin, the operating-lever 3, pivoted in said frame, the brace 12, pivoted at its rear end to the frame and toggled to the upper end of said operating-lever, the link 10, pivoted at its rear end to the forward part of said brace, and at its forward end to said sliding breech-pin, the hammer 9, and the arm 14, pivoted at its forward end to the upper end of said operating-lever, and with its rear part bearing against the hammer, whereby the sliding breech-pin and firing-pin are moved back by said brace and link, and the hammer is cocked by said arm when the operating-lever is moved forward, substantially as described.

2. The combination, with the frame of a fire-arm, of the sliding breech-pin 5, the operating-lever 3, pivoted in said frame, the brace 12, pivoted at its rear end to the frame, and toggled to the upper end of said operating-lever, the link 10, pivoted at its rear end to the forward part of said brace, and at its forward end to said sliding breech-pin, the hammer 9, and the arm 14, pivoted at its forward end to the upper part of said operating-lever, and with its rear end bearing against the hammer, whereby the sliding breech-pin and firing-pin are moved back by said brace and link, and the hammer is cocked by said arm when the operating-lever is moved forward, and a guide-piece, 18, pivoted at one end to said hammer and connected at its other end with said arm, to guide the rear end of said arm always into the same position and prevent noise in the working of the parts, substantially as described.

JAMES H. BULLARD.

Witnesses:
T. A. CURTIS,
CHAS. H. WOOD.

(No Model.)

J. H. BULLARD.

MAGAZINE GUN.

No. 287,229. Patented Oct. 23, 1883.

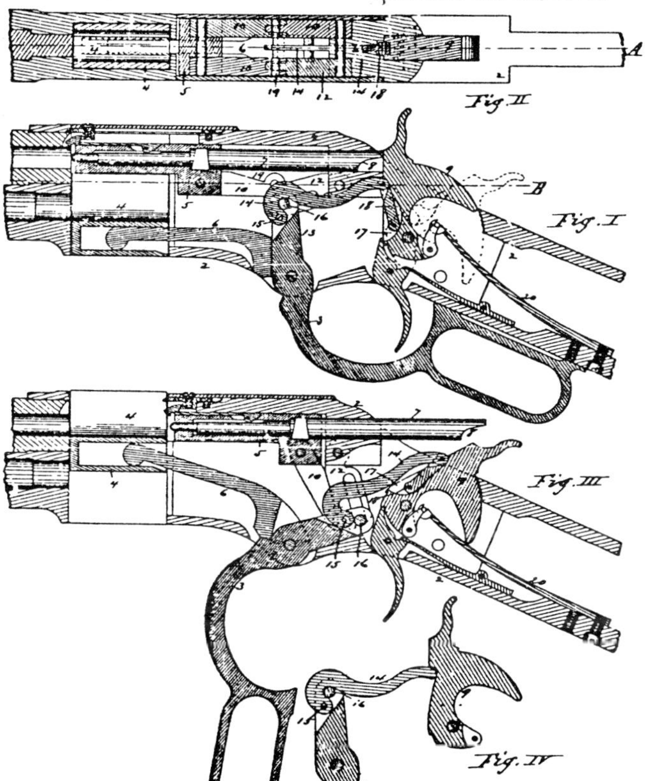

Witnesses.
Chas. H. Wood,
C. E. Curtis,

Inventor,
James H. Bullard,
By T. A. Curtis,
his Atty.

Plate 1-6.

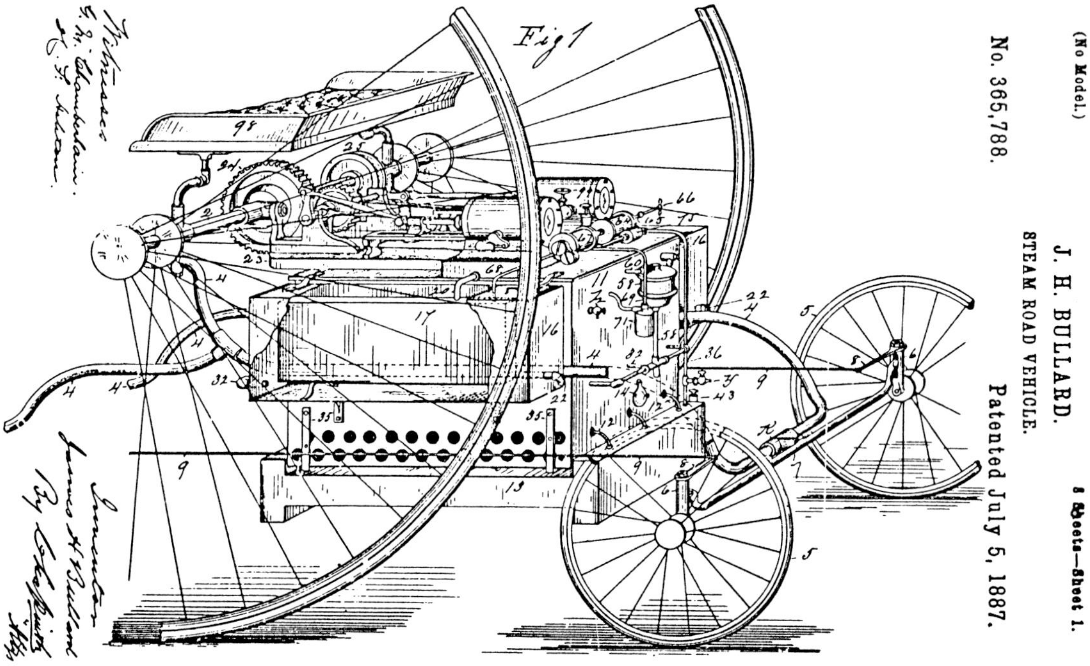

Plate 1-7.

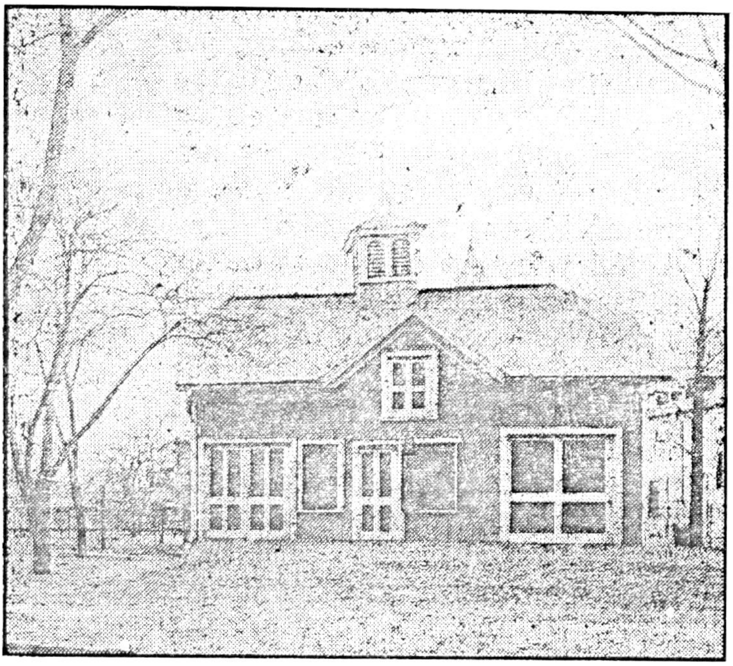

WHERE MR BULLARD BUILT HIS STEAMER

He Made It Out of High Wheel Bicycle Wheels and Pipe. This Is His Workshop on, Hawley Street, Near, State

Plate 1-8.

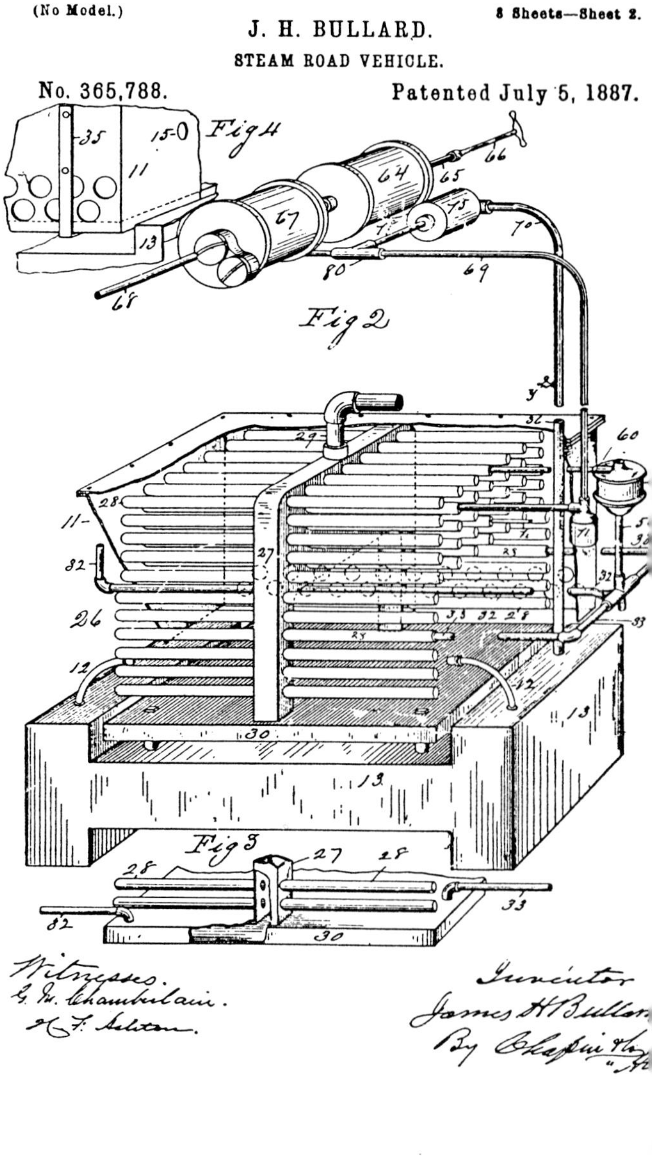

An interesting sidelight to Bullard's arms patents is shown by patent #287229, granted October 23, 1883, for improvements upon the device shown in B.F. Henry's[7] patent drawing of October 16, 1860. Its object was to remove much of the friction in the operation of cocking the hammer and in moving the operating lever forward (Plate 1-6). It's difficult for the average person to fathom the remarkable inventiveness and mechanical aptitude of Mr. Bullard's brain. He never stopped inventing.

As well as being involved in many careers and patent designs, Bullard spent much of his spare time working on a steam car in the shed behind his house on State Street (Plate 1-8). He patented his steam car on July 5, 1887, patent #365788. His vehicle actually ran in Springfield, but not until December 31, 1898, and it is doubtful that more than one was ever built (Plates 1-7, 1-9).

Proof that the vehicle actually ran comes from an article in the Springfield *Republican* dated Wednesday, January 4, 1899, stating:

James H. Bullard of 777 State St. has nearly completed a steam wagon, on which he has been engaged for over 13 years. On Saturday, Mr. Bullard tried the wagon on the road about his home, and even though on that day the going was exceedingly heavy the trial was comparatively satisfactory. However, he has been at work for some time on several additions which he hopes to make soon, and when these are made he thinks the wagon will be one of the most perfect of its kind in the world. The wagon is similar in appearance to those already in the market, run without horsepower. The box is built about eight inches high in front between the seat and the dash, and behind the seat is an enclosed box something over a foot high and covered. The seat is simple in design and upholstered in corduroy. The wheels are regular tandem bicycle wheels with tandem tires running on ordinary ball bearings. The weight of the wagon is about 400 pounds, when empty, and when loaded with fuel and water about 500, so that the pressure on each wheel when in full operation is approximately 150 pounds, or about the ordinary pressure on a tandem wheel. The mechanism by which the wheels are driven is a simple chain sprocket. The steerage is done by the hands by moving a lever which extends down through the bottom of the vehicle and attaches to the two front wheels.

On the front of the dash is fastened a large bell and a lamp, and on the back, facing the operator, are two little dial indicators showing the pressure of steam and air. The wagon is made of the very best material, and its lines of construction are such as to give it a symmetry not always found in our present stylish vehicles.

Beneath the seat, entirely covered and hidden from view, is the mechanism by which it is run. A pair of vertical engines of the marine type, together with the boiler, are here to be found. The boiler will hold from three to four gallons of water, which is a sufficient quantity to run the wagon about 50 miles, the distance varying somewhat with the country traversed. The fuel used to produce the necessary steam can be naptha or oil, as suits the taste of the owner. Reckoning on the cost of locomotion with either of these materials places it at about one-third of a cent a mile or to exhaust one boilerful of water, and hence run the machine a distance of about 50 miles, entails an expense of 16 2/3 cents. Moreover, the boiler is free from any danger of explosion so that even if the water becomes entirely exhausted there is no fear of accident. It has been tested with a pressure of 700 lbs. a square inch, which it easily withstood, while the working pressure is only about 160 pounds.

The movement of the wagon is variable. It is hard to tell what its speed will reach, but it can be run as slowly as a man would ordinarily walk. Experiments have already been made in which it ran a mile in two minutes and 11 seconds. Twenty miles an hour has been done easily, and greater speed can be made. The valve for admitting the steam is at the right hand of the operator, and near it is the lever for reversing the motion. The brake is near the driver's right foot, and by simple pressure engages with the rear axle.

Of course there is nothing new in the idea of a steam wagon. Thirty years ago one was attempted, but it amounted to little, and endeavors have been made ever since, with varying success, but there are certain points about this one which place it above all the rest. Its neatness of appearance with all its mechanism concealed from view, and the ease and rapidity with which it moves, together with the trifling expense for fuel, all speak highly in its favor. There is no doubt about its attaining sufficient speed to gratify all desire for rapid riding.

Of course its speed in any case is dependent upon the load in tow and the road travelled. But the point on which it excels is its safety. As stated above, its boiler is made so that there is no danger of an explosion from the water being used up, and the fact that the working of steam is so much more familiar to most people than that of naptha or explosive engines is a strong point in its favor. Furthermore, when this is completed it will have attachments so that any person can run it with safety. Mr. Bullard feels confident that it is one of the greatest things of its kind ever put on the market and that there is bound to be a demand for wagons.

THE SPRINGFIELD SUNDAY REPUBLICAN, SPRINGFIELD, MASS

Historical Sketches of Springfield... No. 168

BULLARD'S "STEAM VEHICLE"—1886

ROBERT HOLCOMB

Long before the invention of the gasoline automobile, James H. Bullard built in Springfield a steam-operated vehicle, for which he was granted patent No. 365,788 on July 5, 1887. Bullard was the holder of many patents and the manufacturer of the Bullard rifle, which he invented and made at a factory located near Winchester Square. In an article on Bullard, Maxime O. Freniere of this city, who was a designer for the old Knox Automobile Co. in the auto pioneering days, gives the following information about Bullard's steam vehicle: It had two high wheels of the type used on bicycles in that period, and two similar wheels only smaller. The steering arrangement was attached to the rear wheel, pivots and rods leading to the front. The seat was located over the front axle and the engine and boiler were located above and between the two axles. There was a clutch in the main gear that was attached to the driving axle, which engaged or disengaged the engine. The boiler was a horizontal tubular type of special design, heated by a liquid fuel. It had a special water tank, condenser, and water-gauge fuel-controling devices. No smoke pipe was necessary. Bullard's steam vehicle actually ran in Springfield, but whether more than one was ever manufactured is not known.

Plate 1-9. Reprinted from the Springfield *Sunday Republican*, March 20, 1955.

Bullard made several attempts to interest investors in his steam motor carriage, but when he failed he concluded that the time was not ripe for such an invention. The car was allowed to stand in the workshop and was finally sold for scrap metal.

Several articles and newspaper clippings state that the car ran in 1885. This is very unlikely given that the summer of 1885 found Mr. Bullard in England studying much of the technology that he was to incorporate into his vehicle as the design progressed. The car was not patented until 1887.

Other snippets of information on Bullard's car include the following:

Bullard never, in my opinion, received the credit he deserved for his work on the automobile principle," said Charles R. Culver, general manager of the Arrow Automobile Company, one of the few who remember seeing the little car in operation.

"It had the first automatically controlled power plant in an automobile," said Frederick H. Bullard, the inventor's son, a mechanical engineer with offices in the Stearns Building.

"Every essential element of the later successful steam automobile was there and working, and no element of automobile control which it did not possess had ever been added in any design by any builder," the younger Bullard said in 1919.

He said the driving wheels of the car were 50 inches in diameter with solid rubber tires. The steering wheels, also bicycle wheels, were about 14 inches in diameter, with solid rubber tires. The car which had a wheelbase of four feet was steered with a cross-handlebar.

The engine, which had two cylinders, and the boiler were both at the rear of the driving axle. Kerosene was used as fuel, atomized by a process which appears to have been one of Bullard's better known achievements.

Bullard's son said the framework was of iron pipe with screwed and pinned joints. There was a single seat for the driver who was the only passenger. A regulator was attached to the main framework by a horizontal pivot, thus providing for the inequalities of the road. The whole thing was not too heavy to be easily lifted by two men.[8]

Bullard's ideas for development of heat generated by petroleum were so successful that he was led to believe that it was of equal importance for other purposes, and he began to devote his whole time to developing it for heat in metallurgical uses and processes. Between 30 and 40 patents for its application were secured by Bullard over the years. This method of generating heat was used by hundreds of factories around the world.

About 1888 Bullard formed the Aerated Fuel Company, located at 425 Main Street, and became its general manager. Ex-state governor George D. Robinson became its president, Mr. A.B. Wallace of Forbes and Wallace its treasurer, and Charles E. Stickney its secretary. The company was formed to manufacture and sell industrial-type oil burners, the device Bullard had invented for high-pressure steam boilers.

J.H. Bullard had married Martha Bullard (her maiden name) on December 31, 1863, and they had three sons and one daughter: Frederick Herbert, Herbert Augustus, Edwin and Mary Louise.

In the 1888-89 Springfield City Clerk Directories, Bullard was listed as manager of the Aerated Fuel Company, located at 425 Main Street until the year 1906. In 1907 he was listed as a manufacturer of speed recorders (speedometers). No business address is listed so he probably worked from his house. In 1908 his son Frederick joined him, and they remained in this business until 1910, about which time Bullard sold his home at 777 State Street to a neighbor, Col. A.H.H. Goetting, who a few years later converted Bullard's old shop into a garage where the widely advertised cars of later, more successful automakers were sold.

In 1911 Bullard's home address became 45 Warner, but sometime after Bullard died, the name of the street was transferred from one road to another, so if the Bullard-owned home still stands on Warner, it would be difficult to prove. In any event, he only lived in it three or four years. His house on State Street is now a library parking lot. No monuments have been erected to mark the spot.

The 1912 Directory listed Bullard as manager of the Bullard Specialty Company, 45 Warner, and in the 1913 Directory his capacity and address were unchanged.

James Bullard died on March 26, 1914, at the age of 72, in his home on Warner. He was survived by four children and one adopted daughter. His wife, Martha, had died the previous year, on September 20 at 73 years of age. Bullard was buried in the Oak Grove Cemetery near his old employer, D.B. Wesson, but unlike Wesson he left no monument save the steel pin marker #1380 in section 14 (Plate 1-10). The ashes of his son Frederick are also buried in the 20 by 16-foot plot.

Plate 1-10. The author holding the only physical reminder of J.H. Bullard's resting place, a steel cemetery marker with the number 1380 on it.

Just before his death in 1914, Bullard contributed information to the Springfield newspapers outlining the role of two of his cousins, George A. and Silas Bullard, in the capture of Jefferson Davis, President of the Confederacy.

By all accounts James H. Bullard was a well-respected citizen and a member of many groups and societies in Springfield. His greatest failing, if one can call it that, was his inability to remain with one project long enough to see it through to commercial success before venturing on to the next project. His adult life was spent satisfying his inventive mind's pursuit of new inventions and patents, but never fully capitalizing upon the inventions which should have made him wealthy.

James Bullard was the archetypal eccentric Victorian inventor; everything interested him. If one's success is not measured by wealth but by the freedom to do as one wishes, then Mr. Bullard was very wealthy indeed.

The greatest and most lasting legacy of James H. Bullard was his development of the solid-head semi-rimless (rimless) shell for small arms. Perhaps every center-fire cartridge made today should be headstamped with a B in honor of the man who made that technology possible.

FOOTNOTES

1 U.S. Department of Commerce Patent and Trademark Office records of 1878, listed as patent #207010, August 13, 1878.

2 The author has never been able to ascertain what a Shaker Hood was. Can any reader help?

3 The author has never seen or been able to locate a picture of James Bullard, which has led to speculation on the author's part that Mr. Bullard may have been self-conscious about some aspect of his physical appearance.

4 H.H. Bigelow was to provide that finance: "The Bullard Rifle Co. of Springfield, Mass., is indebted to him [Bigelow] for its establishment." *Geneaology of the Bigelow Family*, p. 413.

5 A word of caution when dating employment status from city clerk directories: their dates appear to be almost two years behind actual fact. The 1884-85 directory is the first to place him with the Bullard Co., yet he left it after the summer of 1885.

6 *Bibliographical Review of the Leading Citizens of Hampden County* (Boston: Biographical Review Publishing Company, 1895), pp. 255-56. Most biographical material on J.H. Bullard seems to have emanated from this sketch, and because it was published during Bullard's lifetime, no doubt it is accurate and can be accepted as primary source material.

7 The patent papers read B.F. but, as every firearms historian knows, it was B.T. Henry.

8 Kenneth A. Anderson, "Bullard's 1885 Steam Auto Chugged into Oblivion," Springfield *Union*, January 3, 1967.

2

THE FACTORY AND ITS DEVELOPMENT

The exact founding or completion date of the factory proper has been obscured by time and somewhat muddled by the promotional blurbs extolling the industrial prowess of the region in the 1880s. However, the date the doors closed forever on the Bullard Company is certain.

According to a contemporary account published in June 1891, the following took place: "In January 1891, by the generosity of Mr. H.H. Bigelow of Worcester, the owner of the Bullard plant, an offer of that building, valued at $50,000, was made to the school for one-half that sum. This property includes a most substantial brick factory, containing a first-class engine and boiler, with main shafting, 23,000 square feet of floor space and 64,000 square feet of surrounding land situated on the New England railroad, between the Wilbraham and Boston roads."[1]

By 1882 James Bullard was busy perfecting his rifle design and the machinery necessary to place it into full-scale production.

Mr. H.H. Bigelow of Worcester, Massachusetts, a wealthy entrepreneur and philanthropist, was one of several men who were to play a part in supplying the necessary capital to purchase land and machinery for the soon to be erected Bullard factory (Plate 2-2). Sketchy business records of the early 1882 period seem to indicate that Bullard had a company known simply as J.H. Bullard and Co. Records also indicate that by May 6, 1882, the company intended to call itself the Bullard Repeating Arms Association as soon as it was established.[2] Bullard Repeating Arms Association was in existence by 1883.

On June 28, 1883, two separate legal documents secured the six parcels of land needed for the new factory.[3] The land was acquired for the sum total of $3,000. The land was purchased in Bigelow and Bullard's names, not the company's name. The author has not been able to locate any documents that show that the property was ever transferred to the company.

On October 11, 1883, the company was reorganized and capitalized with $500,000. It was now a corporation registered in Maine with a new name, The Bullard Repeating Arms Company.[4] This was to be the company name until its official demise in January 1891.

The question arises as to whether or not there was more than one factory. The new factory could not have been finished much before the end of 1883, as the land was only secured in June 1883. It may not have even been completed before early 1884, yet some of the earliest known Bullard advertising occurs in the October 20, 1883, issue of *American Field*. This, combined with the fact that the earliest or lowest serial numbered repeaters are stamped Bullard Repeating Arms Association, leads to the conclusion that some small-scale manufacturing of firearms must have taken place before the name change from Association to Company. This will be discussed more fully later in the chapter, with the inclusion of contemporary accounts.

A legal document dated October 27, 1884, indicates that the Bullard Company

Plate 2-1. The Bullard factory as it looked about 1891 after its acquisition by the Christian Industrial and Technical School. This view was taken from the rear of the building.

Plate 2-2. Horace Holly Bigelow, 1827-1911.

had recently taken over the premises.[5] Just prior to this undertaking, litigation of some sort took place between the Pratt and Whitney Company and the Bullard Company over the cost of machinery. It was eventually resolved, but the legal document mentioned above may have been a form of legal maneuvering, as it gave the land the factory was situated on to one George H. Ball. One of the main terms in the document directed towards George H. Ball was his promise "to allow the Bullard Repeating Arms Company a corporation by law duly established and now in the occupation of the premises to occupy the same for its manufacturing purposes in such way and manner as it shall elect fully and without let or hindrance or liability for the payment of rent at any time . . . ."

This document now meant that George H. Ball was president, Horace Holly Bigelow was treasurer and James H. Bullard manager of the factory.[6]

After reviewing the agreement between Ball and Bigelow, and then recalling that it was Bigelow who sold the Bullard plant and its surrounding property in 1891, it would seem that all of the above legalities were taken to either fend off further lawsuits or reduce taxation and liabilities, for in the end Bigelow still owned it all.

To further complicate the business activities of the Bullard Company during its infancy, it was involved in the police courts of Springfield. During June and August 1884, the personnel of the Bullard Company were in court for three separate cases involving six court days (Plate 2-3). The records do not show the reason for the frequent court appearances, but they may have been due in part to the Pratt and Whitney litigation.

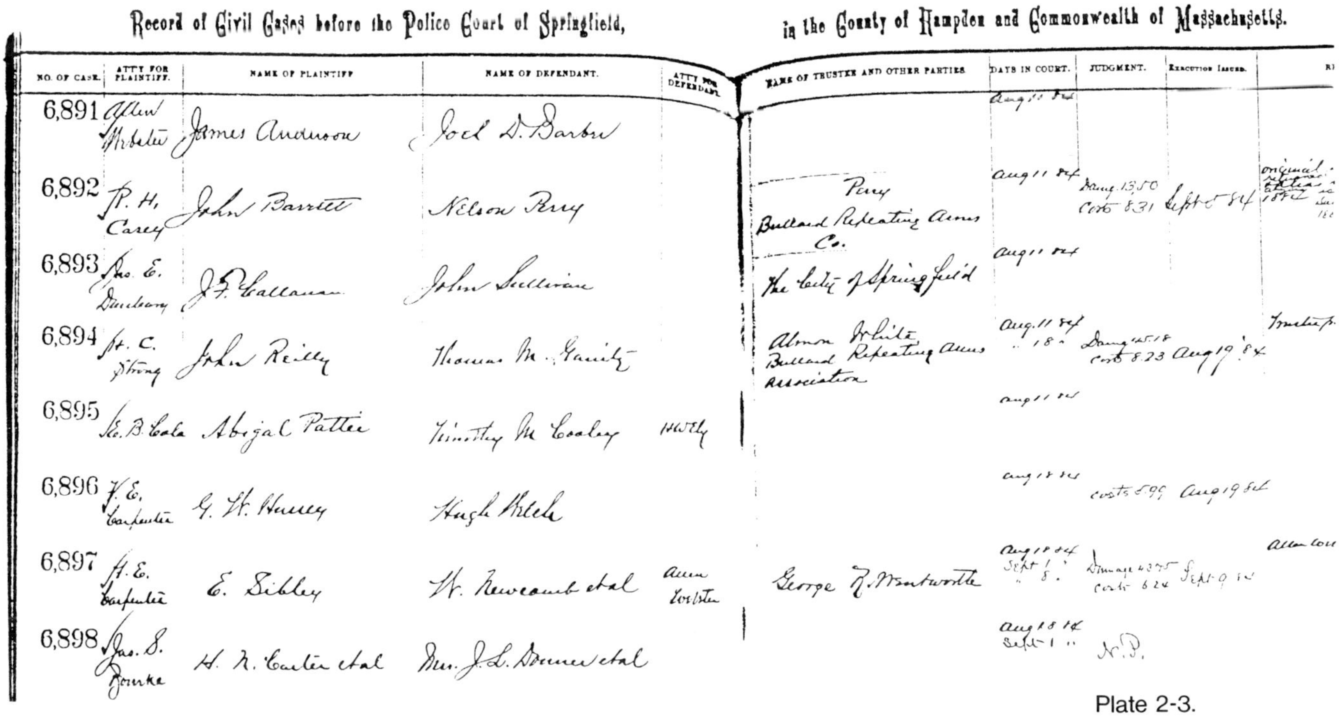

Record of Civil Cases before the Police Court of Springfield, in the County of Hampden and Commonwealth of Massachusetts.

No. of Case	Atty for Plaintiff	Name of Plaintiff	Name of Defendant	Atty for Defendant	Name of Trustee and Other Parties	Days in Court	Judgment	Execution Issued	[illegible]
6,891	Allen Webster	James Anderson	Joel D. Barber			Aug 11 '84			
6,892	R. H. Carey	John Barrett	Nelson Perry		Perry Bullard Repeating Arms Co.	Aug 11 '84	Dam. 13.50 Costs 8.31	Sept 5 '84	Original [illegible] 1884
6,893	Jas. E. [illegible]	J. F. Callanan	John Sullivan		The City of Springfield	Aug 11 '84			
6,894	H. C. Strong	John Reilly	Thomas McGarity		Almon White Bullard Repeating Arms Association	Aug. 11 '84 " 18 "	Dam. 45.18 Costs 8.23	Aug 19 '84	Trustee [illegible]
6,895	Geo. B. Cole	Abigail Patter	Timothy McCooley	H.W. [illegible]		Aug 11 '84			
6,896	F. E. Carpenter	G. H. Hussey	Hugh Welch			Aug 18 '84	Costs 5.99	Aug 19 '84	
6,897	F. E. Carpenter	E. Sibley	H. Newcomb et al	[illegible]	George R. Wentworth	Aug 18 '84 Sept 1 " " 8 "	Damages 43.75 Costs 6.24	Sept 9 '84	Allen [illegible]
6,898	Jas. S. Rourke	H. N. Carter et al	Mrs. J. L. [illegible] et al			Aug 18 '84 Sept 1 "	N.P.		

Plate 2-3.

At the end of summer 1885, upon his return from England, J. H. Bullard severed his connections with the company he had formed and his brother William A. Bullard became plant superintendent for the years 1885-86. (The years mentioned for Wm. A. Bullard's superintendency come from the city directory and may also be suspect.)

It has been shown that the Bullard plant as it now exists was occupied by middle to late 1884. (The transfer of property to Ball on October 27, 1884, made that clear.)

But was there in existence an earlier, much smaller establishment where small-scale manufacture of Bullard rifles under the style of Bullard Repeating Arms Association took place? Certain records seem to indicate the existence of two factories.

According to an article in *Springfield and Environs: Commerce, Manufactures and Resources*, published in June 1883, there was:

> The Bullard Repeating Arms Association was organized in 1882 with a capital stock of $500,000 and immediately commenced the manufacture of firearms in a two-story brick building 30x60 feet in dimensions, furnishing employment to about twenty-five skilled artisans, until arrangements could be completed for the erection of a plant of their own. These arrangements are now completed, and the company have procured a site and are about erecting a four-story brick building, 169 feet long and 40 feet wide, with trimmings of Longmeadow stone and Monson granite. There will be four machine rooms, each 40x125 feet and the building when finished will accommodate about two hundred workmen. Ultimately large additions will be made to the plant, if the business warrants. The factory will be provided with all modern facilities, including an engine of 100 horse-power, electric lights etc., and the mechanical equipment is now being manufactured under contract at a cost of about $30,000.

A second article, in the same vein, published in 1884, claims: [7]

> The Bullard Repeating Arms Company is one of the new industries; but it has started on a noteworthy scale, under favorable auspices, superior management, with large capital and unsurpassed facilities . . . It starts with a capital of five hundred thousand dollars, and with George H. Ball of Worcester as president, and Horace H. Bigelow of Worcester as treasurer. The manager, James H. Bullard, is the inventor of the Bullard rifles. The manufacturing establishment is worthy of special mention. The buildings are unusually well constructed, and compare favorably with any manufacturing establishment in this country. They were built all of brick, from the plans of Mr. Bullard, who had in mind every demand of the business . . . The main building is three storys high, 165 feet long, and 40 feet wide. The equipment of the finest of metal and wood-working machinery is ingeniously arranged . . . There is good working room for 200 men, and about 50 guns a day can be made . . . The situation is most delightful; being at the head of State Street, at the crossing of the New York and New England Rail road. The company's property includes two acres, and provision has been made for future enlargement.

A Springfield catalog article from 1887 had this to say about the Bullard Company: [8]

> This company has taken a high position rapidly in furnishing the world with a superior class of repeaters, carbines, hunting and target rifles for military, target and sporting use. The business of this concern was originated in 1882 under the style of the Bullard Repeating Arms Association, and in 1883 the present company was organized with the title of the Bullard Repeating Arms Company, and was incorporated with a capital of $500,000. Mr. George H. Ball, of Boston, is the president, and Mr. H.H. Bigelow, of Worcester, is the secretary and treasurer, Mr. S.K. Hindley, manager. The company's works are located contiguous to the New York and New England R.R. and near State Street, and comprise a three-story brick building, 30x120 feet, annex 40x60 feet in dimensions, with separate buildings for engine house and boiler room. The factory is equipped with the finest class of machinery, which is operated by a 150 horse power steam engine. From seventy-five to one hundred hands have been employed in the various departments . . .

Similarly, a Springfield trade article of 1890 had these words of praise for the Bullard Company:[9]

> The works of the company rank with the most complete and extensive private plants of the kind in this country. They are situated at Winchester Park, near State Street, and adjoining the New York and New England railroad, with which they are connected by a switch. The shops, equipped with a comprehensive and costly array of special machinery and employing from fifty to two hundred and fifty hands as the exigencies of trade require, are contained in a massive brick structure of three stories 40x125 feet, with a 40x60 foot wing, the engine and boiler houses being detached. The capacity is about fifty finished arms per day and the policy of producing only in response to the demand having been adopted, no accumulation of stock occurs . . . The trade with Canada, England, and the antipodes is particularly good . . .
>
> The company was organized with a capital of $500,000 in 1882, and commences manufacturing in 1884. President Ball resides in Boston, Secretary and Treasurer Bigelow in Worcester, and Manager Hindley in Springfield.

An 1891 description of the interior of the factory after its takeover by the Elektron Manufacturing Company and the Christian Industrial and Technical School had these comments:[10] (see Plate 2-4)

The works are situated on Wilbraham Street, beside the New York and New England Railroad, from which a side track will soon be running. The main building is of brick, forty feet by fifty-five feet, and four stories high. It has a long three-story wing 125 feet by forty feet. In this wing the workrooms are located. A brick engine and boilerhouse, forty-five feet square, is situated north of the main building, with a fifteen-foot passageway between. The sixteen-inch leather belt, by which power is transmitted from the engine to the main shafting, passes above this driveway in a covered bridge.

In the main building the first floor is at about the level of the street. The main entrance to the works is by a stairway leading from the street to the second floor. The office occupies a space in the main building about thirty-five feet by thirty feet, near the entrance. Here may be found the desks of the general manager, superintendent and electrician . . .

From the descriptions given it's quite possible and plausible that a smaller building elsewhere in Springfield housed the Bullard Company until its factory was completed and ready for production. From the varying descriptions and dimensions of the factory, it's doubtful that some of the pundits penning those articles extolling the virtues of the company ever laid eyes upon it.

Plate 2-4. The factory as it looked when occupied by the Elektron Manufacturing Company circa early 1890s. The railroad switch is very much in evidence.

The author would caution readers that due to publishers' lead times and the fact that the quoted articles were intended to advertise the companies mentioned, many details given are suspect and the information should be viewed in that light. The author has used the foregoing articles to support his two-factory theory because it is felt that the main thrust of the articles is correct. The articles bog down in detail and hyperbole.

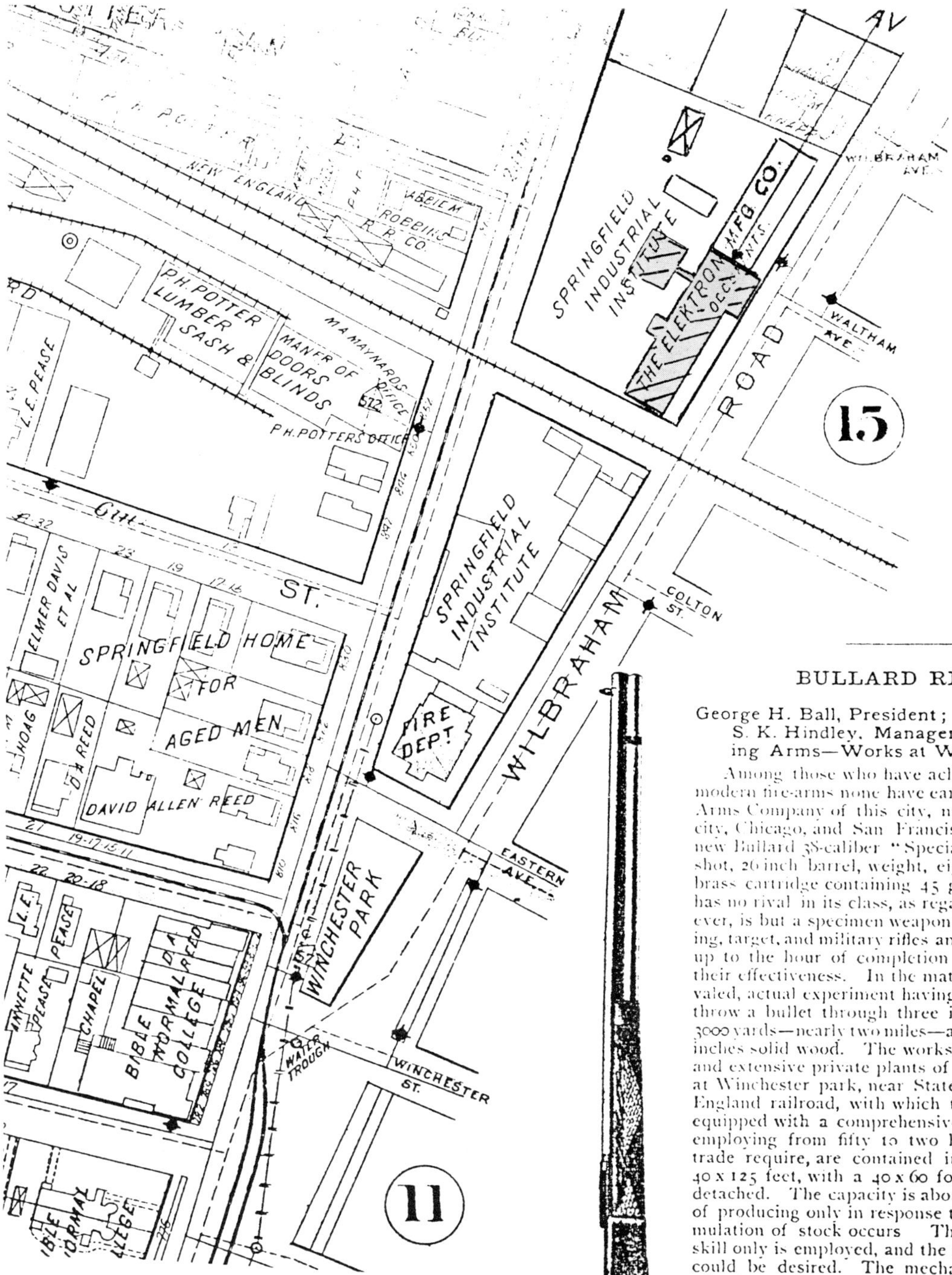

Plate 2-6. 1899 Springfield Survey showing factory extension of 1896. Hatch marks represent original factory dimensions.

BULLARD REPEATING ARMS CO.

George H. Ball, President; H. H. Bigelow, Treasurer and Secretary; S. K. Hindley, Manager—Manufacturers of the Bullard Repeating Arms—Works at Winchester Park.

Among those who have achieved marked success in the improvement of modern fire-arms none have earned a higher place than the Bullard Repeating Arms Company of this city, maintaining wholesale branches at New York city, Chicago, and San Francisco. The cut printed herewith represents the new Bullard 38-caliber "Special" rifle, with globe and peep sights, eleven-shot, 26 inch barrel, weight, eight and a quarter pounds. This arm uses a brass cartridge containing 45 grains of powder and 190 grains of lead, and has no rival in its class, as regards force, rapidity, and accuracy. This, however, is but a specimen weapon from a long list that includes repeating, sporting, target, and military rifles and carbines in all desirable styles and improved up to the hour of completion with every valuable novelty that can add to their effectiveness. In the matter of penetration alone these arms are unrivaled, actual experiment having proved that a Bullard rifle of 45 caliber will throw a bullet through three inches of spruce and eight inches of sand at 3000 yards—nearly two miles—and has a penetration at fifty yards of twenty-two inches solid wood. The works of the company rank with the most complete and extensive private plants of the kind in this country. They are situated at Winchester park, near State street, and adjoining the New York & New England railroad, with which they are connected by a switch. The shops, equipped with a comprehensive and costly array of special machinery and employing from fifty to two hundred and fifty hands as the exigencies of trade require, are contained in a massive brick structure of three stories, 40 x 125 feet, with a 40 x 60 foot wing, the engine and boiler houses being detached. The capacity is about fifty finished arms per day, and, the policy of producing only in response to the demand having been adopted, no accumulation of stock occurs The highest grade of technical and mechanical skill only is employed, and the material, workmanship, and finish are all that could be desired. The mechanism, while simple and strong, is singularly effective, acting directly without the intervention of springs; the piece may be loaded from above or below after the magazine is filled, thus securing an extra shot in case of emergency—twelve in all—which may be discharged in the unexampled short space of five seconds. All of the parts are of fine forged steel and gun metal, and the nicest adjustment characterizes the machine throughout, especial care being given to the rifling, which is superior to any similar work of the kind ever performed, thus assuring almost absolute accuracy of aim and execution.

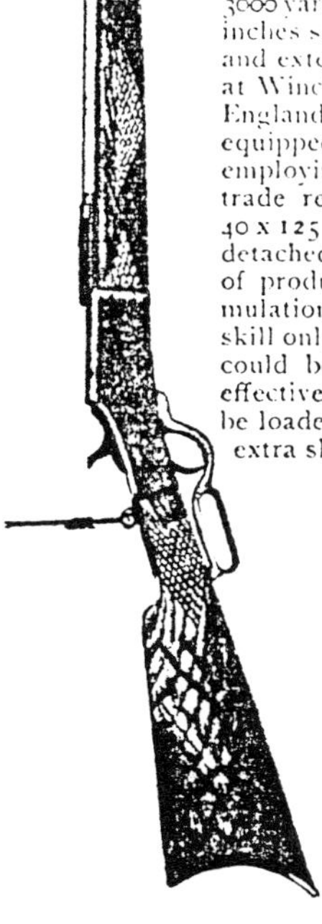

As might be supposed, the Bullard repeating arms are fast superseding those of older design, and are being rapidly adopted by hunters of large and small game, many of them going out to India and others to Africa, where the lion, the tiger, and the elephant are becoming acquainted with them to their sorrow. The American frontiersman has quickly appreciated their merits. The trade with Canada, England, and the antipodes is particularly good.

The company was organized with a capital of $500,000 in 188[illegible], and commenced manufacturing in 1884. President Ball resides in Boston, Secretary and Treasurer Bigelow in Worcester, and Manager Hindley in Springfield.

Plate 2-5. Reprinted from *Inland Massachusetts Illustrated*, 1890.

After Bullard's departure from the factory in mid-1885, the vacuum he left was filled more and more by Solomon K. Hindley. Original correspondence from May 1885 has him signing the letters as secretary of the Bullard Company. With William A. Bullard's departure in 1886, the way was clear for Hindley to assume the position of manager of the works and to become a rifle inventor in his own right. By mid-1886 Hindley was submitting his own patents, and by 1888 his own rifle designs, all under the auspices of the Bullard Company.

The Bullard factory should have flourished. It had a fine product, the best machinery, excellent financing and a modern plant. What went wrong? No one crisis seems to have been responsible for the factory's failure in 1891. Examination of its business methods must be studied in order to better understand the complex reasons for its failure.

The Bullard Company lost few opportunities to promote themselves, both in the printed media and on the target range (as with the hiring of such notable shooters as W. Milton Farrow). The Bullard Company offered its rifles in any caliber a purchaser could desire. The first known catalog offered the .45-70 and .50-95 almost exclusively because of the popularity of those calibers. The company was also to develop its own line of proprietary cartridges, but *never* to the exclusion of the more popular Winchester and Marlin loads of the day. As has been said many times, the factory's finances were such that any project that they may have wished to develop was well within their grasp. The rifles they manufactured were second to none, both in their design and in the quality of materials used. The fitting of these parts, steel to wood, was superb. The mechanism was the smoothest lever action ever made, with little friction and wear on the parts. The Bullard people set up an aggressive network of dealers, both in the United States and Canada. They exhibited at shows in Europe. A future President of the United States was a Bullard rifle owner. The price of a Bullard rifle was higher than that of its competitors, but it was better made, with more hand finishing. The price factor alone could not have been the cause of failure, for many small manufacturers of firearms have survived for a hundred years by catering to those with expensive tastes. Did the introduction of the Model '86 Winchester, with its very fine action, spell the end for Bullard? There may be some reason to suspect that the introduction of this rifle caused some consternation in the Bullard camp, as very suddenly Bullard began stamping their repeaters with the legend "Model 1886." But they may have overreacted, for Marlin repeaters survived very nicely and Savage had yet to even commence with its lever-action design, but was to do so knowing full well the competition it faced.

With their financial clout, Bullard should have been able to weather any of the many mini-recessional storms of the 1880s and come out even more successful. On the face of it, how could this company have failed.

The reasons the Bullard Company failed are many and varied. They are underlying, and combined with the failure of the company to update their designs in keeping with the many changes taking place in ballistics and public taste, were to grow insurmountable.

When J.H. Bullard left the company, he took with him its technical expertise. True, S.K. Hindley was also a designer and inventor, but he only built on existing Bullard designs, never developing anything radically better or different in concept. The Bullard management at this point should have been on the lookout to hire another inventor — a Burgess, a Browning — but the author has seen nothing to indicate that this was even attempted.

The upper level management often did not participate in the day-to-day affairs of the company. Bigelow resided in Worcester and Ball in Boston, and the everyday running of the company fell upon Hindley. To be blunt, upper management became lethargic and failed the company by not providing direction or plans for the future. No efforts were made to build up stocks and distribute them widely to the dealers, and because no large military orders were ever obtained, the work force in the plant was allowed to rise and fall accordingly, which may account for some of the oddities that occur on the arms themselves. Also, Mr. Bigelow was a rather successful and wealthy businessman from Worcester who doubtless had little time for the affairs of the Bullard Company, and by all accounts the business languished and died not with a bang but with a whimper.

The Bullard Company, like many others its size, employed other specialty companies to manufacture their drop-forgings.[11] The Bullard Company used the Billings & Spencer Company of Hartford, Connecticut. In this way it was possible for a smaller company to produce arms for a price much lower than if they had to invest in the drop-forging equipment themselves. A letter from the Billings & Spencer Company relates the amount of work done for the Bullard Company:[12]

> Hartford, Conn., Nov 25
>
> "Dear sir, — Yours of the 22nd at hand, and we herewith submit the information requested. For the Bullard Repeating Arms Co. of Springfield Mass., we make drop-forgings for the rifle parts as follows: Receivers, levers, lock-frames, carrier levers, main springs, trigger springs, extractors, guard lever latch, guard lever links, brace, breech block, gear link, brace link, racks, bolts, hammers, triggers, sliding covers, side plates, loading-slides, stirrups, tips, front-tube studs, tube tenants . . . ."
>
> The Billings & Spencer Co.,
> H.E. Billings

Nothing has been seen by the author to indicate what method was used to dispose of the Bullard Company's machinery. Correspondence to all of the Bullard Company's competitors still in business to this day has turned up no evidence that one of them may have bought out the Bullard Company or its portable assets.

Following the Bullard Repeating Arms Company's failure and sale of the factory to the Christian Industrial and Technical School in January 1891, its facilities were shared with the Elektron Manufacturing Company, the company trained apprentices in return for their cheap labor. About 1895 the Springfield Industrial Institute succeeded the Christian Industrial and Technical School and a new building was erected to house it, with Elektron still occupying the ex-Bullard plant.

The 1899 Springfield Survey shows the Springfield Industrial Institute and the Bullard building still occupied by the Elektron Company, but it also seems to be part of the Springfield Industrial Institute and probably still had the same arrangements in regards to labor and training as it had with the earlier school. The map also shows the 1896 expansion to the building.

Sometime in 1911 the ex-Bullard plant became part of George M. Hendee's motorcycle-manufacturing empire. The Hendee Manufacturing Company, manufacturers of the famous Indian motorcycle, became the leading motorcycle producer in the United States. The building remained in that capacity until 1953, when the last Springfield Indian left the factory.

The original Bullard building underwent many changes to appear as it does today. The building was extended past the eastern tower end, and again in 1911 when Hendee added two storeys to the building at a cost of $84,000. The now massive Hendee works have been basically empty since 1953, with the exception of small craft shops and clothing remnant merchants. As of June 1985, no glass remained in the windows, its facade was hidden by trees and underbrush, and its loading dock was burned and charred. The western end of the Hendee complex has been torn down after one of innumerable fires finally made it unsafe to remain standing. The oldest structure in the complex, the Bullard factory, still remains. It has been nominated for inclusion in the National Register of Historic Places and will probably have its insides gutted. It may again find a use, housing inner-city families of Springfield (Plates 2-6, 2-7).

Plate 2-7. Bullard factory, photo taken in 1985.

FOOTNOTES

1 *Progressive Springfield*, Springfield, Mass., October 1891, No. 5, p. 22.

2 The records that the author refers to here are unfortunately not available for publication purposes due to legal stipulations set out by the sources. However, these records are available for research purposes by writing to the Baker Library, Harvard Business School, Boston, Mass., 02163, and asking for copies of documents marked Mass. V.42 p. 414; V.43 p. 151 part a; part b; part c; Mass. V.43 p. 236; Mass. V.44 p.32.

3 Hampden County Registry of Deeds, Book 402, pp. 296-297, and Book 399, p. 543.

4 In order to ascertain the logic behind a Massachusetts company incorporating under another state's laws, the author called upon the Maine Historical Society. Mr. Stephen Trent Seames was kind enough to provide a plausible answer: "Very often companies formed under Maine laws due to the small assessment of taxation required by this state. Perhaps the only connection is that an attorney for the company was the resident agent."

5 Hampden County Registry of Deeds, Book 408, pp. 529-531.

6 Hampden County Registry of Deeds, Book 408, pp. 529-531. "That we Horace H. Bigelow of Worcester in the county of Worcester and James H. Bullard of Springfield in the county of Hampden both in the Commonwealth of Massachusetts for and in consideration of one dollar and other good and valuable considerations to us paid by George H. Ball of Worcester aforesaid Esquire. The receipt whereof we do hereby acknowledge do hereby give grant bargain sell and convey unto the said George H. Ball and his heirs and assigns forever a certain tract or parcel of land situated in Springfield aforesaid bounded and described as follows . . .

7 Moses King, *King's Handbook of Springfield* (Springfield, Mass.: James D. Gill, 1884), p. 366.

8 *The Massachusetts Industries Leading Manufacturers and Merchants* (Springfield, Mass. 1887), p. 221.

9 *Inland Massachusetts Illustrated*, (Springfield, Mass. 1890).

10 *Progressive Springfield*, Springfield, Mass. October 1891, No. 5, p. 132.

11 Drop-forging is the process whereby hot metal is shaped by the weight of a heavy hammer dropping against a die, and by doing so fashioning the shape desired. The hammer weight contains the other half of the die. The next time you pick up a wrench that is stamped "drop-forged" note the seam that runs around the outline of the wrench. That is produced by the two halves of the dies meeting.

12 A.C. Gould, *Modern American Rifles*, (Plantersville, South Carolina: Thomas G. Samworth, 1892; republished 1946), p. 29.

BACKGROUND TO CHAPTERS THREE AND FOUR

This background information is most important to the understanding of the way in which the Bullard Company serial-numbered their arms. The repeaters were split into two groups, **large-frame** and **small-frame**. The large-frame repeaters included all calibers from .40 on up to the .50s. The small-frames included all calibers below .40 but **not** including .40. The only exception to this rule was that .44 caliber was offered in the small-frame version, although the author has yet to see a Bullard in .44 caliber. Both the large- and small-frames differed quite substantially, not only in sheer size, but also in much of the detail (which will be elaborated on further in the following chapters). Even within their own groupings, there are substantial detail differences, especially among the large-frames, in the areas of levers, triggers, lever latches and lever pin design, to name but a few.

These variances again point to the fact that, although the Bullard plant was substantial, their output was not. They were never to achieve mass-production status, but sold arms in small lots as the demand warranted. This may help account for the many differences in the models. The lack of a permanent factory work force completely familiar with the arms could also have been a contributing factor accounting for the many model variations.

Many of the sights described in Chapters Three through Seven were identified by Ed Iwanchuk.

The repeater has a standard Winchester rear sight. Single-shot has a Lyman tang sight. *Courtesy Verne W.L. MacRoberts/Photo by Don Reid*

3

LARGE-FRAME REPEATERS

The patent covering the large-frame repeater was granted to James Herbert Bullard on August 16, 1881. It was number 245700 (Plate 3-1).[1] After describing his invention in the Letters Patent, he claimed as new —

> 1. In a fire-arm, the combination of the lever H, the spring-pawl h, pivoted thereto, the link I, pinion f, and stationary rack f'', as a means of moving the bolt "a" to and fro in loading the arm and ejecting the shell, substantially as described.
> 2. The combination, in a fire-arm, of the lever H, with its spring-pawl h, and the arm L, as a means of moving the brace into and from its place beneath the block, substantially as set forth.
> 3. In a fire-arm, the combination of the bolt a, the block E, and the brace N, substantially as set forth.
> 4. In a fire-arm, the combination of the lever H, with its spring-pawl h, the link I, pinion f, bolt a, the arm L, brace N, and block E, operating substantially as set forth.

This same patent covered the small-frame repeater as well. Bullard was also granted a British patent on September 23, 1881, for his rifle design. Mr. H.H. Lake was Bullard's representative[2] (Plate 3-5). The patent number was 4110.

The mechanism of the repeater is shown very clearly in the drawings rendered by R. Bruce McDowell (Plates 3-7, 3-8) and show the inherent strength of the action with its brace N securely holding the breechblock E. In his description of his rifle, Bullard called the breechblock a vertically swinging breech-block. Whatever term is given to this breechblock design, it was very strong; probably the strongest repeater action then in existence. It was not superseded in strength until the advent of the Model '86 Winchester with its double-lug locking system. Author Ken Waters states that the Bullard rifle has a medium-strong action for purposes of cartridge reloading.[3] This author agrees with Mr. Waters' assessment of the Bullard action. The rack-and-pinion design of the action not only made the Bullard the smoothest and quietest of all lever actions, but also one of the best at seating and ejecting cartridges. This ability was the result of the mechanical advantage given the under lever by the pinion, which moved the breechbolt to and fro in a very positive and direct manner. Added to this was the fact that the breechbolt moved twice as fast as the pinion, which gave the rifleman the advantage of a portable arbor press in the field. This action could seat and eject slightly oversized cartridges with ease because of its rack-and-pinion action, which gave the guard lever its tremendous leverage. Nowhere in the action does one part moving against another cause any undue wear, abrasion or friction. All these factors were eliminated by pinning the moving parts together so that they moved flawlessly in relation to the movement of the pinion. The pinning of all moving parts eliminated

J. H. BULLARD.

MAGAZINE FIRE ARM.

No. 245,700. Patented Aug. 16, 1881.

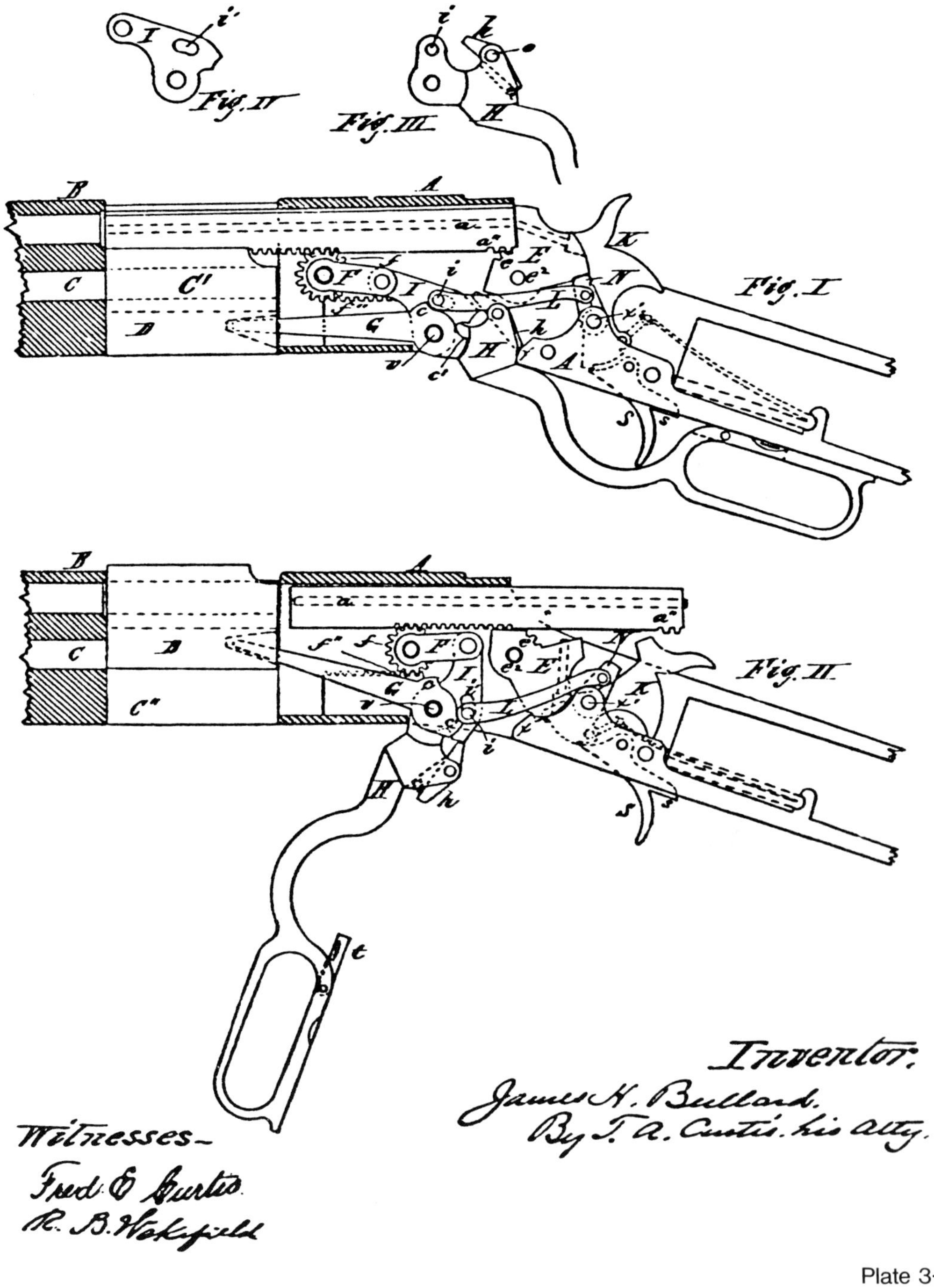

Plate 3-1.

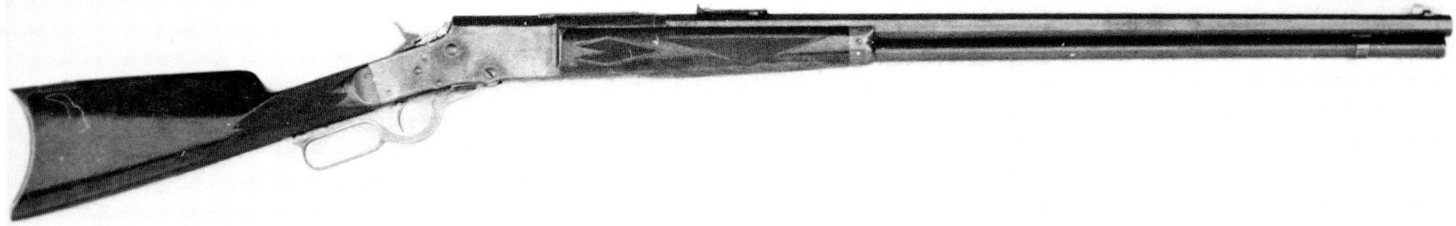

Plate 3-2. Deluxe with checking, crescent buttplate, full octagonal barrel. The style of checking shown appears to be the most common coverage and design used by the Bullard craftsmen on their repeaters. It has a sporting rear sight. *Courtesy Harold R. Jonah*

A.

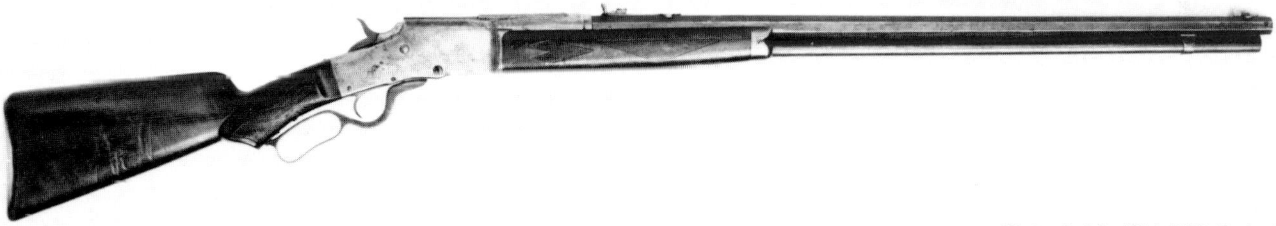

Plate 3-3A. (SN 585 Cal. .40)
Courtesy E.R. "Ted" Cotton

B.

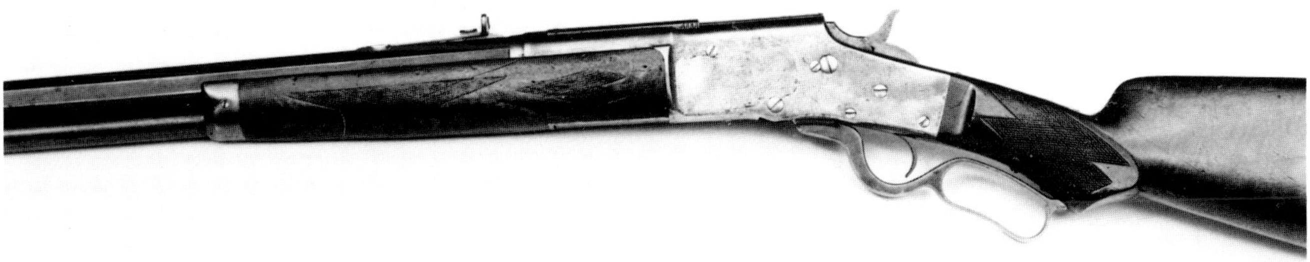

Plate 3-3B. (SN 585 cal. .40) Twenty-eight-inch octagonal barrel, no lever pivot pin lockscrew, 7-screw receiver. Style 'A', dustcover serrations, elk motif on buttplate, shotgun butt. Rocky Mountain front sight, Buckhorn rear sight. *Courtesy E.R. "Ted" Cotton*

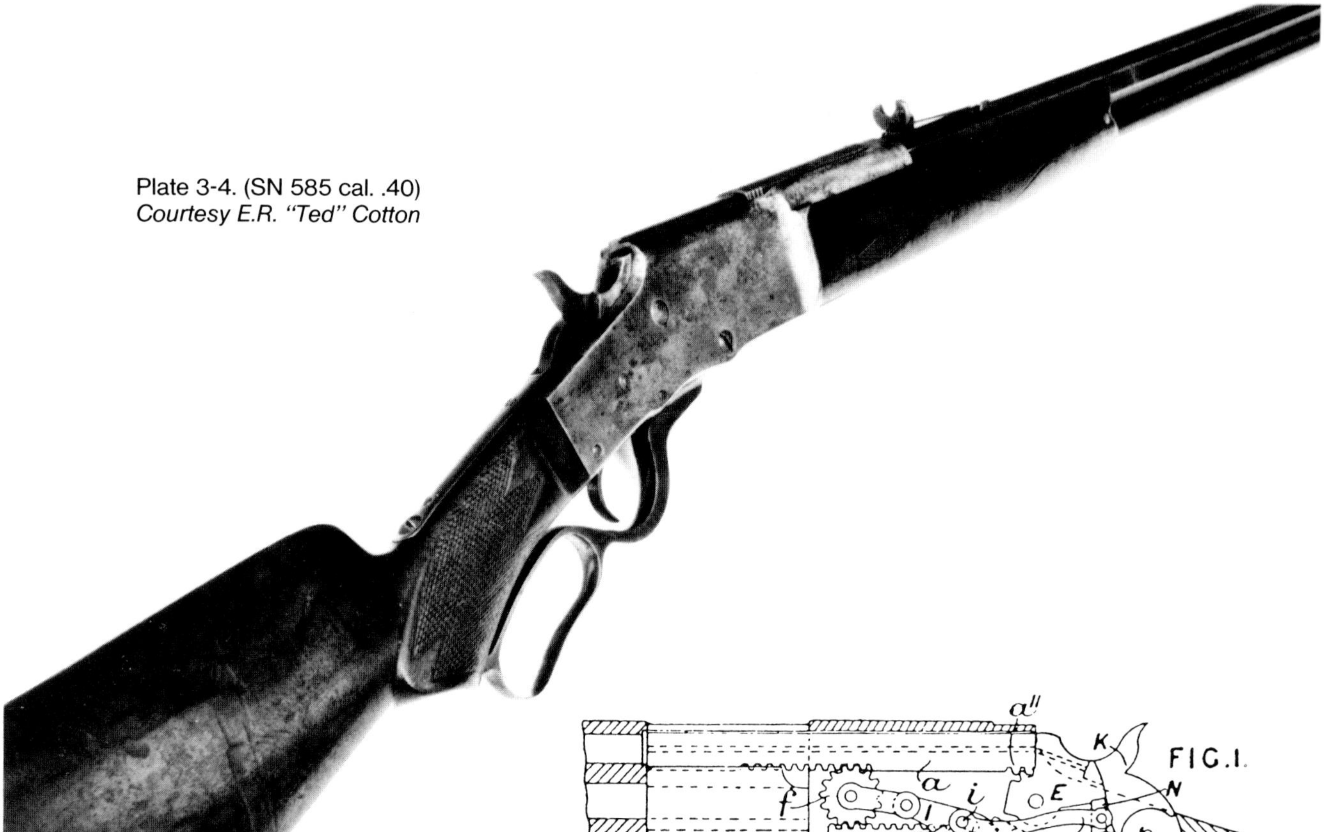

Plate 3-4. (SN 585 cal. .40)
Courtesy E.R. "Ted" Cotton

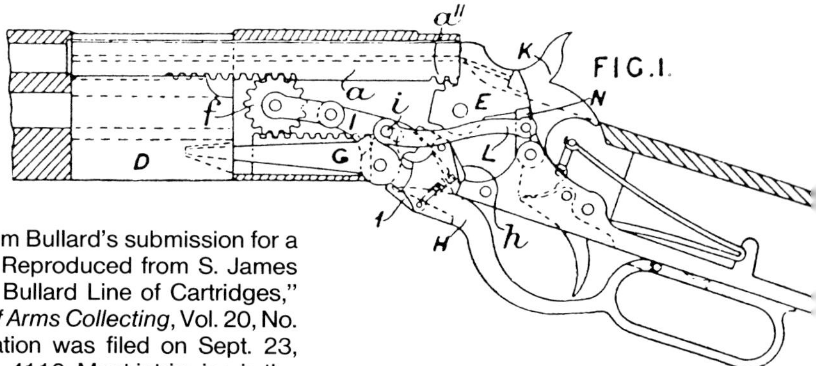

Plate 3-5. Illustration from Bullard's submission for a British patent in 1881. Reproduced from S. James Gooding's article "The Bullard Line of Cartridges," *The Canadian Journal of Arms Collecting*, Vol. 20, No. 4. The patent specification was filed on Sept. 23, 1881, and assigned No. 4110. Most intriguing is the use of a magazine cutoff device built into the front of the lever (1). The American patent drawings do not show this device. The mag cutoffs made by Bullard were buttons mounted on the left side plate cover (see Plates 7-21 & 7-22). It would be most interesting to see if any Bullard's ever turn up with this style of magazine cutoff.

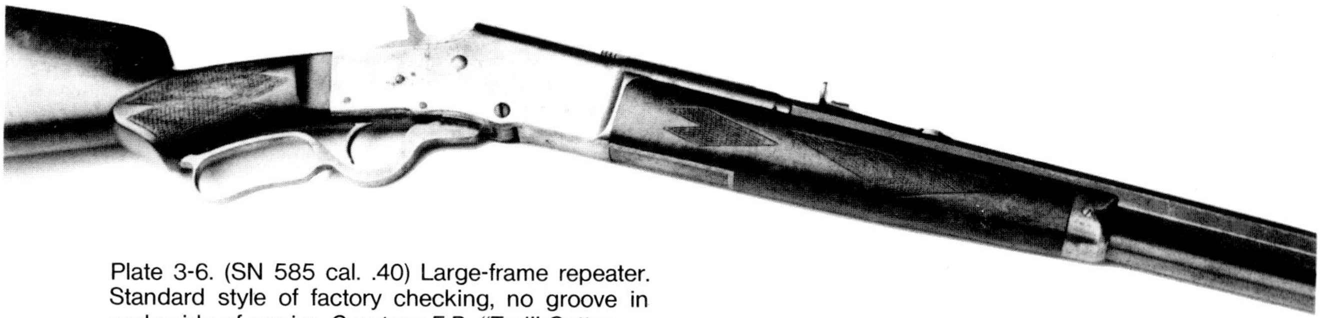

Plate 3-6. (SN 585 cal. .40) Large-frame repeater. Standard style of factory checking, no groove in underside of carrier. *Courtesy E.R. "Ted" Cotton*

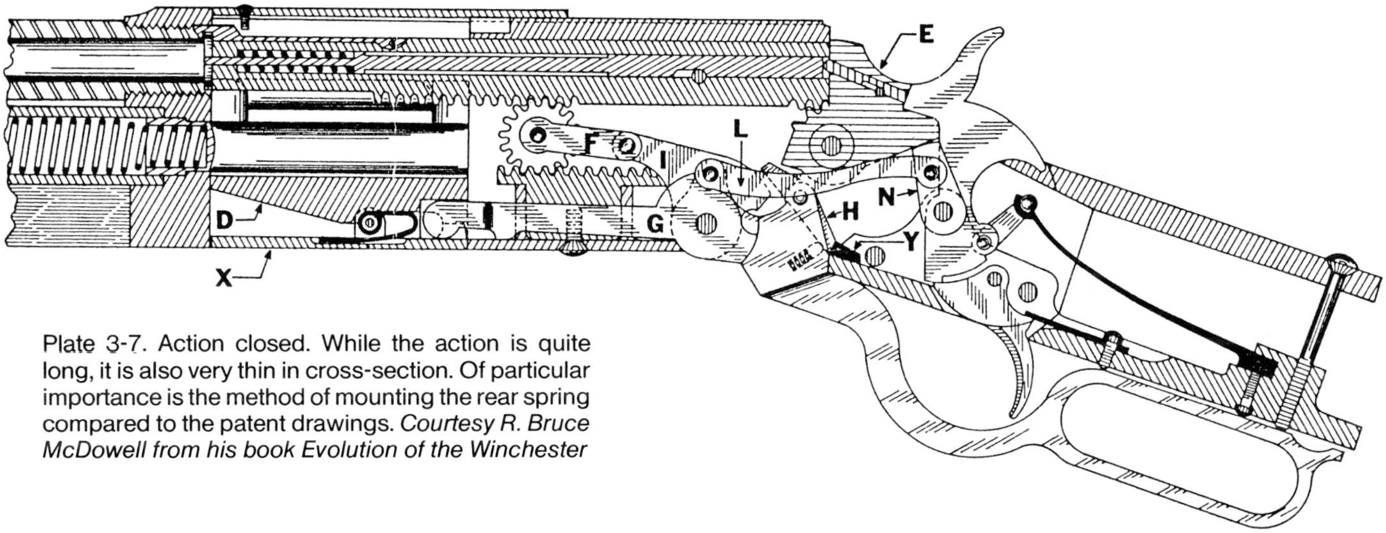

Plate 3-7. Action closed. While the action is quite long, it is also very thin in cross-section. Of particular importance is the method of mounting the rear spring compared to the patent drawings. *Courtesy R. Bruce McDowell from his book Evolution of the Winchester*

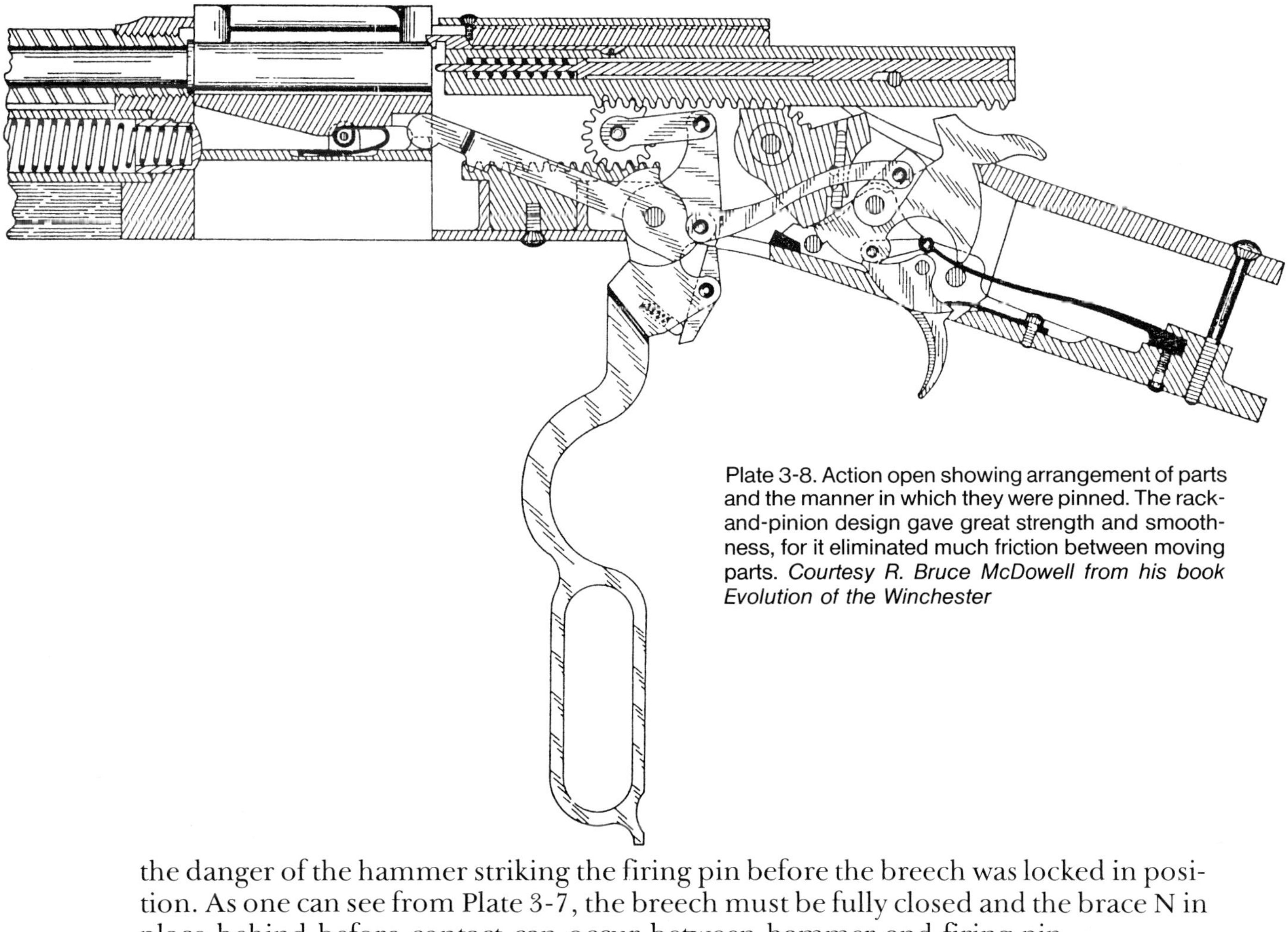

Plate 3-8. Action open showing arrangement of parts and the manner in which they were pinned. The rack-and-pinion design gave great strength and smoothness, for it eliminated much friction between moving parts. *Courtesy R. Bruce McDowell from his book Evolution of the Winchester*

the danger of the hammer striking the firing pin before the breech was locked in position. As one can see from Plate 3-7, the breech must be fully closed and the brace N in place behind before contact can occur between hammer and firing pin.

In this author's opinion the Bullard action was one of the finest made. It incorporated the very best in design from both engineering and safety standpoints. It was made from the finest materials then available and to precise tolerances. If you have the pleasure of owning a Bullard, then you have already experienced what this author is trying to convey.

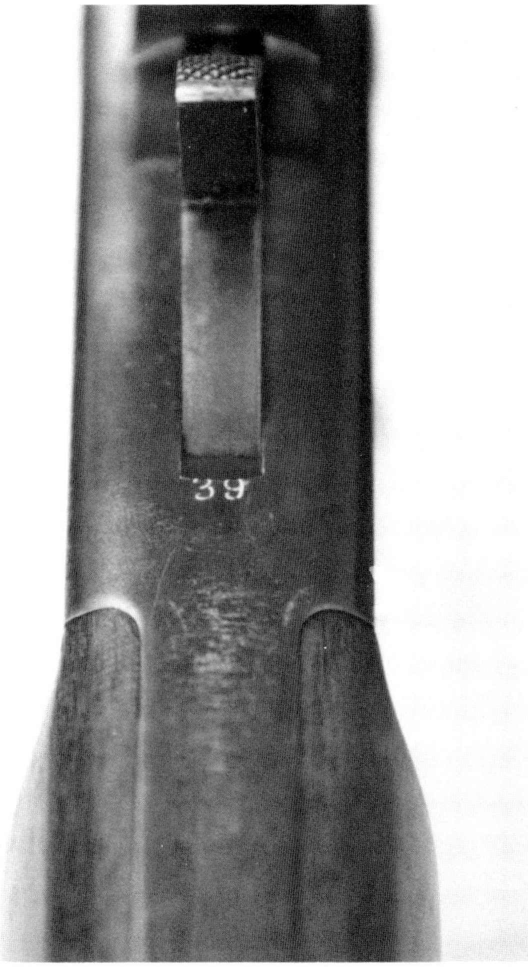

Plate 3-9. (SN 39 cal .45) Large-frame repeater showing serial number location. Standard position on all Bullard arms seen. *Courtesy O.L. Peacock, Jr.*

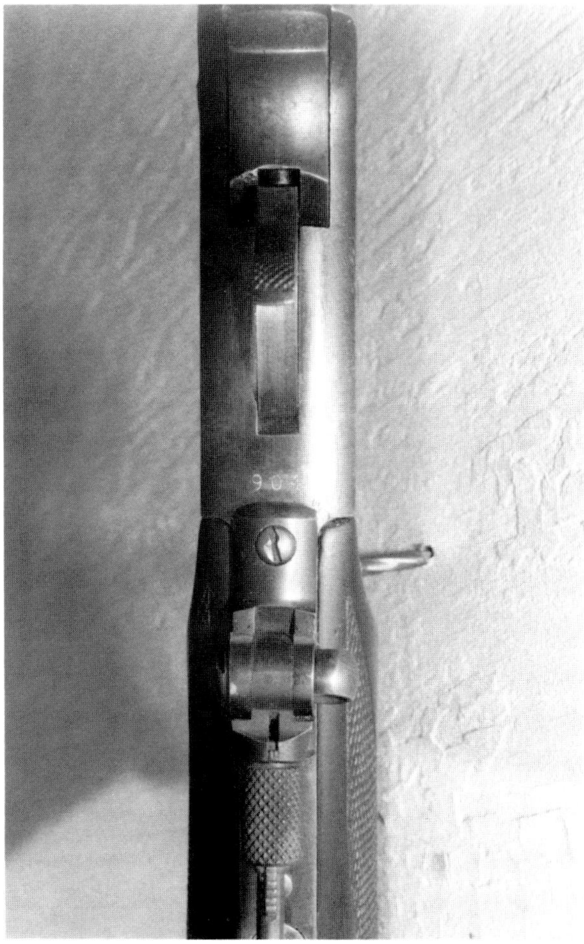

Plate 3-10. Showing location of serial number behind hammer, SN 904. All serial numbers the author has seen have been very lightly stamped, as in this photograph. Early Lyman tang sight. *Courtesy Harold R. Jonah*

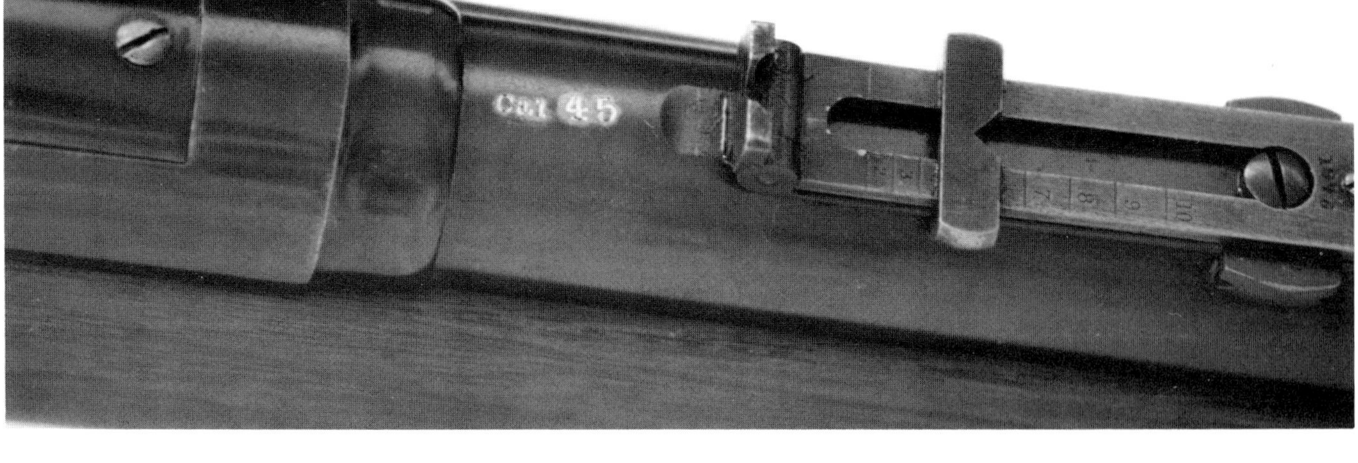

Plate 3-11. (SN 39 cal .45) Early style and location of caliber markings corresponding to 'A' position in tables at back. Standard Winchester 1876 carbine elevator sight. *Courtesy O.L. Peacock, Jr.*

SERIAL NUMBER NOMENCLATURE

All Bullard rifles were serial-numbered using blocks or groups of numbers assigned to specific models. The large-frame models were assigned the serial number blocks of 1 to 1500 and from 2000 to 3500. There are some discrepancies in this second block of numbers (see Tables). Since the assigning of a block of numbers to a particular model does not in itself prove they were produced, the author has spent many years tracking down serial numbers in order to prove or disprove the production totals. No repeating rifle with a serial number higher than 2684 has appeared and all models, with few exceptions, have fallen within their assigned blocks of numbers. All known Bullard serial numbers appear on the upper tang directly behind the hammer in rather small lettering (Plates 3-9, 3-10). The chart at the back of this chapter will illustrate fully all of the documented Bullard large-frames of which the author is aware. The chart well shows that no gaps exist in the 1 to 1500 number blocks, but that from 2000 to 2500 only one repeater has been documented. From 2684, no more repeaters of either size are to be seen. This could lead to the well-founded conclusion that no repeaters were produced beyond that range of serial numbers.

Serial numbers on the large-frames also appear on the breechblock, inside surface of lower tang, concave side of breechblock brace, underside of breechbolt and on the front of the buttstock where it mates with the rear face of the receiver, to list but a few locations.

CALIBER MARKINGS, LOCATIONS AND CALIBER CHOICES

Repeater caliber markings and locations were at best haphazard and lacking detail. The repeaters marked 1 through 85 that the author has seen have either had no visible caliber markings or were stamped on barrel. The location of the caliber markings varied widely from serial number 1 through to about 500, when the location became somewhat standardized. The first markings were usually on the barrel just forward of the breech flare, running in line with the bore (Plate 3-11). Others were on the barrel ahead of the rear sight. Even after serial number 500 there are still discrepancies. Serial number 874 has its caliber marked both on the barrel in front of the rear sight and on the breech ahead of the dustcover. Still some others are marked on the beveled area of the breech where it narrows to meet the barrel. The position on the breech immediately in front of the dustcover, though, was to become the predominant location for the caliber stamping (Plate 3-13).

The earliest repeaters in the range of serial numbers 1 through 85 are either not marked as to caliber, but accept .45-70 government or .40 cal cartridges, or are marked cal .40 or cal 45. This seems to agree with the offerings from what may be the earliest known Bullard catalog, in which the standard cartridges of the day, all manufactured by Winchester, are advertised. They were .45-75, .45-60, .40-60, U.S. government .45-70, and an express version the .50-95.

As time went on, Bullard offered a complete range of cartridges then in popular use, as well as a proprietary range of cartridges of his own design for the large-frame repeaters. These will be described extensively in the chapter on cartridges.

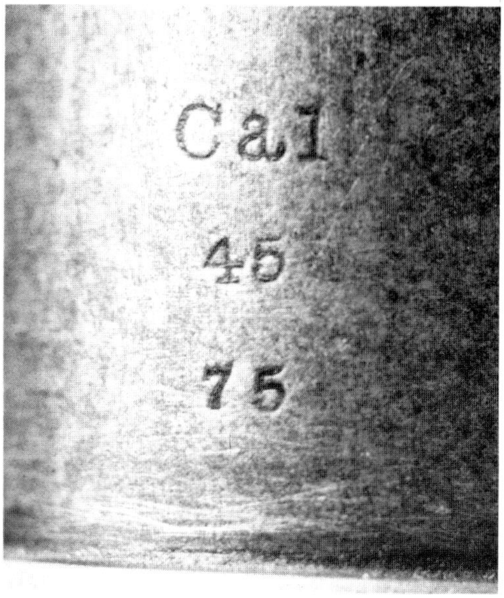

Plate 3-12. Caliber markings on repeater SN 203 on barrel just ahead of breech and behind sight. An 'A'-position caliber stamping but stamped at right angles to bore, not in line, as on SN 39 (Plate 3-11). See tables for frequency of 'A' position. *Courtesy Richard A. Failes/photo by Ted Jez*

Plate 3-13. (SN 585 cal .40) This plate clearly shows the caliber script style and location, corresponding to the 'C' location referred to in the tables. The location shown appears to be the one most often encountered. The ramp for the rear sight is mounted backwards. More comments about the location of the rear sights will be made later. *Courtesy E.R. "Ted" Cotton*

COMPANY ADDRESS TYPES

All documented repeaters are stamped with the company address and patent dates on the left side of the receiver just below the opening for the cartridge elevator. There are three documented and one undocumented styles for the large-frame repeaters. The first is only used on rifles from serial number 1 through 85, but may go up to about serial number 100, when the style changes. Serial number 118 has the word "company" in the address, instead of association. (This address is shown in Plate 3-14):

1) BULLARD REPEATING ARMS ASSOCIATION
SPRINGFIELD, MASS. U.S.A. PAT. AUG. 16, 1881

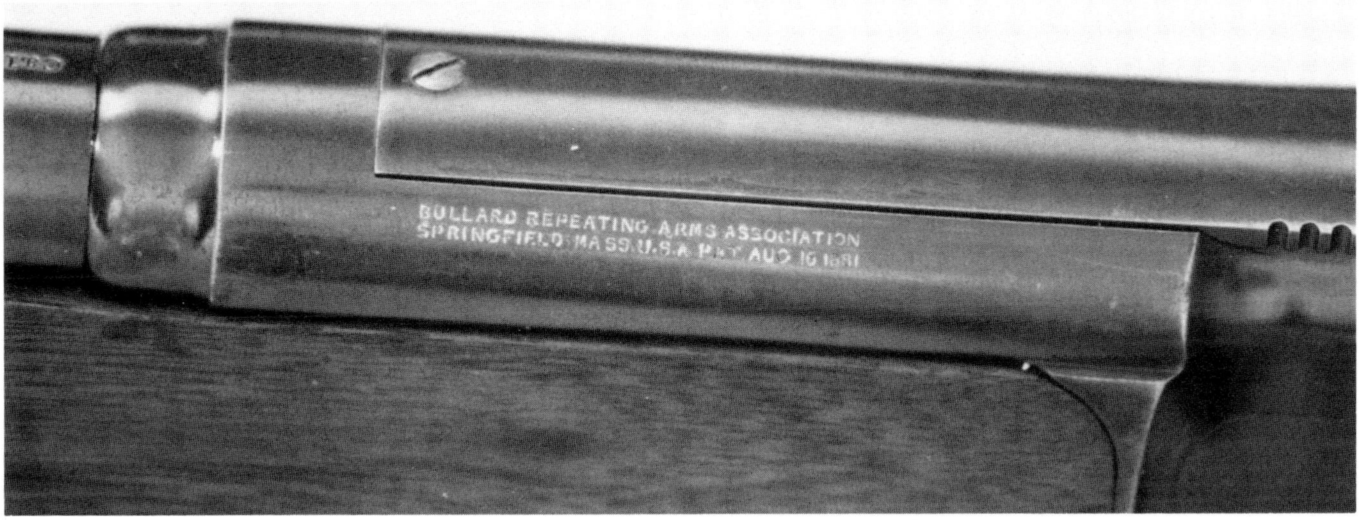

Plate 3-14. Approximately the first 100 large-frame repeaters were stamped with the number 1 style of address. This rifle has 'A'-style dustcover finger serrations. *Courtesy O.L. Peacock, Jr.*

The second and, at this point, most commonly seen address type is.

2) BULLARD REPEATING ARMS COMPANY
SPRINGFIELD, MASS. U.S.A. PAT. AUG. 16, 1881

Plate 3-15. (SN 203) The number 2 style of address. See tables for usage. *Courtesy Richard A. Failes/ photo by Ted Jez*

Plate 3-16. (cal .45) This photograph shows the number 2 style of breech address and the 'C' location for the caliber markings. As mentioned on Plate 3-13, one can see the impossibility of raising the rear sight to its fullest extent due to the stepped ramp coming into contact with the breech flare. This is a common occurrence on the large-frames. The ramp in Plate 3-13 was probably reversed in order to gain more elevation. Thanks to Ed Iwanchuk for pointing out this error or oversight on the factory's part. *Courtesy Verle W. Hunt.*

The few rifles made under the Association name could lead one to speculate that about 100 large-frames were made at a smaller factory before the Bullard Association changed its name to Company on October 11, 1883.

A third style of address may also be used, but less frequently than number 2. Because no photographic evidence has been forthcoming, the author has been forced to label this style of address as unknown in the charts at the rear of the book. It is included here only to allow the readers to see what it may look like and in the hope that someone will contact the author with photographs of this style:

BULLARD REPEATING ARMS COMPANY
SPRINGFIELD, MASS. U.S. PAT. AUG. 16, 1881

Notice the only difference is that the A in U.S.A. may be deleted. Until photographic proof is supplied, this address will have to remain on the unknown list.

As can be seen in many of the photographs, the upper portion of the A was almost illegible and may account for the occurrence of some of these address styles. Unless a magnifying glass is used, it is sometimes very difficult to discern whether the A is there or not. Photographic evidence is the proof needed to clarify this problem.

A third documented style of address also appears on later large-frames (Plate 3-17). It has two lines spaced further apart, possibly to accommodate the extractor access hole. In this way, less of the address is obliterated by the hole. Also note that the letters U S A are dropped from the second line. At this stage of research it seems to be the least seen address style.

3) BULLARD REPEATING ARMS COMPANY
SPRINGFIELD, MASS. PAT AUG 16, 1881

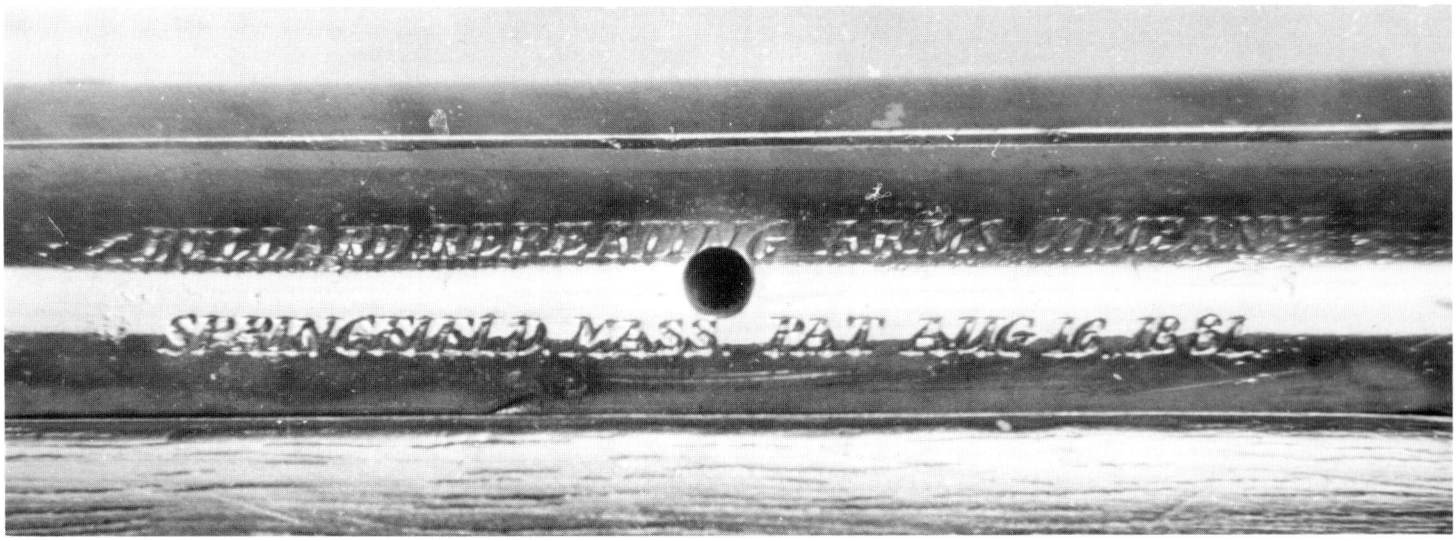

Plate 3-17. (SN 2590 cal .45) Number 3 style of address. This style has a larger spacing between the lines and larger, deeper etched lettering. By spacing the lines, less of the address was obliterated by the extractor access hole. The USA has been completely dropped. *Courtesy A.J. Newberry*

The tables at the back of the book will show the usage of each type of address where that information is known.

VARIATIONS IN DESIGN

The large-frame repeaters have an inordinate number of variations in their design. This has led to speculation on the author's part that two or possibly three different sets of drop-forging dies were used over the years by the Billings & Spencer Company on behalf of the Bullard concern. The tremendous number of differences found in the large-frame repeaters are much less noticeable on the other lines of Bullard rifles.

These differences are found in (a) dustcover finger grooves or serrations, (b) guard levers, spring-pawls and pinion links, (c) number of receiver screws used, (d) trigger styles, (e) cartridge elevators, (f) magazine tube and tenant location, (g) use of extractor removal access holes, (h) breech design, (i) Model 1886 designations, (j) .50-95 Winchester receiver design, (k) receiver contour lines, (l) buttstock flares. No doubt there are other, as yet undiscovered differences in the large-frames.

(A) Dustcover Finger Serrations or Grooves: There are two distinct designs used, and the author has taken the liberty of naming each in order that they may be identified more easily. The names chosen were types "A" and "B". As can be seen from the tables, "A" style is more prevalent at the beginning or lowest serial numbers, but "B" is never totally unused. By later serial numbers, "A" is still more prevalent in use, but neither style totally predominates on the large-frames. Plates 3-18 and 3-19 clearly show style "A", which has four protruding ridges and four grooves with a protruding ridge on the rear extremity of the cover. Style "B" has only three protruding ridges and four grooves, with the last groove closest to the rear extremity of the cover. Plate 3-20 shows this well. A possible explanation for these two styles may be that two distinct forging dies were used or different craftsmen at different times may have removed more material from the end of the dustcover than was necessary, thereby making a "B" cover from an "A". However, this is only speculation. Of the documented repeaters, both large- and small-frame, 34 use the "A" style and 13 use the "B" style of dustcover serrations.

(B) Guard Lever, Spring-Pawls and Pinion Links: This variation became apparent when Mr. Paul Simkin and the author were physically comparing rifles in their collections. There are at least two different lever, spring-pawl and pinion link designs used in the large-frames. Plate 3-21 shows the external lever differences. The external differences are purely one of shape rather than function. Study closely the curvature of the lower surface of the lever between the trigger guard and the finger loop. The upper rifle or background rifle has a more abrupt inverted V shape to the space between the trigger guard and the finger loop. The rifle in the foreground has a much gentler curve to that same space. The pawl differences can be seen in Plates 3-22 through 3-27. The spring-pawl is the device used on the Bullard rifle to hold the lever in the closed or firing position without the use of an external lever latch as used on the early models of Winchesters. The two designs noted thus far consist of a T-shaped pawl supported by two thin, cross-sectioned arms or supports. The second pawl is a flat design of thin cross-section supported by rather thick arms. This design has less surface area on the face that contacts the pinion arm, whereas the T-shaped pawl has a much greater surface area in which to contact the pinion arm. The T-shaped pawl should give a lower rate of wear due to its larger, less stressed area of contact. This difference was discovered too late for the author to do a great deal of research into it, but if the photographs of the underside of the receiver are clear, the author can tell which type of spring-pawl was used. In order to understand this, compare the lever pivot area in Plate 3-38 with Plate 3-27. You will note that when the thin pawl is used (3-26) the center section of the pivot area of the lever has a thin section in the middle, but the

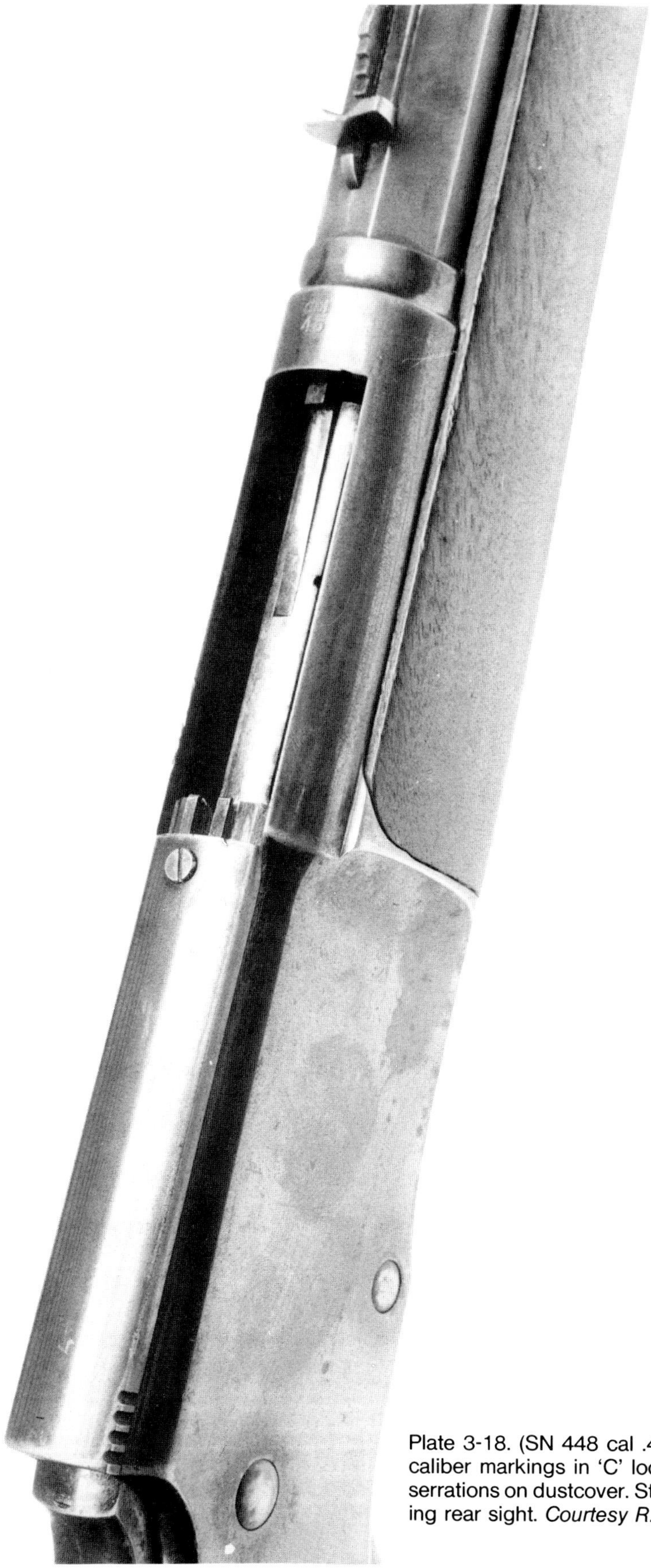

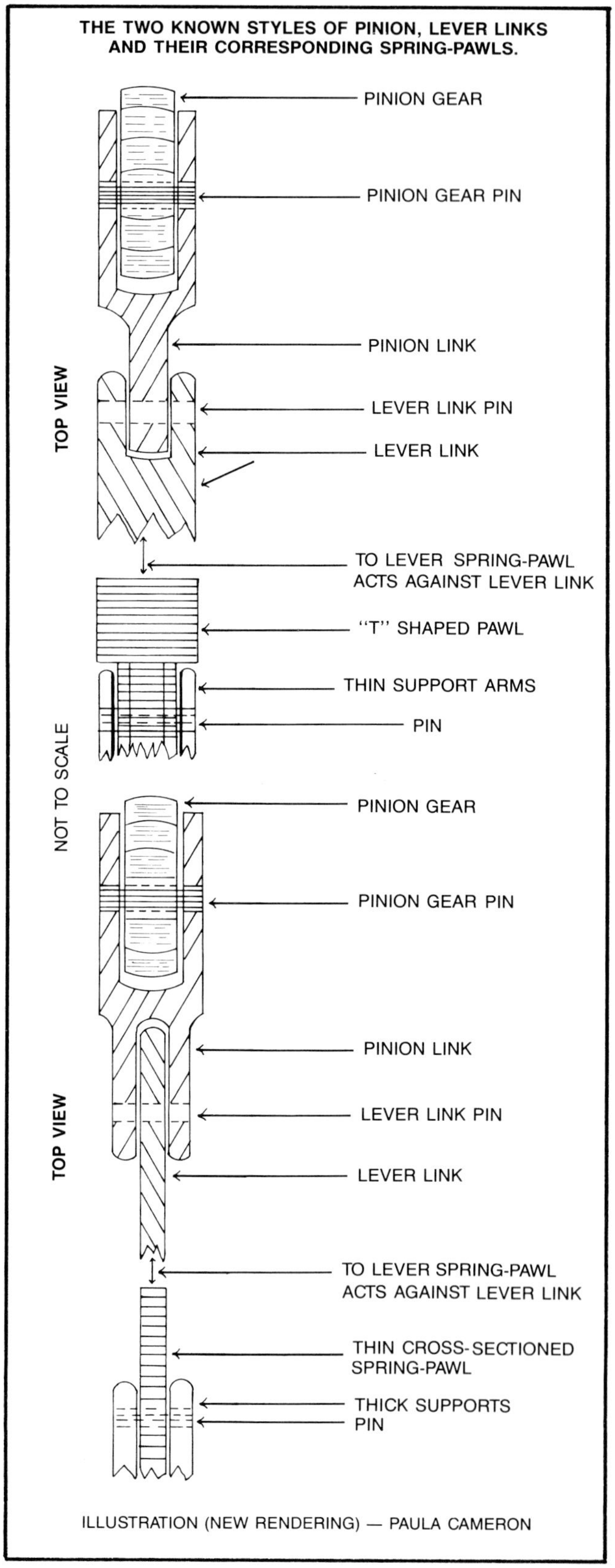

Plate 3-18. (SN 448 cal .45) Large-frame. Detail of caliber markings in 'C' location and 'A'-style finger serrations on dustcover. Standard Winchester sporting rear sight. *Courtesy R. Bruce McDowell*

T-shaped pawl lever has a thick-sectioned middle (Plate 3-38). Internally the T-shaped pawl works against a thick or heavy proportioned lever link, with the pinion link having a single arm. The thin pawl acts against a thin-sectioned lever link that is pinned to a pinion link with two arms (see diagrams).

Be aware that the shape of the curvature of the lever discussed earlier is *not* an indicator of which pawl was used, for both the gentle curve and inverted V levers have been seen with both styles of pawl.

(C) **Number of Receiver Screws Used:** This particular variation has so far only shown up on the large-frame repeaters and involves the number of slot-headed pins and screws used when viewing the receiver from the *left* side only. The receivers are either 7- or 8-headed. The extra or eighth slot head is used to lock the lever pivot pin in place. On the 8-head receiver the lever pivot pin is just that, a pin with no threads. The small slot-head screw just forward of it is threaded to lock the pivot pin in place.

On the 7-screw or pin-receiver, a different method is used to retain the lever pivot pin in place. This involves using a larger diameter, hollow, internally threaded pin (this portion allows the lever to swing or pivot on it) and a thin externally threaded male screw which mates with its counterpart to lock solidly. Both types of receivers can be discerned from either side of the receiver, for the 7-screw receiver has a slot head on both sides of the lever pivot pin, while the 8-screw receiver is slotted on the left side only. The tables indicate that generally the 8-screw receiver was more prevalent at first, with the 7-screw receiver becoming dominant in about the 500 serial number range. Note that serial number 36 is a 7-screw receiver. Plates 3-19, 3-20 and 3-34 show both 7- and 8-screw receivers. Plates 3-73A & B show the difference very clearly.

D) **Trigger Styles:** The large-frame repeaters come with two styles of trigger, one with a slight curvature and generally found on straight-stocked rifles. The second style has much more curvature and is generally found on rifles stocked with pistol grips. This is not a hard and fast rule, though, and again points up the nature of Bullard manufacturing — much handfitting and little true mass-production or steady work force (Plates 3-35, 3-73).

E) **Cartridge Elevators:** The variation now discussed concerns whether or not the bottom *external* spring-loaded plate in the cartridge carrier has a groove milled in it (Plates 3-36, 3-38). This groove, when it appears, is not even of consistent dimensions in many instances. The first theory about the grooves' use was that it was meant to facilitate loading of the bottleneck cartridges by relieving the flat surface of the plate and allowing easier access to the magazine for the bulkier cartridge. This theory was not long in being dispelled. The grooved and non-grooved carrier was used rather indiscriminately throughout the caliber ranges (see Tables).

F) **Magazine Tubes and Tenants:** When full-length magazines are used they are usually secured to the barrel by means of a tenant encircling the tube and dovetailed into a slot cut into the underside of the barrel. The full-length tubes run the spectrum from being almost flush with the muzzle to being some ½" or more shorter than the muzzle (Plates 3-39 through 3-42). Most full-length tubes used the same system of spring retainer plate at the muzzle end of the magazine tube, that being a small slotted flat plate with a protrusion on one edge that fitted a slot cut in the underside of the barrel and a screw to lock the plate on the underside of the tube (Plate 3-41B).

The most standard distance from the muzzle to first edge of the bevel slot in the barrel for retaining the magazine band was from 3⅛ to 3¼ inches. However, rifle serial number 461 has a full-length magazine with a tenant only 2 inches from the muzzle of the barrel (Plates 3-40, 3-41A and B). Of course magazine tubes came in all lengths to accommodate the buyer. The tables show which rifles were equipped with what length of tube.

G) Use of Extractor Removal Access Holes: When used, these holes were generally drilled through the left side of the breech area, through the address and patent dates. Some small numbers of rifles were drilled from the right side only. These were probably done by gunsmiths, not by the factory. Some few rifles have this same hole drilled right through both sides of the receiver (Plates 3-17, 3-43, 3-44). Plate 3-47 shows how this access holes was used. The purpose of this hole was to allow a broken extractor to be removed without disassembling the whole action, which would be the case if no hole was drilled. It is not until the 2000 range of serial numbers that the hole seems to be a standard feature. How many holes were drilled by the factory and how many by outside gunsmiths is unknown. As indicated by the tables, little pattern of use is observed until the 2000 range of serial numbers. In one of the few instances that Bullard rifles are mentioned in contemporary firearms' literature, a broken extractor is the complaint. Surely the factory was aware of this problem, which accounts for the almost universal appearance of this hole in the 2000 serial number range.

H) Breech Design: In another development too late to document fully, it has been noted that serial number 521 has a slightly different design at the breech where the extractor rests when the breechbolt is closed. Other rifles may also have this larger cutaway, but further study will be necessary.

I) Model 1886 Designations: The use of this designation by the Bullard Company is limited to the large-frame and small-frame repeaters. It is stamped on the right-hand side of the receiver opposite the barrel address in letters ⅛-inch high. This should give an indication as to production figures before and after 1886 (if not to production, then to sales), since it is still the author's belief that the drop-forgings were made in two or three lots and stamped with their serial numbers at the time of assembly. They were assembled as demand required, with no accumulation of stocks. In other words, they were sold out of serial number sequence. For example, serial numbers 461 and 770 are not marked Model 1886, but serial number 521 is! The tables give a complete rundown of the Model 1886 usage. Most repeaters in the 2000 range are marked Model 1886, but were they made after 1886 or only sold after 1886? Obviously the advent of the Model 1886 Winchester had something to do with Bullard stamping their rifles in such a manner, since if they were to be labelled as any model, it should have been Model 1881. It's also a possibility that repeaters sent back to the factory for repairs were also stamped Model 1886 after that year as a matter of factory policy. This may account for some of the haphazard use of the designation. But in the absence of any factory records, it's only conjecture (Plates 3-48, 4- 5B).

J) .50-95 Winchester Receiver Design: Thanks to Mr. Stuart W. McCallum for noticing the size difference between the receiver on the .50-95 Win. and the other large-frame calibers. While disassembling his rifle, he noticed that the carrier, carrier well or mortise and the breechbolt were all ¼ inch *shorter* than either of his .40 or .50-115 caliber Bullards. Research has not found any other size variation on the large-frames thus far.

The .50-95 Win. load, being of a stubby nature (2.26 inches long), was indeed shorter than the other common calibers by ¼ inch. The .40-75 Bul. was 2.54 inches long, the .40-90, Bul. was 2.55 inches long and the .50-115 Bul. was 2.56 inches long. Why the factory went to such lengths to accommodate a cartridge loading that never gained much popularity is a mystery, but it does show the extraordinary way in which the Bullard Company gave their customers the widest choice possible in their rifles (Plates 3-55A & B).

K) Receiver Contour Lines: Two or three receivers in the 2000 range have been found to have a very distinct line on the receiver running from the front of the receiver flat to the breechblock. On all the other known receivers, the receiver flows in a gentle

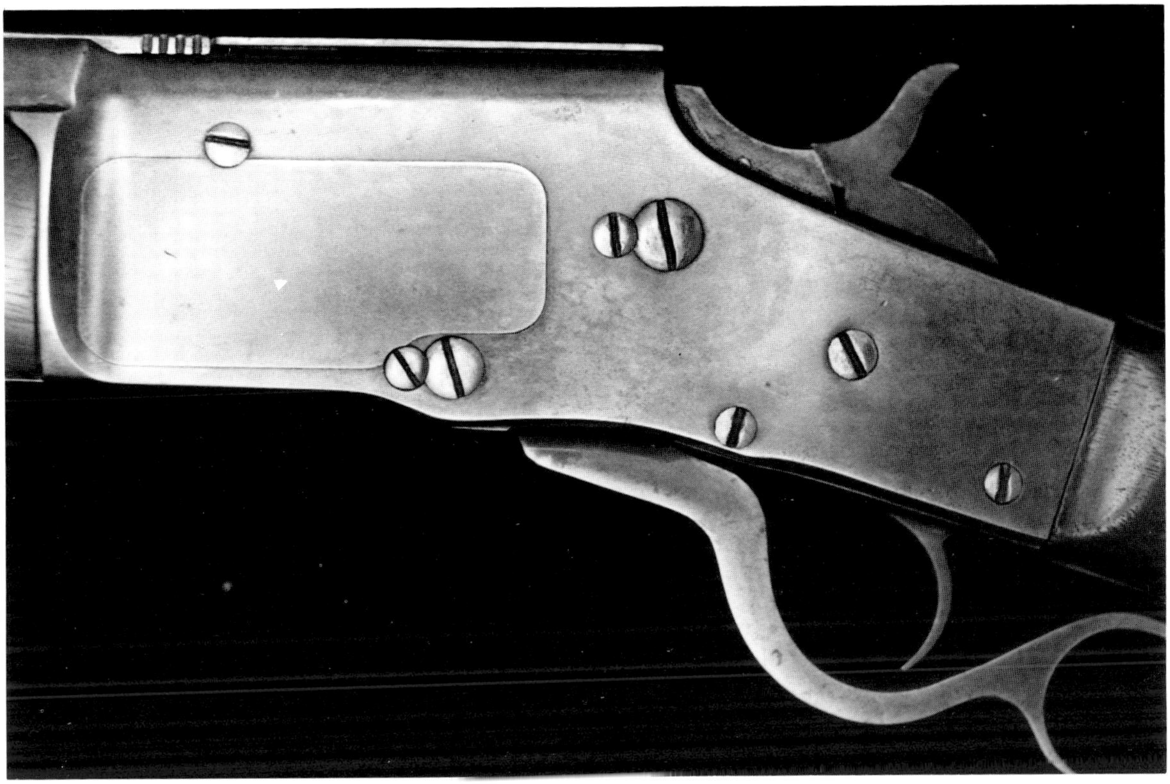

Plate 3-19. Large-frame .45 clearly showing finger notches or serrations of style 'A' on the dustcover with protrusion closest to hammer end of dustcover. Eight-screw receiver type. *Courtesy Verle W. Hunt*

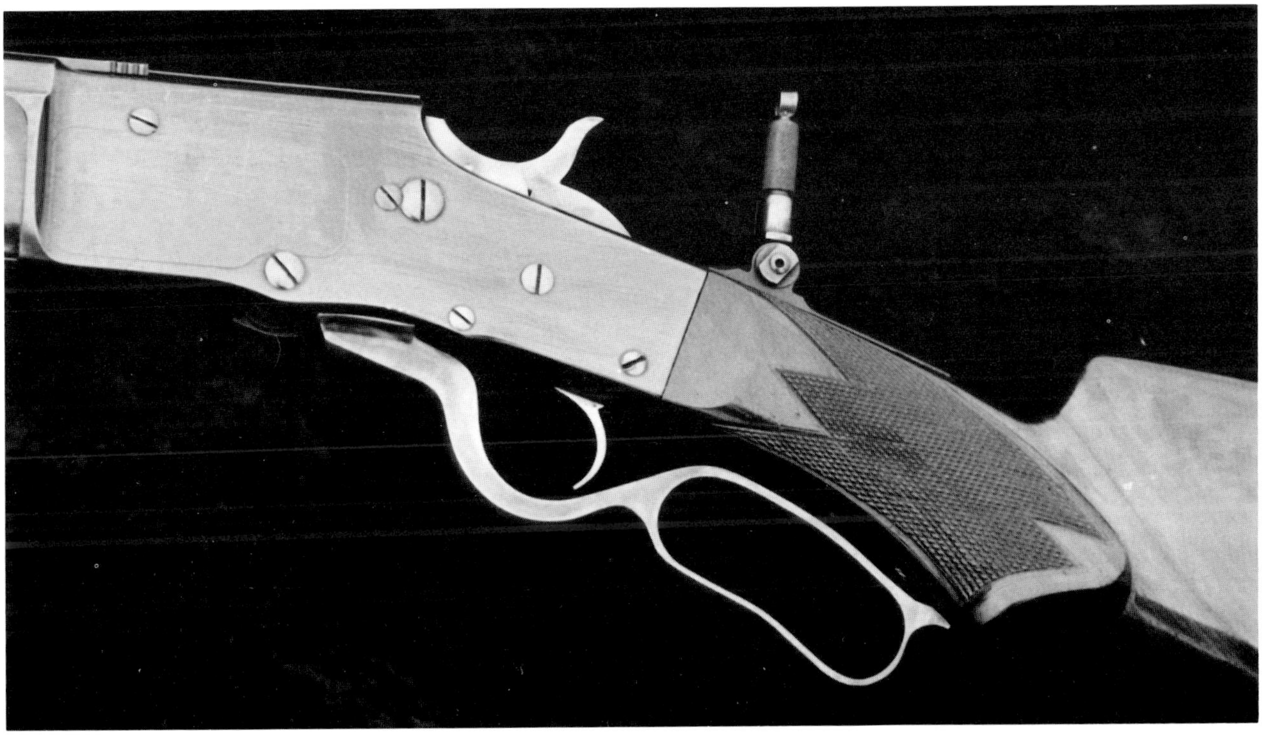

Plate 3-20. (SN 1142 cal .40) Large-frame. Clearly shows 'B'-style finger serrations of dustcover. Early Lyman tang sight. *Courtesy Joel M. Tastad*

slope from the flat to the top of the receiver, to the dustcover track. The receivers in Plates 3-29, 3-49, 3-59, 3-77A and 3-77B have a very distinct line of demarcation between the flat and the top part of the receiver. This may have been the third batch of receivers forged with a different die, as mentioned previously, *or* as pointed out by Ed Iwanchuk, the three rifles in question may have been refinished leaving a distinct line between the flat and the top of the receiver.

L) **Buttstock Flares:** This refers to the flare of the wood stock where the butt meets the receiver. Two styles have been noted. One style has a rather shallow half-moon, while the other has a broader crescent much more pronounced (Plates 3-49, 3-50, 3-51). Again this may indicate manufacture at two different periods of time. The difference is slight, but it is noticeable and therefore has been mentioned.

COMMON CHARACTERISTICS OF THE LARGE-FRAMES AND GENERAL INFORMATION

Rifling	.40-70	1 twist in 20 inches
	.40-75	1 twist in 20 inches
	.40-90	1 twist in 18 inches
	.50-115	1 twist in 72 inches

All widths on the rear barrel sight slots we 24/64 inch and the width of the front sight slots was also 24/64 inch. A.C. Gould had this to say about tang sights in his book *Modern American Rifles*: "The Bullard Arms Co. endeavored to accommodate those using its rifles by boring a series of screw-holes in the tangs of its rifles that they might take any of the several varieties of rear tang sights."

FOOTNOTES

1 Note in Plate 3-1 that there is a small pivoted trip t on the lever H which may be extended to strike projection s on the trigger, thereby allowing the rifle to be discharged with the closing of the lever only. The author has seen no rifle containing this device.

2 S. James Gooding, "The Bullard Line of Cartridges," *The Canadian Journal of Arms Collecting*, Vol. 20, No. 4, pp. 130-133.

3 Ken Waters, "Loading the Old Ones," *Gun Digest Treasury* (Chicago: Follett Publishing Company, 1972), pp. 268-277.

Plate 3-22. Background: (SN 461) Note groove in underside of magazine carrier. Foreground: (SN 521) No groove in underside of magazine carrier. Examine closely the lever latch pawl mechanism. The pawl itself is thinner in cross-section than the pawl of SN 461, but the pawl pin supports are much thicker in comparison to SN 461 supports. See Plate 3-23 for more detail. The pawl on SN 461 also forms a 'T' shape when looked at from the butt. So far research has indicated that the mechanism on SN 521 is the rarer of the two. *Courtesy Paul Simkin and G. Scott Jamieson* ⟶

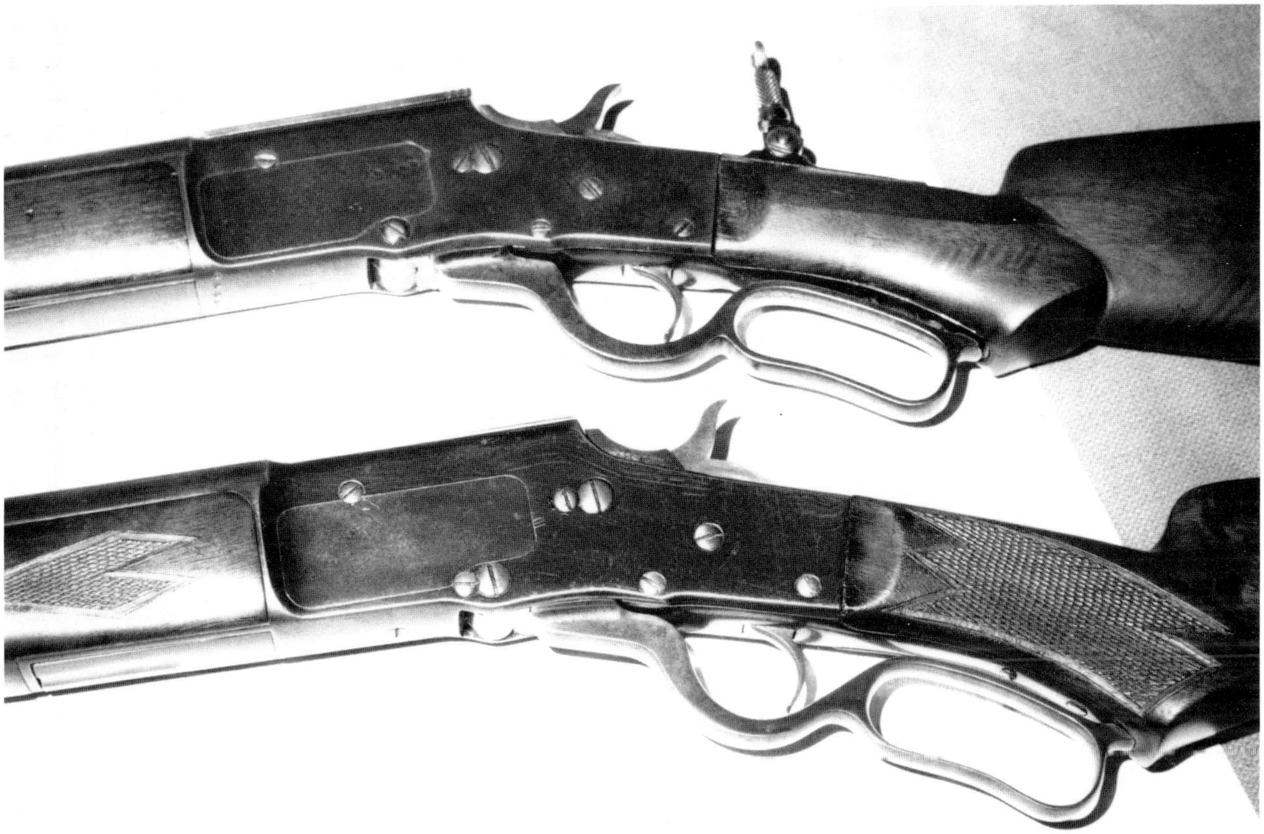

Plate 3-21. Background: (SN 521 cal .40) Large-frame marked 'model 1886'. 'B'-style dustcover serrations. Note that the curve on the underside of the lever between the triggerguard and the lever loop forms a distinct V. No lockscrew on lever pivot pin. Foreground: (SN 461 cal. .40-90 Bul caliber) Large-frame not marked 'model 1886.' No dustcover (missing). The curve on the underside of the lever between the triggerguard and the lever loop forms a gentle arch. Lockscrew fitted to lever pivot pin. Eight-screw receiver. *Courtesy Paul Simkin and G. Scott Jamieson*

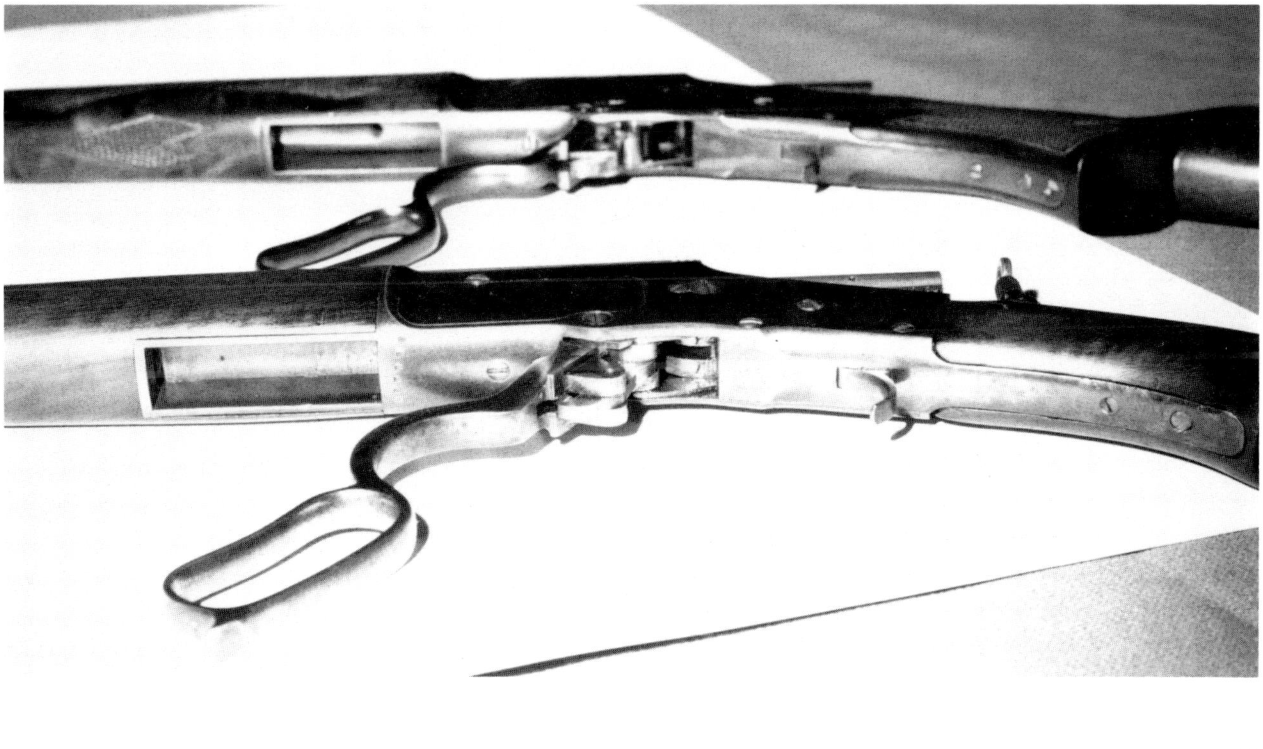

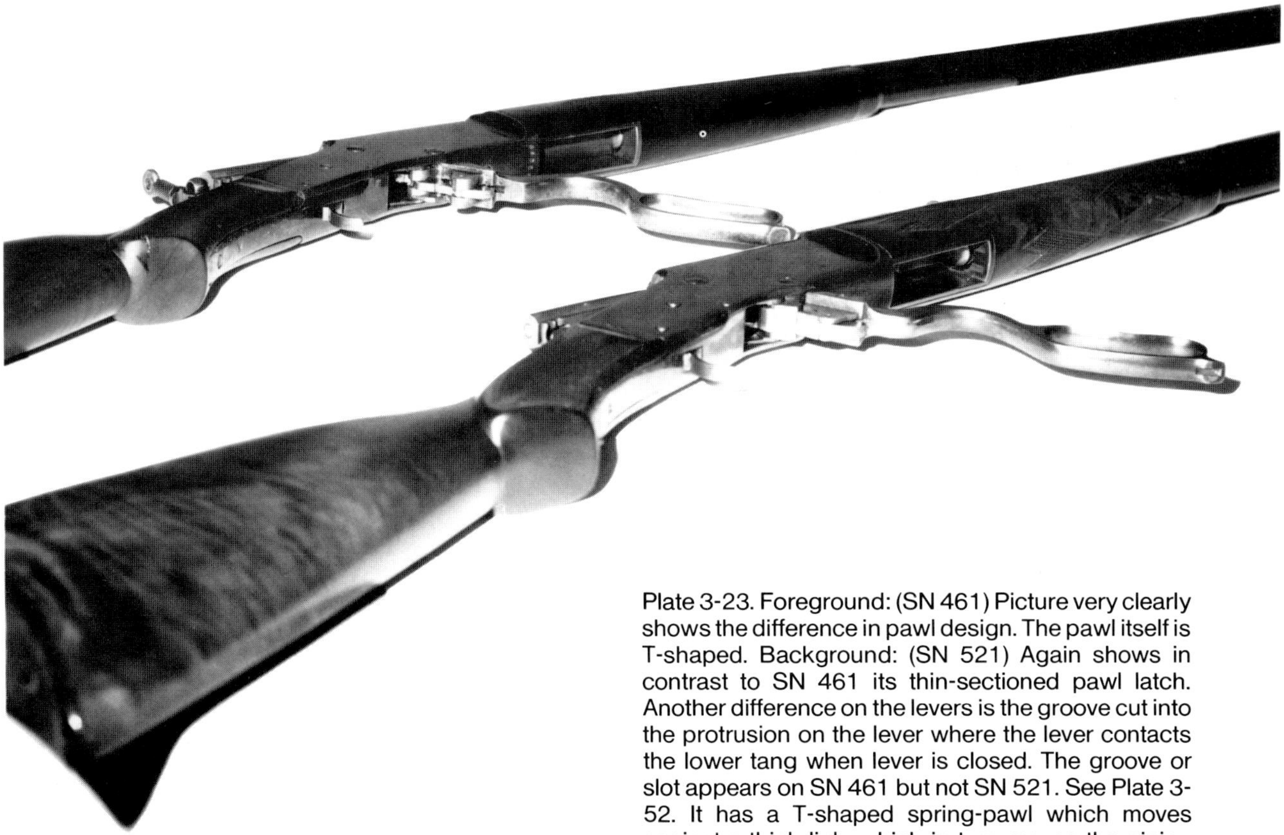

Plate 3-23. Foreground: (SN 461) Picture very clearly shows the difference in pawl design. The pawl itself is T-shaped. Background: (SN 521) Again shows in contrast to SN 461 its thin-sectioned pawl latch. Another difference on the levers is the groove cut into the protrusion on the lever where the lever contacts the lower tang when lever is closed. The groove or slot appears on SN 461 but not SN 521. See Plate 3-52. It has a T-shaped spring-pawl which moves against a thick link, which in turn moves the pinion and its support link. Notice that the link holding the pinion is fastened or pinned to the lever link (which the pawl manipulates) and that the lever link arms fit around the outside of the pinion link. The diagrams illustrate the difference. *Courtesy Paul Simkin and G. Scott Jamieson*

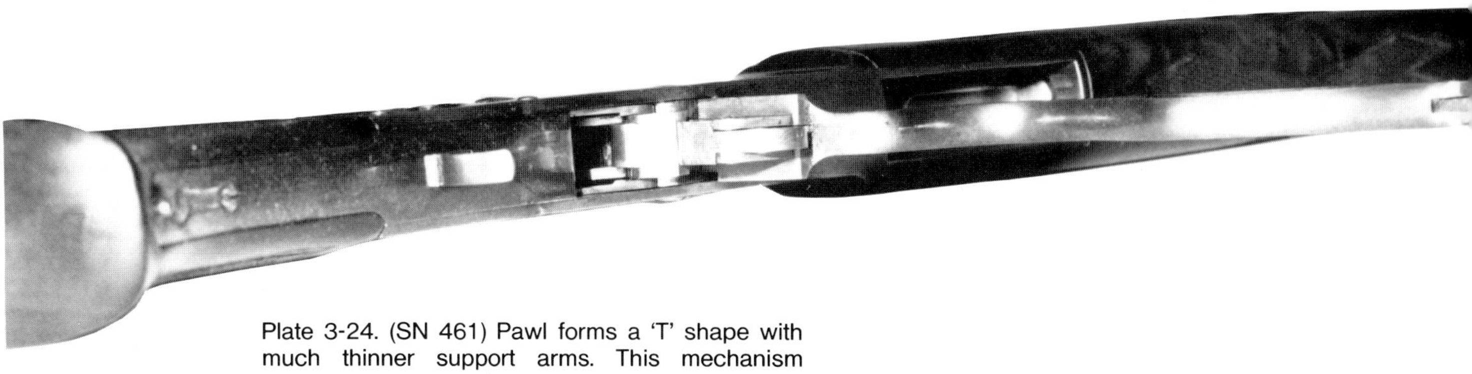

Plate 3-24. (SN 461) Pawl forms a 'T' shape with much thinner support arms. This mechanism appears much more frequently than the one on SN 521. Small slot cut on end of lever is very much in evidence. What purpose this slot served is unclear. As well, examine Plates 3-23 and 3-25 closely and you will see that the T-shaped or thick pawl mates to a thick-sectioned arm, which in turn is pivoted and moves the pinion. The thin pawl pushes against a thin counterpart, therefore there must be at least two different styles of pinion support arms. (Link 'I' in patent drawings.) *Courtesy G. Scott Jamieson*

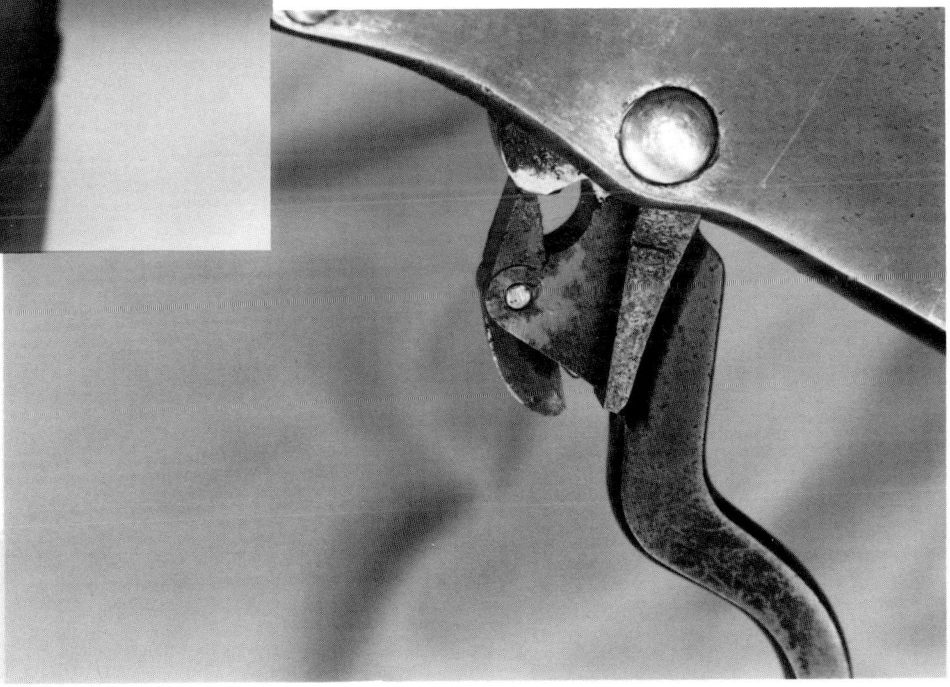

Plate 3-25. (SN 825 cal .45) Large-frame repeater. Both shots show detail of T-shaped pawl and lever. *Courtesy Kenneth Howell*

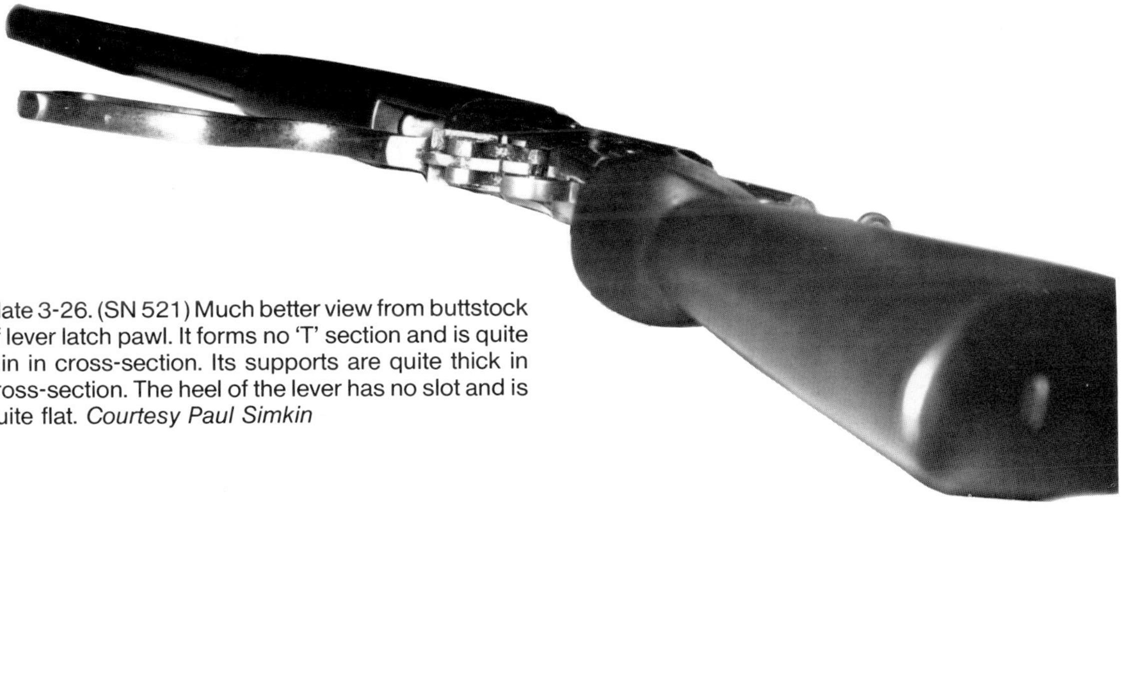

Plate 3-26. (SN 521) Much better view from buttstock of lever latch pawl. It forms no 'T' section and is quite thin in cross-section. Its supports are quite thick in cross-section. The heel of the lever has no slot and is quite flat. *Courtesy Paul Simkin*

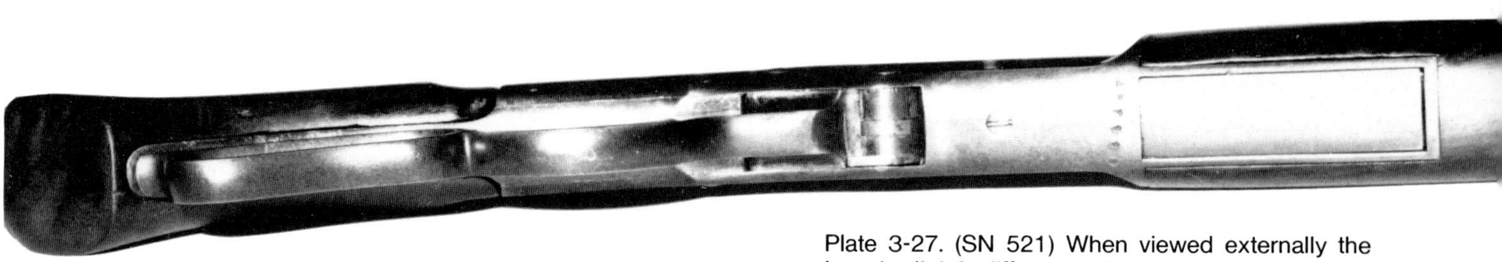

Plate 3-27. (SN 521) When viewed externally the lever is slightly different, at its pivot point, from standard, with a thin cross-section arm to manipulate the pinion arm internally. this large-frame also carries a six-digit serial number stamped on the underside of the receiver behind the cartridge carrier. It is not a factory stamping. *Courtesy Paul Simkin*

Plate 3-28. (SN 448 cal .45) Large-frame repeater. Shows very clearly the more common T-shaped pawl and lever design. This photo shows how the upper edge of the spring-pawl pushes against the pinion link until the lower sliding surface meets with lower side of opening causing it to move away from the link. *Courtesy R. Bruce McDowell*

Plate 3-29. (SN 2590 cal .45-85) Large-frame repeater. The type of finger serrations shown on the dustcover of this rifle has led the author to speculate that rather than a type 'C', it is a type 'A' that was machined more than normal, leaving the fourth protrusion rather thin, almost nonexistent. Note case-hardening and lack of lockscrew at lever pivot pin. Seven-screw receiver. Close scrutiny of the photograph will reveal a distinct line running the length of the receiver just above the screw that retains the inspection cover. *Courtesy A.J. Newberry*

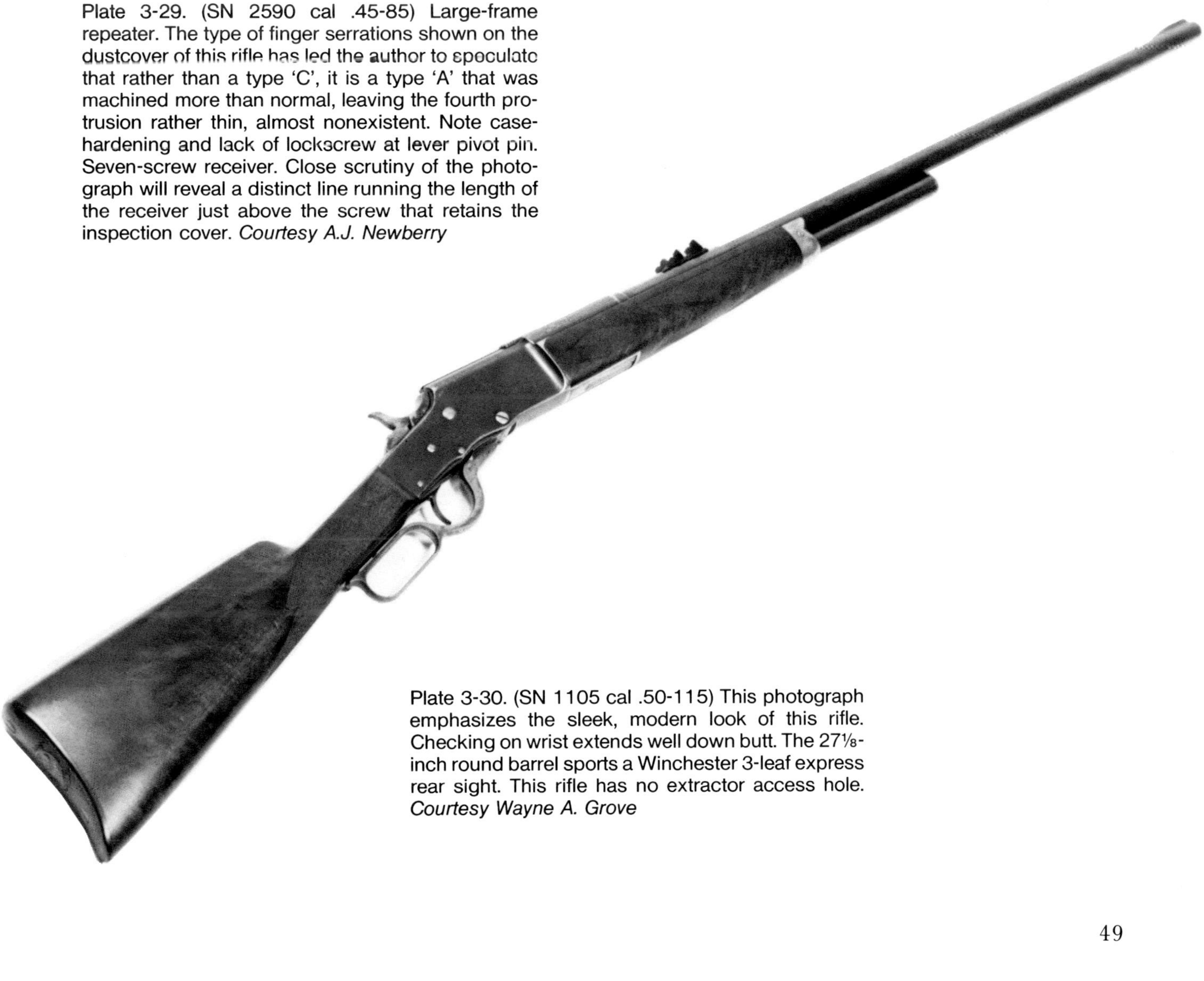

Plate 3-30. (SN 1105 cal .50-115) This photograph emphasizes the sleek, modern look of this rifle. Checking on wrist extends well down butt. The 27⅛-inch round barrel sports a Winchester 3-leaf express rear sight. This rifle has no extractor access hole. *Courtesy Wayne A. Grove*

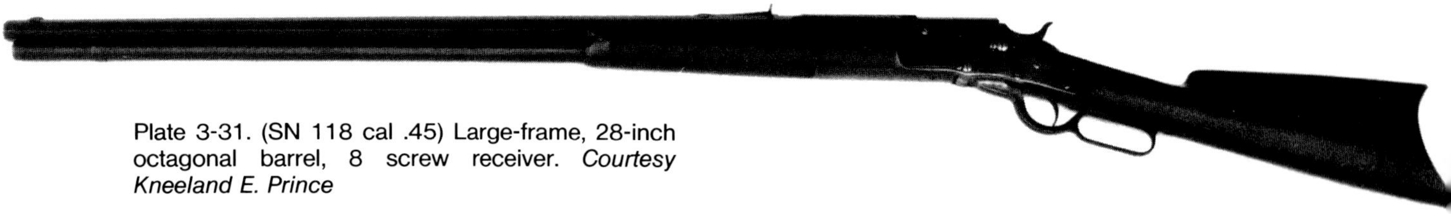

Plate 3-31. (SN 118 cal .45) Large-frame, 28-inch octagonal barrel, 8 screw receiver. *Courtesy Kneeland E. Prince*

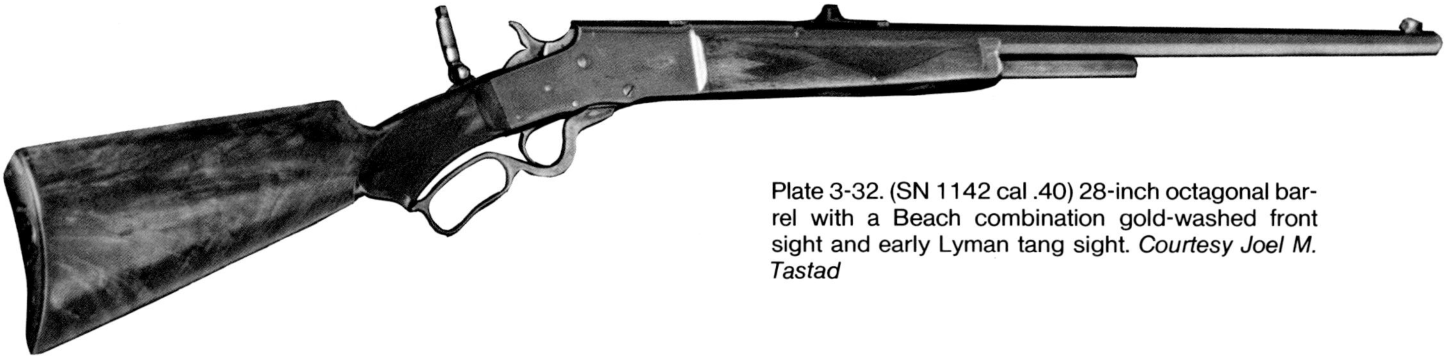

Plate 3-32. (SN 1142 cal .40) 28-inch octagonal barrel with a Beach combination gold-washed front sight and early Lyman tang sight. *Courtesy Joel M. Tastad*

Plate 3-33. Shows, from left to right, four styles of Bullard buttplates. 1. (SN 1649) Small-frame, shotgun butt hard rubber. 2. (SN 1532) Small-frame, Swiss-style. 3. (SN 582) Large-frame, crescent steel (rifle). 4. (SN 1804) Small-frame, military style (crescent). *Courtesy Wayne A. Grove*

Plate 3-34. (SN 118 cal .45) Eight-screw receiver. This rifle is the first documented one to use the word 'Company' in its address. *Courtesy Kneeland E. Prince*

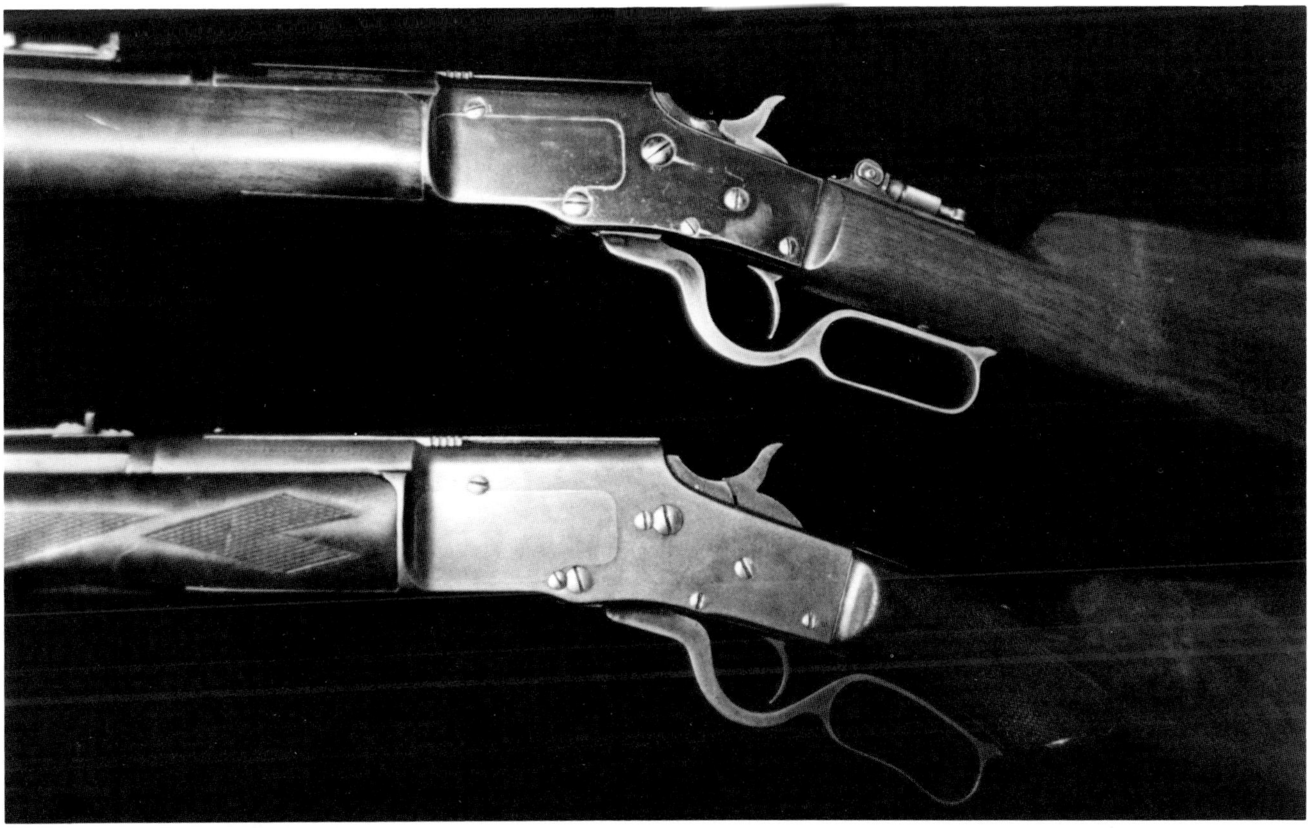

Plate 3-35. Trigger styles. This photograph also shows the difference in size between a small-frame (top) and large-frame (bottom) repeater. To the author's knowledge, the small-frame repeaters never used a small lockscrew to retain the breechblock pin, but the large-frame always did — at least all rifles seen to date have. Top rifle: Lyman tang sight for 'model 1873' Winchester. Rear barrel sight '1873' carbine elevator. Bottom rifle: standard Winchester sporting sight. *Courtesy Verle W. Hunt*

Plate 3-36. (SN 203 cal .45-75) Showing groove in external underside of cartridge carrier. *Courtesy Richard A. Failes*

Plate 3-37. (SN 203) Showing clearly the 'A' finger grip design of the dustcover and the 8-slot head pins on left side of receiver (extra screw at lever pivot pin). *Courtesy Richard A. Failes*

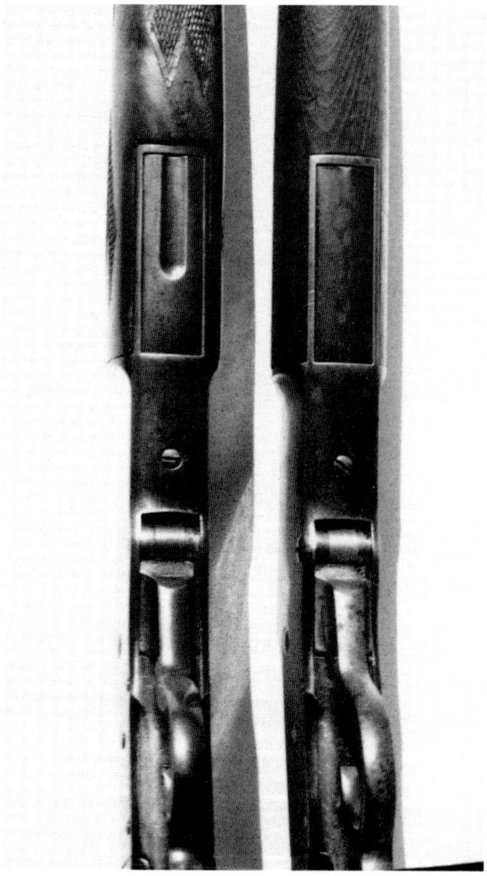

Plate 3-38. (Left SN 461 .40-90 Bullard. Right SN 770 cal .45) 1½-inch-long slot on SN 461. Compare to slot down in Plate 3-36, which is only 1⅜ inches long. The slot length on the repeater shown in the chapter header is 2 inches. In this plate, compare lever pivot point area to Plate 3-37 and see that the two rifles shown here have a much thicker center section than does SN 521 shown in 3-27. These two rifles both have T-shaped pawls. *Photos by Karen Callaghan*

Plate 3-39A. (SN 23) Standard magazine tube retainer mounted in the most common Bullard position, about 3⅛ inches from the muzzle.

Plate 3-39B. (SN 2331) Detail of ½ magazine tube mounting and forearm cap. *Courtesy Robert L. Bullard*

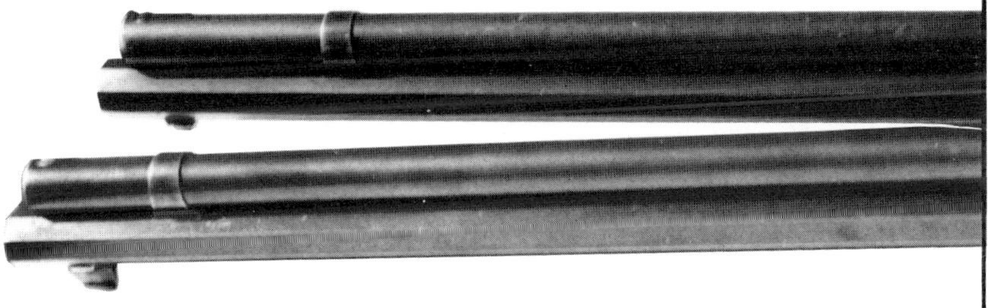

Plate 3-40. (Foreground SN 461) (Background SN 770) Tube tenants. *Photos by Karen Callaghan*

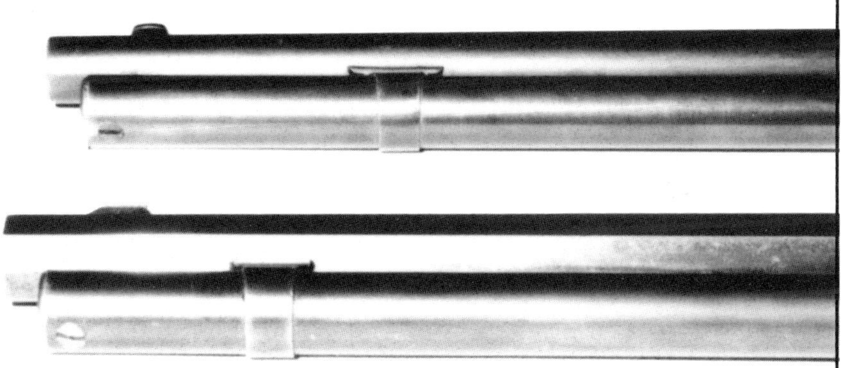

Plate 3-41A. (Foreground SN 461) Least common positioning of front magazine tube tenant or retainer, only 2 inches from muzzle. (Background SN 521) Mounted much further from muzzle; also wider slot than most common dovetail for magazine tube retainers or tenants. *Courtesy Paul Simkin*

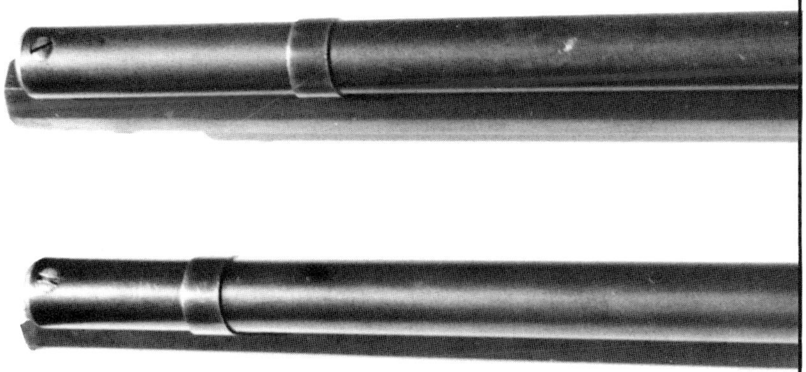

Plate 3-41B. Tenant differences between SN 461 in foreground and SN 770 in background. Serial number 770 has the more common position for Bullard tube tenants. *Photo by Karen Callaghan*

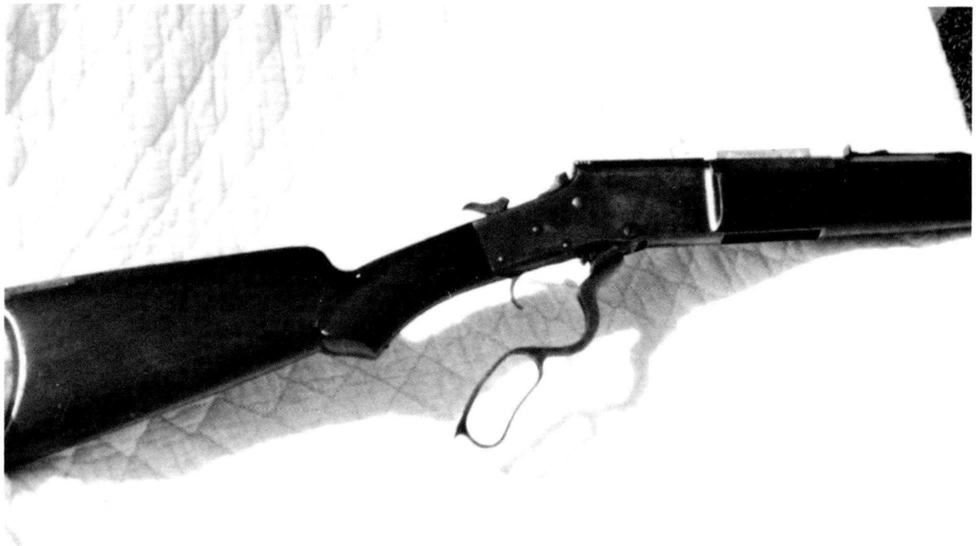

A.

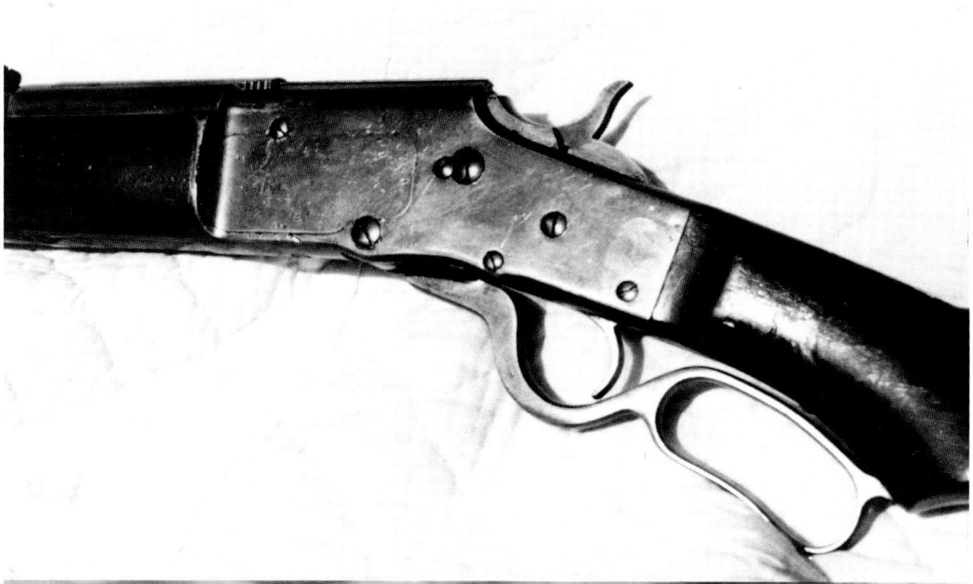

B.

C.

Plate 3-42. (SN 602 cal .45) Large-frame repeater with 20¼-inch octagonal barrel, 'B' style finger serrations. Of particular interest is the method of pinning magazine tube to barrel on carbines. (Standard tube retainer is not used. Sheard front sight.) *Courtesy Chuck Peterson*

Plate 3-43. (SN 974 cal .40) Large-frame. Frame still shows much case-hardening. Pistol grip but no checking on pistol grip or forearm. Has 7-screw receiver with 'B'-type finger serrations. The extractor removal hole is drilled just below address rather than right through it, as is often the case. The trigger has a more than normal curvature to it, even for a pistol grip model. *Courtesy Doyle N. Shaw*

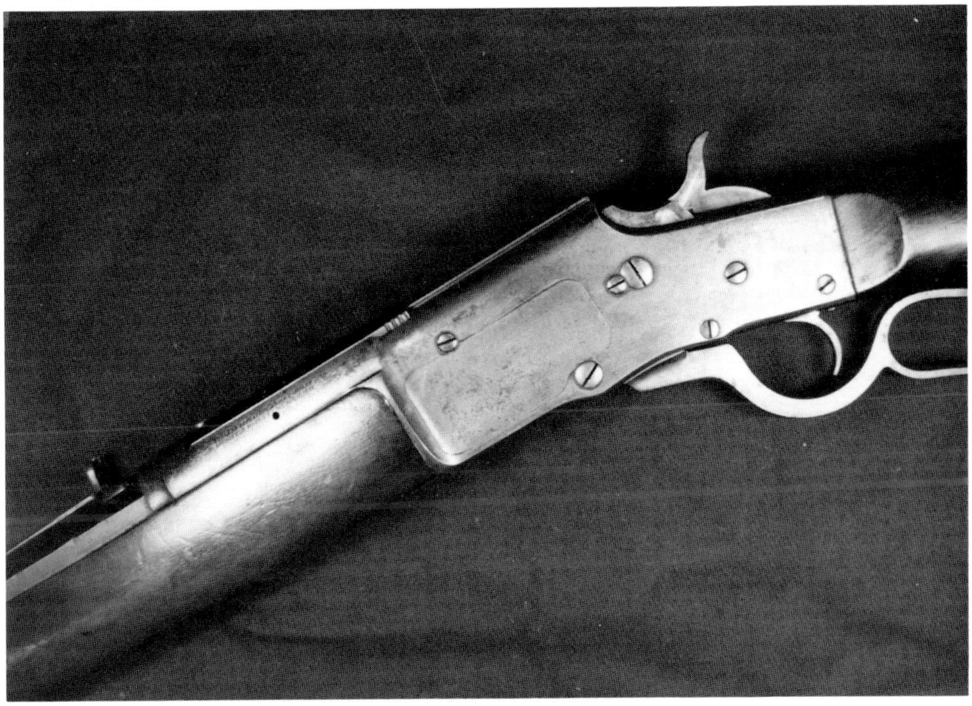

Plate 3-44. (SN 874 cal .40) Shows access hole for easy removal of broken extractor pin. *Courtesy R.P. Howard*

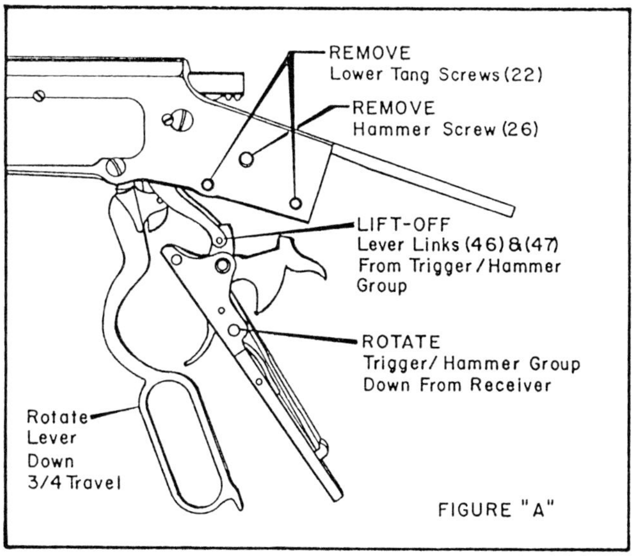

FIGURE "A"

Plate 3-45.

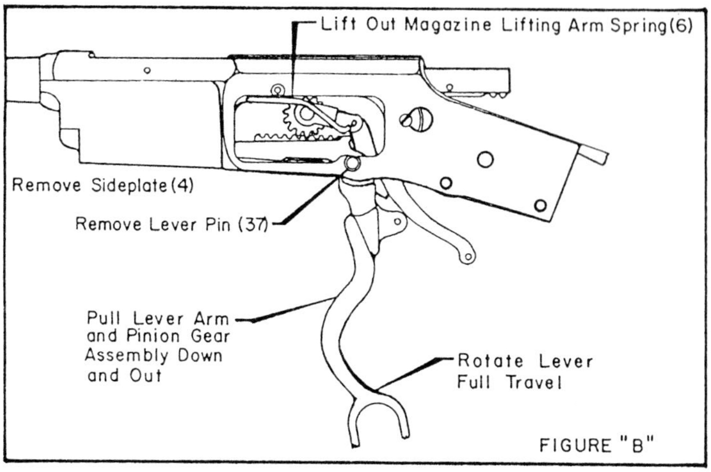

FIGURE "B"

Plate 3-46.

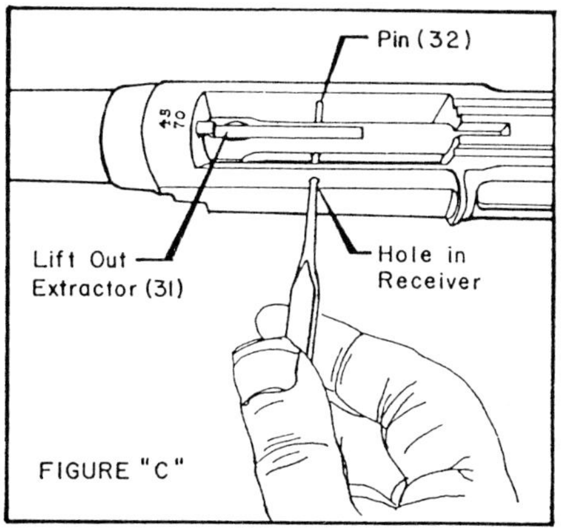

FIGURE "C"

Plate-3-47.

Drawings by Edward A. Tolosky.

Disassembly Instructions

1. Check the rifle to insure it is unloaded. Remove the buttstock by removing the rear tang screw (10). Gently tap buttstock from upper and lower receiver tangs. Cock the hammer (16) and pull the lever (36) down approximately 3/4 of its total travel. Remove the hammer screw (26) and lower tang screws (22) (see **Fig. A**). Rotate the assembled trigger (11) and hammer down and out of the receiver. Lift off the ends of the lever arms (46 and 47) from the lever link pin (21) on the lever link (20).

2. Remove the sideplate screw (5), gently tap walls of receiver (1) and sideplate (4) with a plastic hammer to loosen the sideplate, and gently pry the sideplate from the receiver. Lift out the magazine lifting spring (6), NOTE: Check orientation of this spring with respect to the cartridge carrier lifting arm (48) before removing. Remove the lever pin (37) and lever pin screw (38). Pull the lever and pinion gear assembly (45) down and out from frame of receiver (see **Fig. B**). Disconnect the lever from pinion gear assembly. Lift out the cartridge carrier lifting arm. Push the bolt (30) forward and remove the breechblock (27). Remove the dustcover (63) by removing its retaining screw (65). Align the hole in the bolt with the access hole in the receiver (see **Fig. C**). Insert a drift punch into the hole and push the extractor pin (32) out of the bolt. Remove the pin punch and lift out the extractor (31). NOTE: Extractor pin will not completely clear the bolt body, but will move sufficiently to allow removal of the extractor. Once the extractor is removed, push the extractor pin back.into the bolt, then pull the bolt from the receiver. This will allow the pinion gear assembly to drop out through the bottom of the receiver.

3. Remove the magazine cap screw (58). Then lift the magazine cap (57) out of the magazine tube (54). Pull out the magazine spring (56) and follower (55). Push down on the cartridge carrier (49) and remove it from the receiver. Remove the lower gear rack screw (8) and the lower gear rack (7).

4. To disassemble the trigger and hammer, cock the hammer, place a clamp on the mainspring (15), press the trigger (11) and release the hammer. Loosen the mainspring tension screw (25), remove the clamp and lift off the mainspring. Push the hammer pin sleeve (19) out of the lower tang (9), lift out the hammer. Remaining parts can be removed by driving out the appropriate retaining pins.

All other sub-assemblies are readily disassembled. Make sure to note the orientation of all component parts. To avoid mis-matching parts, work on individual sub-assemblies, cleaning and reassembling each one before proceeding with the next unit.

Assemble in reverse order. **DO NOT** install the extractor until after the pinion gear, bolt, and lower pinion gear rack are installed and aligned. When assembling the pinion gear to the body of the bolt and lower pinion gear rack, the bolt must be in the **CLOSED** position. The teeth of the pinion gear must be engaged in both the lower gear rack and the gear teeth machined on the underside of the bolt body. Misalignment of the pinion gear will cause a change in the relationship of the bolt, breechblock, and lever, in their closed position. When assembling the trigger and hammer, the hammer must be in the forward position. Install the hammer pin screw and lower tang screws. **DO NOT TIGHTEN** until all three screws are started in all threaded holes in the receiver. ■

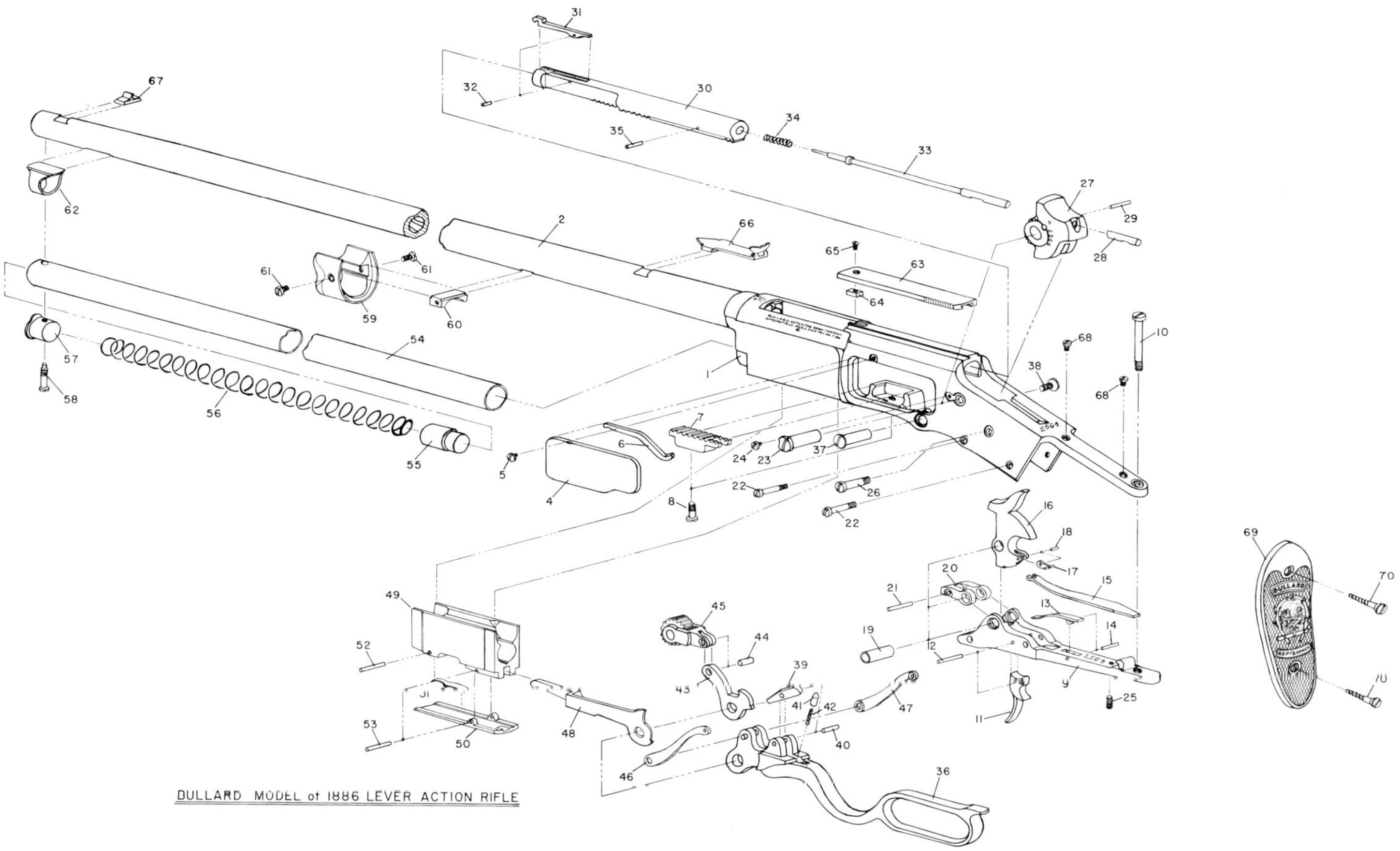

Exploded view of Bullard Model of 1886 lever action rifle. *The title "Exploded Views" belongs to the National Rifle Association and the American Rifleman.*

Parts Legend

1. Receiver
2. Barrel
3. Buttstock (not shown)

3a. Forestock (not shown)

4. Sideplate
5. Sideplate Screw
6. Magazine lifting spring
7. Lower gear rack
8. Lower gear rack screw
9. Lower tang
10. Rear tang screw
11. Trigger
12. Trigger pin
13. Trigger spring
14. Trigger spring pin
15. Mainspring
16. Hammer
17. Stirrup
18. Stirrup pin
19. Hammer pin sleeve
20. Lever link
21. Lever link pin
22. Lower tang screws (2)
23. Breechblock pin
24. Breechblock retainer screw
25. Mainspring screw tension
26. Hammer screw
27. Breechblock
28. Striker pin
29. Striker retainer pin
30. Bolt
31. Extractor
32. Extractor pin
33. Firing pin
34. Firing pin spring
35. Firing pin retaining pin
36. Lever
37. Lever pin
38. Lever pin screw
39. Pinion gear link retainer
40. Retainer pin
41. Retainer plunger
42. Retainer plunger spring
43. Pinion gear link
44. Pinion gear assembly pin
45. Pinion gear assembly
46. Lever arm, left hand
47. Lever arm, right hand
48. Cartridge carrier lifter arm
49. Cartridge carrier
50. Floorplate
51. Floorplate spring
52. Floorplate spring pin
53. Floorplate pin
54. Magazine tube
55. Magazine follower
56. Magazine spring
57. Magazine cap
58. Magazine cap screw
59. Nose cap
60. Nose cap base
61. Nose cap screws (2)
62. Front magazine tube ring
63. Dustcover
64. Dustcover retainer
65. Dustcover retainer screw
66. Rear sight
67. Front sight
68. Blank tang sight mounting screws (2)
69. Buttplate
70. Buttplate screws (2)

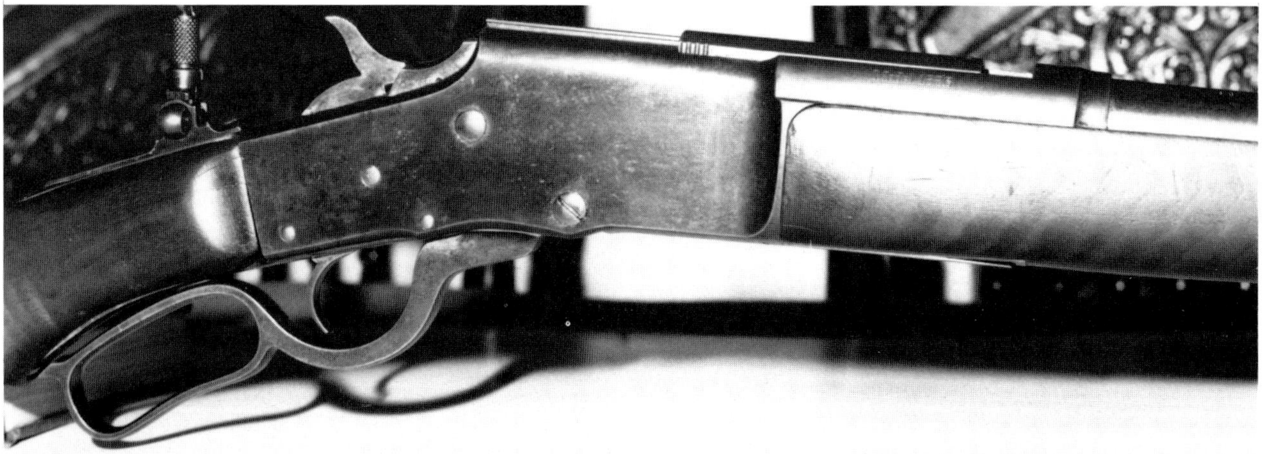

Plate 3-48. (SN 521) Stamped 'model 1886' on right side of breech. This stamping may indicate that Bullards were removed from factory stock in no particular order except to fill a customer's needs. Neither of the author's rifles, SN 461 or SN 770, are stamped 'model 1886.' Tables at back list those marked and not marked. This anomaly has made the dating of Bullard rifles difficult in the extreme. This particular rifle also has a drilled extractor access hole, but neither of the author's have. The Lyman tang sight carries the patent dates of Jan 8/89 and Feb 18/? *Courtesy Paul Simkin*

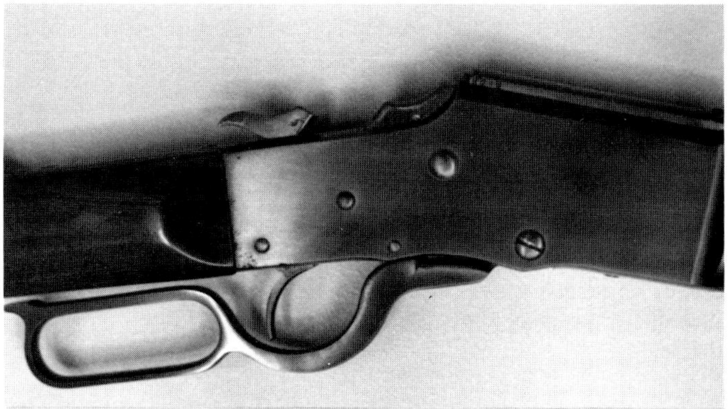

Plate 3-49. (SN 2331 cal .50-95) This rifle has been refinished, which might account for the sharp receiver line.

Plate 3-50. (SN 2331 cal .50-95) Long or fuller buttstock flare.

Note in Plates 3-49 and 3-50 that there is a distinct line or sharp ridge running the length of the receiver just under the dustcover. Plate 3-29 (SN 2590) also has this ridge or line. Most receivers look like Plate 3-51, with a gentle rounded curve on top of the receiver leading to dustcover. *Courtesy Robert L. Bullard*

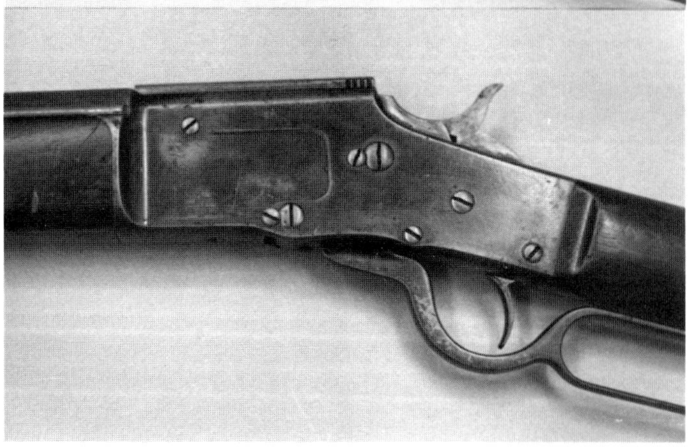

Plate 3-51. (SN 23 accepts .45-70) Shallow or short buttstock flare.

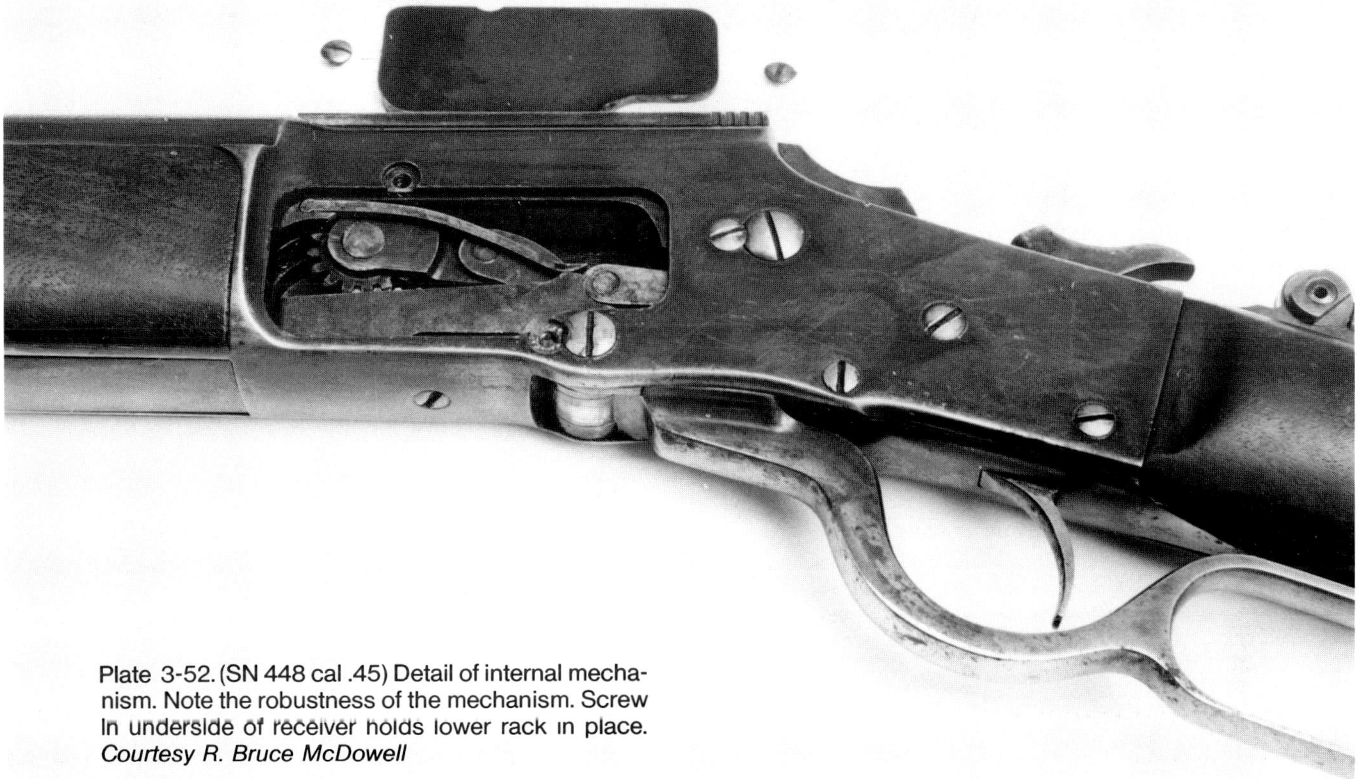

Plate 3-52. (SN 448 cal .45) Detail of internal mechanism. Note the robustness of the mechanism. Screw in underside of receiver holds lower rack in place. *Courtesy R. Bruce McDowell*

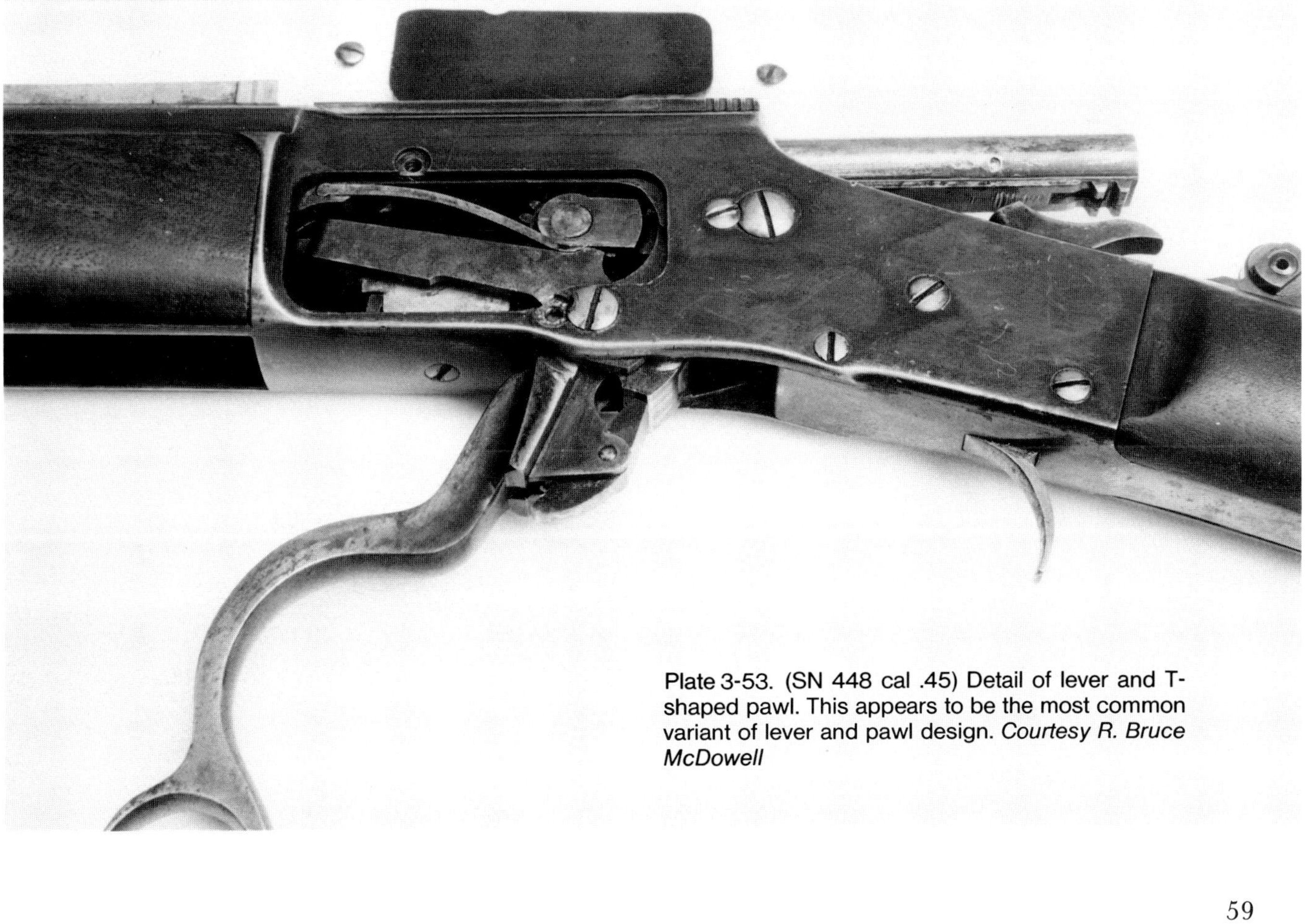

Plate 3-53. (SN 448 cal .45) Detail of lever and T-shaped pawl. This appears to be the most common variant of lever and pawl design. *Courtesy R. Bruce McDowell*

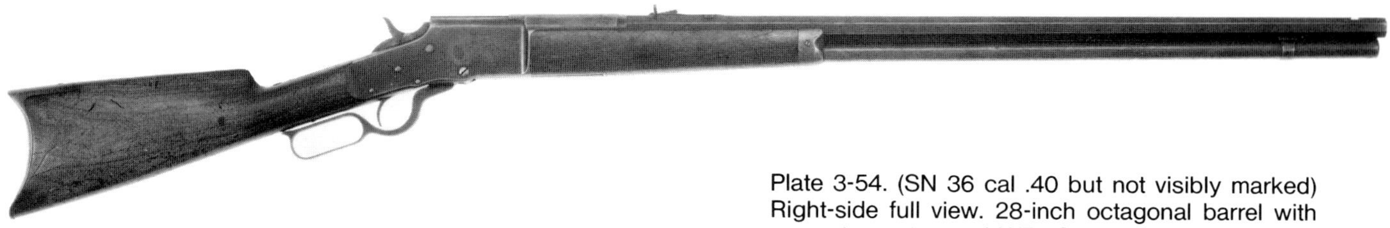

Plate 3-54. (SN 36 cal .40 but not visibly marked) Right-side full view. 28-inch octagonal barrel with upper tang stamped WRACo. The slotted head on the lever pivot pin indicates that this is a 7-screw receiver. Always view the large-frame repeaters from the left side to determine whether they are 7- or 8-screw receivers. *Courtesy The Winchester Arms Museum, Wyoming*

A.

B.

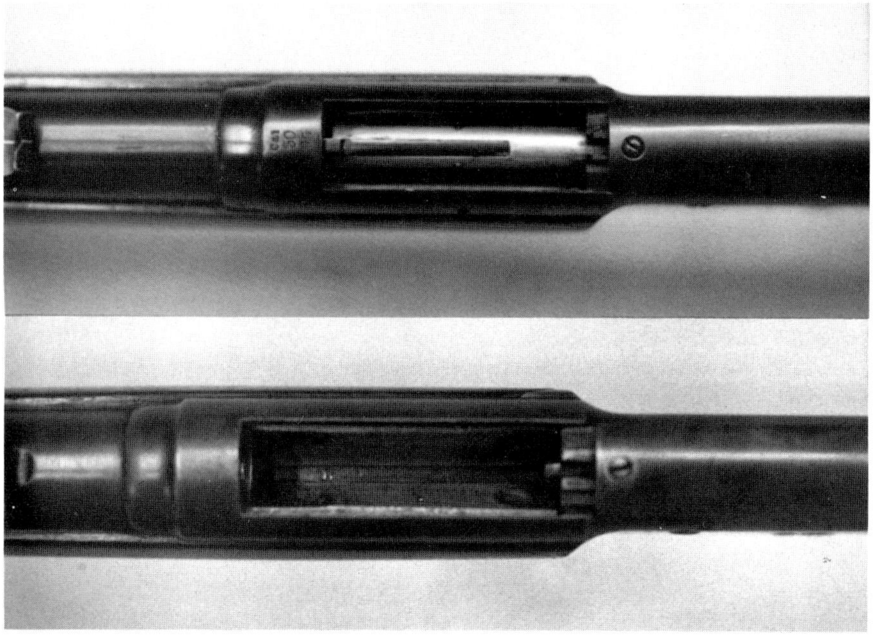

Plate 3-55. (Top: SN 2331 cal .50-95; Bottom: SN 23) When comparing top and bottom photographs, and allowing for the slight difference in distance between the rifles and the camera, it is apparent that the top rifle has a slightly shorter breech opening (¼ inch to be precise) as mentioned in this chapter. The top photograph shows the caliber markings clearly. The bottom shows the absence of any visible caliber markings. *Courtesy Robert L. Bullard*

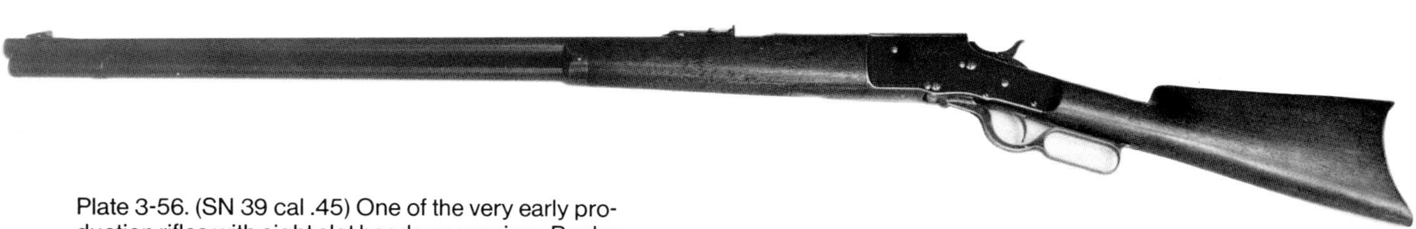

Plate 3-56. (SN 39 cal .45) One of the very early production rifles with eight slot heads on receiver. Rocky Mountain front sight. Three-leaf Winchester express rear sight. *Courtesy O.L. Peacock, Jr.*

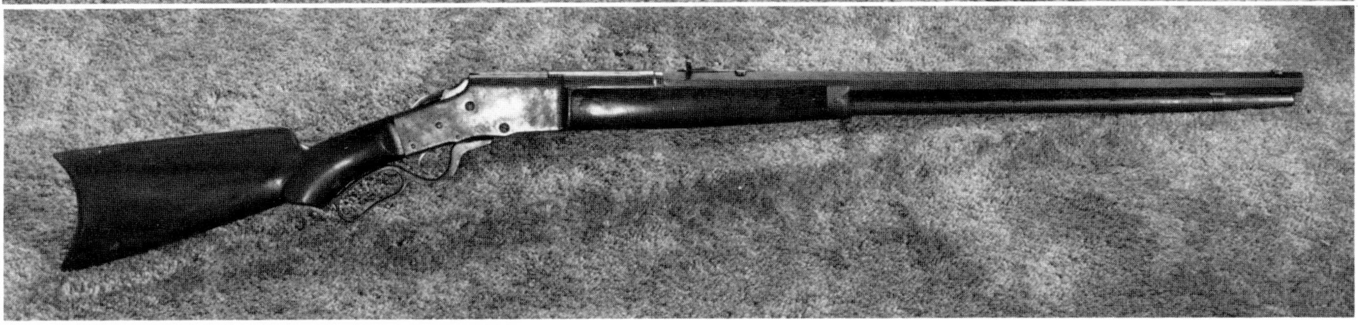

Plate 3-57. (SN 974 cal .40) Large-frame. Left view of rifle pictured in Plate 3-58. Rocky Mountain front sight. Rear sight standard Buckhorn. Has extractor access hole. *Courtesy Doyle N. Shaw*

Plate 3-58. (SN 974) 26-inch barrel, case-hardening. *Courtesy Doyle N. Shaw*

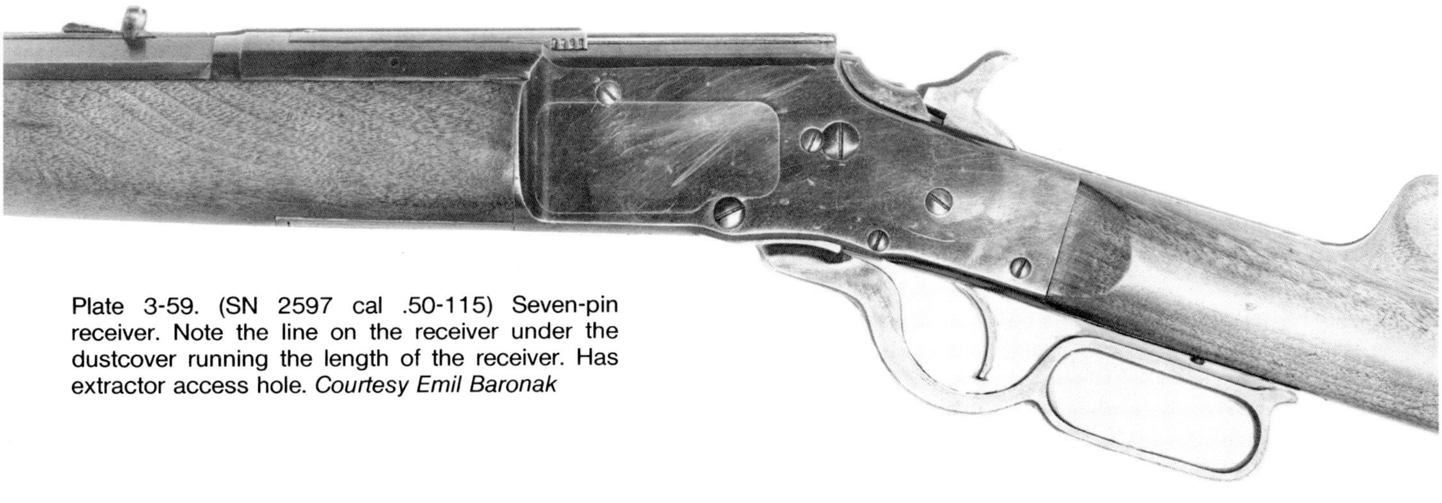

Plate 3-59. (SN 2597 cal .50-115) Seven-pin receiver. Note the line on the receiver under the dustcover running the length of the receiver. Has extractor access hole. *Courtesy Emil Baronak*

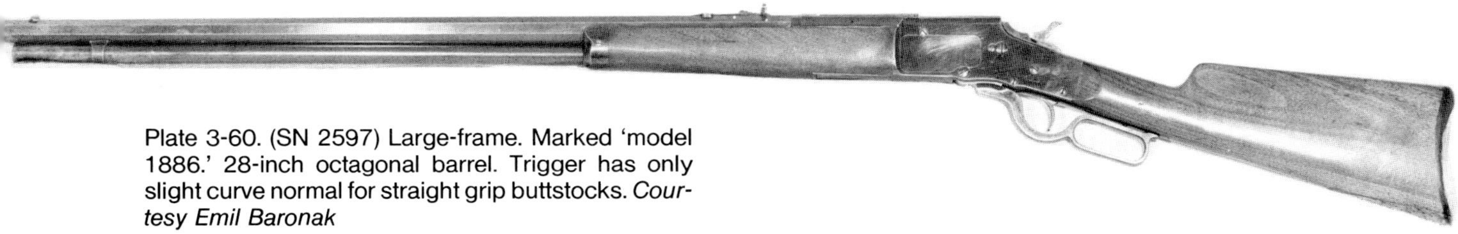

Plate 3-60. (SN 2597) Large-frame. Marked 'model 1886.' 28-inch octagonal barrel. Trigger has only slight curve normal for straight grip buttstocks. *Courtesy Emil Baronak*

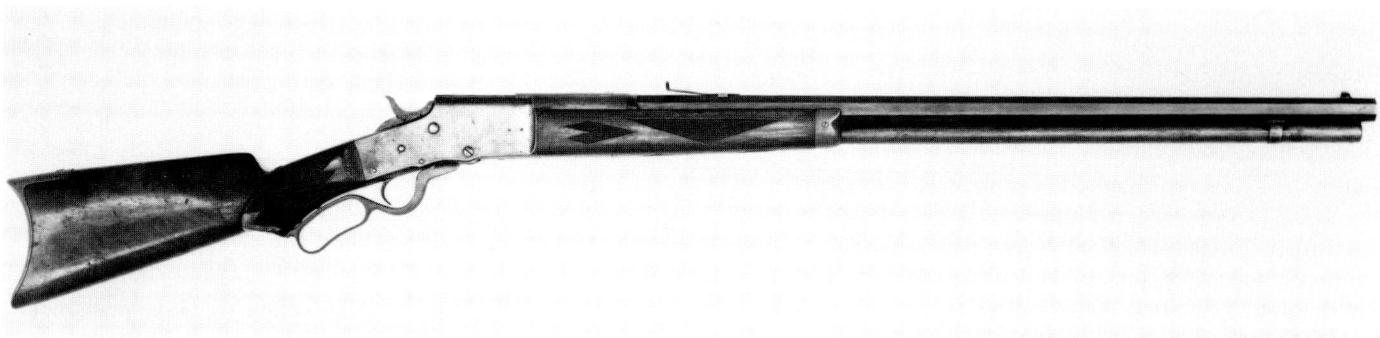

Plate 3-61. (SN 1070 cal .40) Large-frame deluxe. 26-inch octagonal barrel. 'B'-style finger serrations. Note magazine tube ends well before muzzle, a very common occurrence on the repeaters. *Courtesy the Milwaukee Public Museum*

Plate 3-62. (SN 1349 cal .40) 28-inch full octagonal barrel marked 'model' 1886.' *Courtesy of Traders' Den and Doug Mitchell*

Plate 3-63. (SN 1299 cal .40) 23½-inch octagonal barrel. *Courtesy Matthew A. Bailey*

Plate 3-64. (SN 231 cal .40) 26-inch barrel, part octagonal, part round. *Courtesy Matthew A. Bailey*

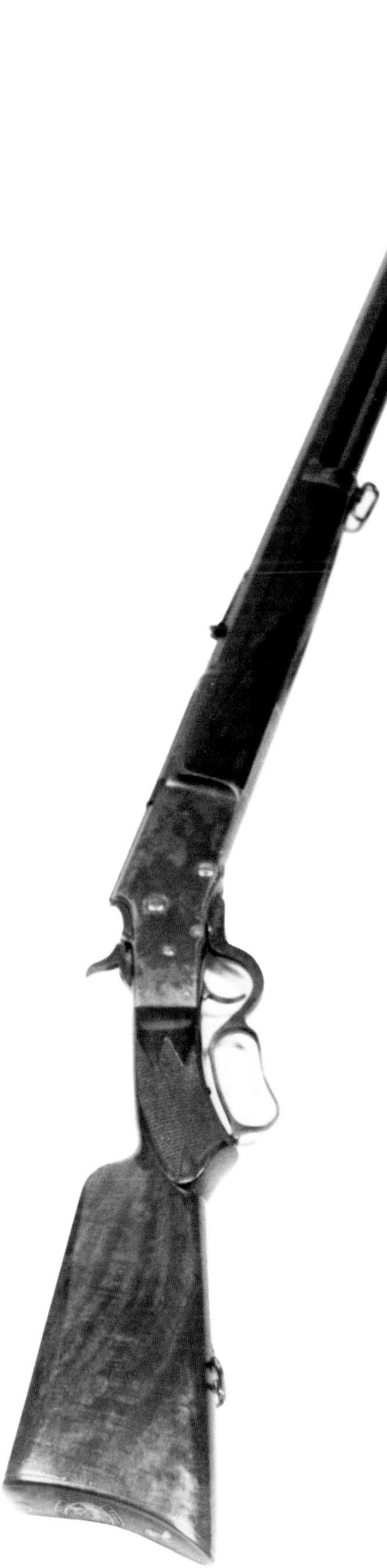

Plate 3-66. (SN 555 cal .45) Large-frame. Elk buttplate motif. 26⅛-inch octagonal barrel. No extractor hole. Has sling swivels and standard large-frame checking. *Courtesy Wayne A. Grove*

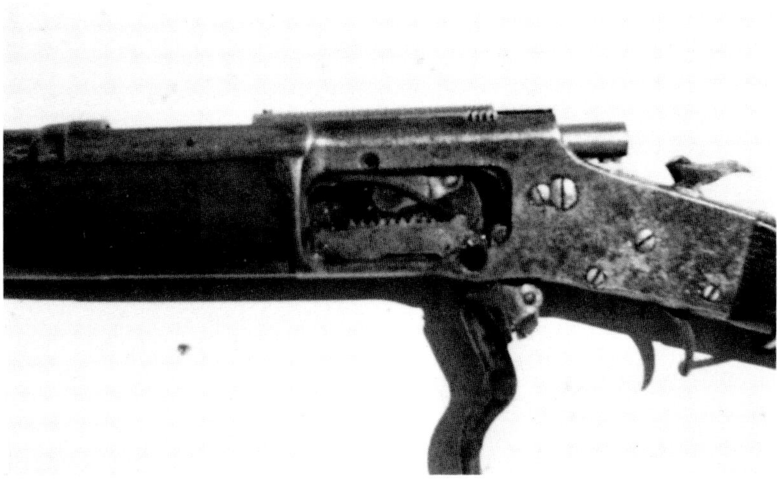

A.

B.

Plate 3-65. (SN 2673 cal .50-115) Large-frame. This rifle was captured from a Boer farmer during the Boer War and from there it was 'liberated' to its new home in South America, where it remains today. *Courtesy O'Brien Forbes Ffrench*

Plate 3-67. (SN 23) 'A'-style finger serrations. *Courtesy Robert L. Bullard*

Plate 3-68. (SN 23) Earliest known Bullard rifle. Caliber not marked but will accept .45-70. It has an 8-screw receiver and type 'A' dustcover. It is marked Bullard Repeating Arms Association. Round 28-inch barrel. It also has T-shaped pawl. *Courtesy Robert L. Bullard*

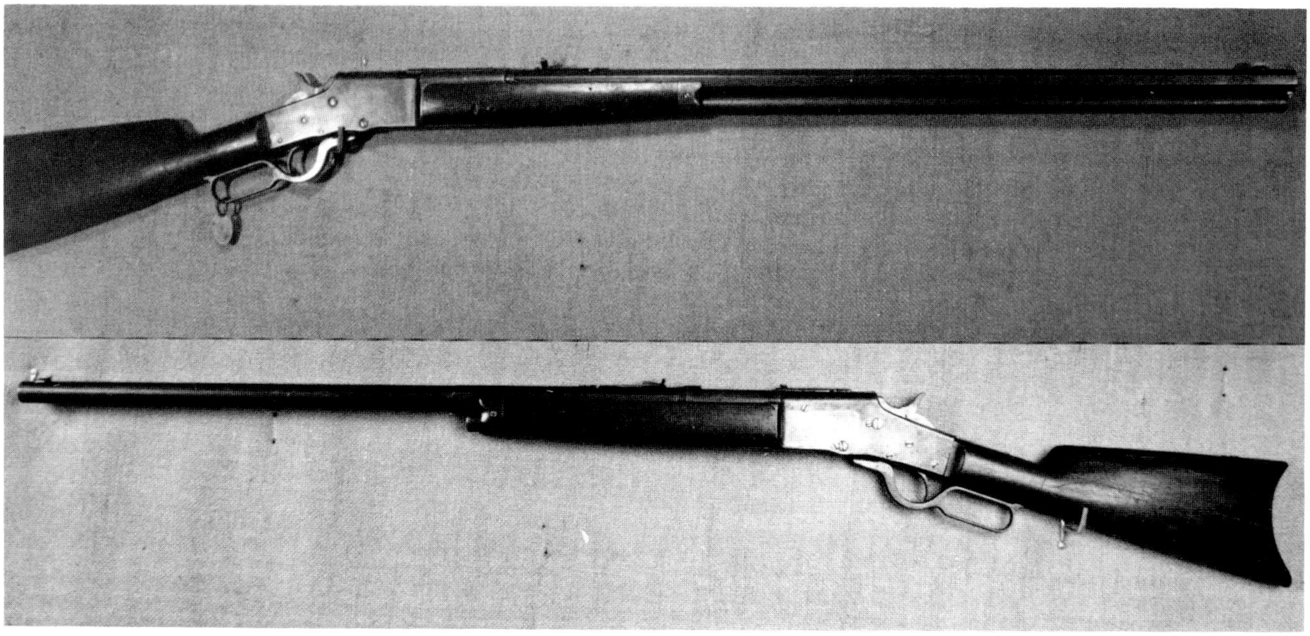

Plate 3-69. (Top: SN 1305 cal .40; Bottom: SN 331 cal .45) Both large-frames. Top rifle has 28-inch octagonal barrel. Seven-screw receiver. Bottom has 24-inch, part round, part octagonal barrel. Button magazine. Lever and hammer case-hardened. This rifle appears to have a non-standard or homemade dustcover, as it is shorter than the factory dustcover. *Courtesy J.M. Davis Gun Museum, Oklahoma. (Accession No. 7017 & 885)*

A.

B.

Plate 3-70. (Top: SN 23; Bottom: SN 2331) SN 23 has no visible caliber markings. Rear sight is a standard Winchester carbine elevator sight. SN 2331 rear sight is a 3-leaf express on sporting base. *Courtesy Robert L. Bullard*

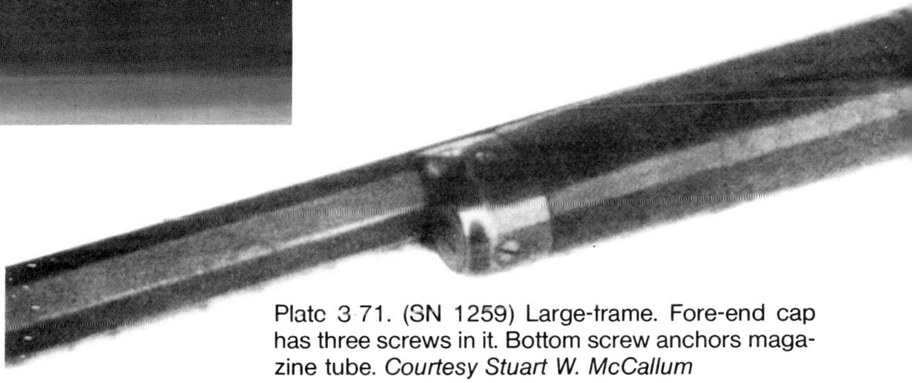

Plate 3-71. (SN 1259) Large-frame. Fore-end cap has three screws in it. Bottom screw anchors magazine tube. *Courtesy Stuart W. McCallum*

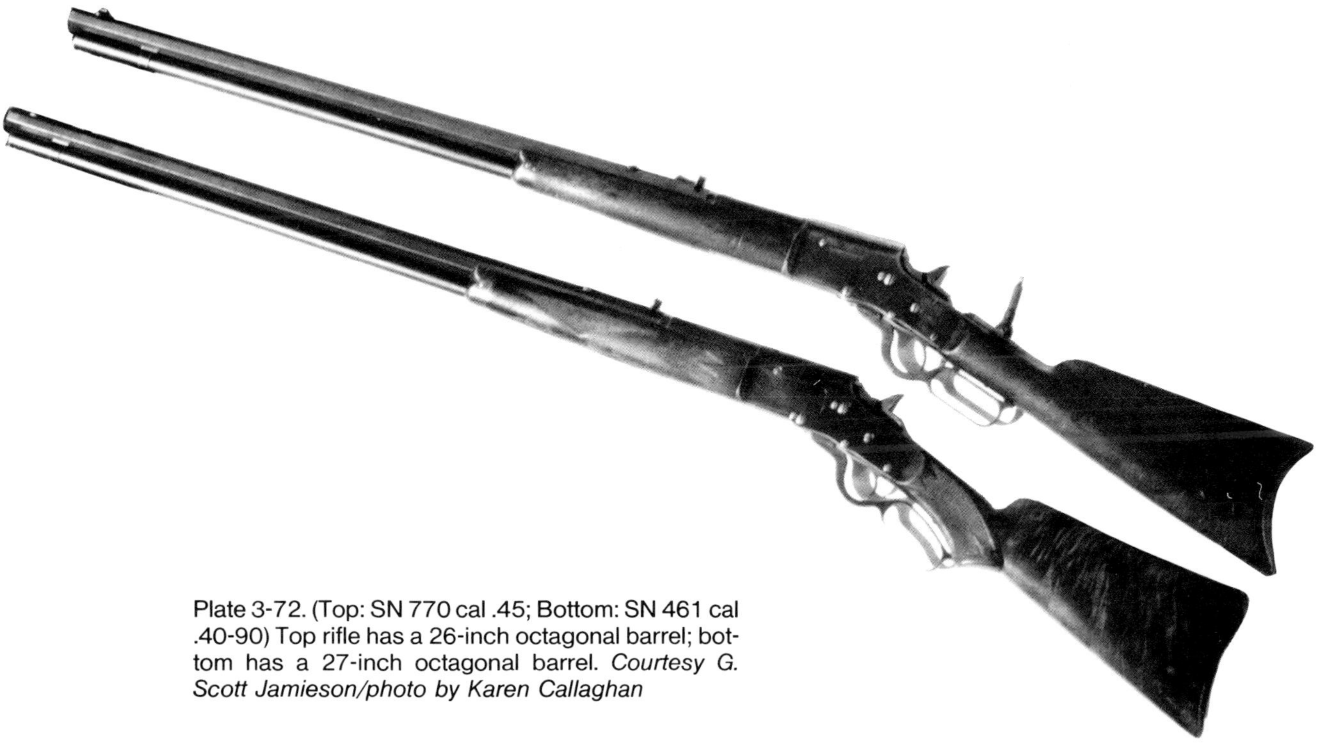

Plate 3-72. (Top: SN 770 cal .45; Bottom: SN 461 cal .40-90) Top rifle has a 26-inch octagonal barrel; bottom has a 27-inch octagonal barrel. *Courtesy G. Scott Jamieson/photo by Karen Callaghan*

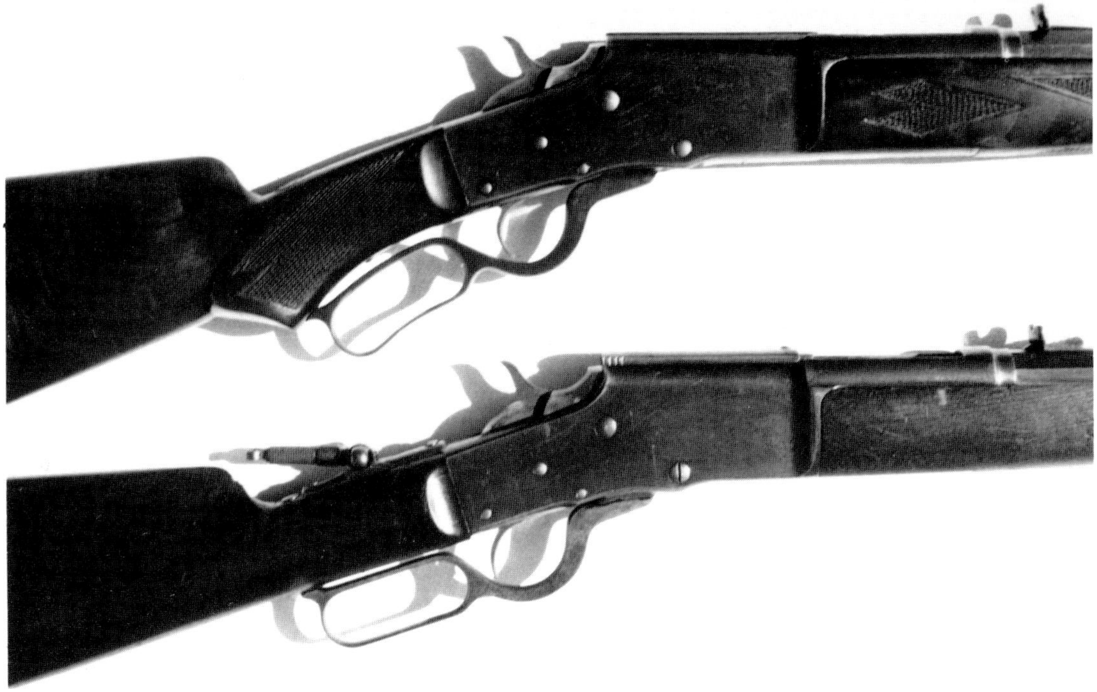

A.

Plate 3-73A. Top: (SN 461) Note pistol grip model has slightly more trigger curvature. Bottom: (SN 770) Straight grip has less trigger curvature. *Courtesy G. Scott Jamieson / Photos by Karen Callaghan*

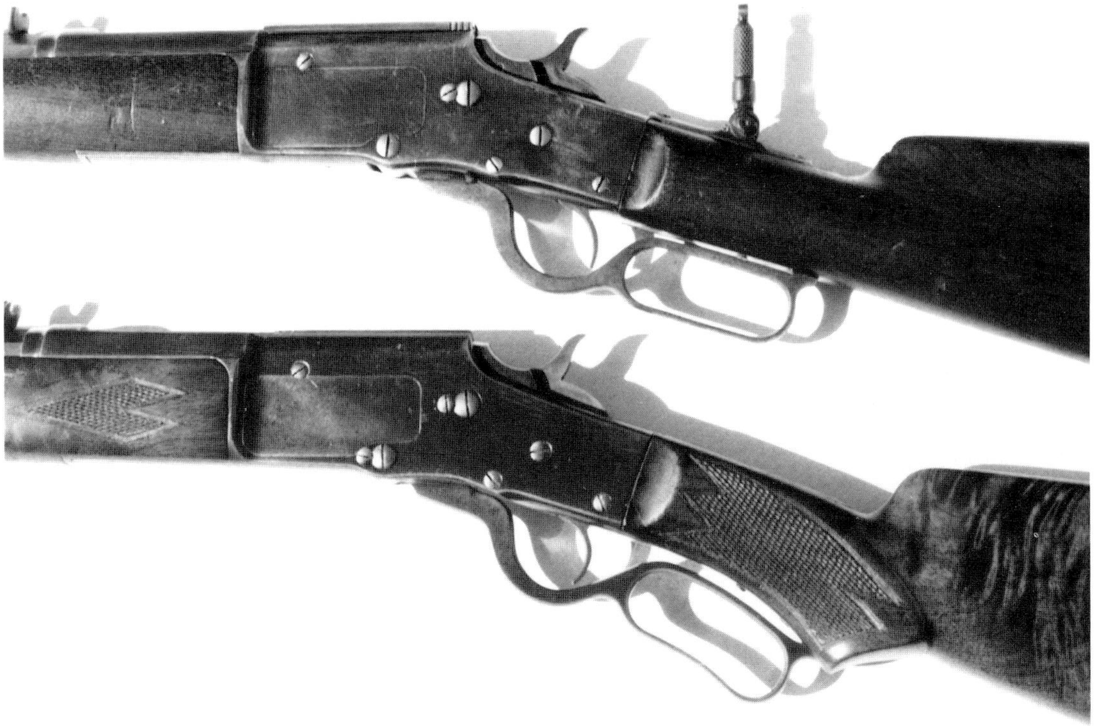

B.

Plate 3-73B. Top: (SN 770). Bottom: (SN 461). Left side details of 7 and 8 screw receivers.

A.

Plate 3-74A. Top: (SN 3856 cal .32-40) 'S' or Serpentine lever (single-shot detachable-interchangeable barrel model). Bottom: (SN 203 cal .45-75). *Courtesy Richard A. Failes*

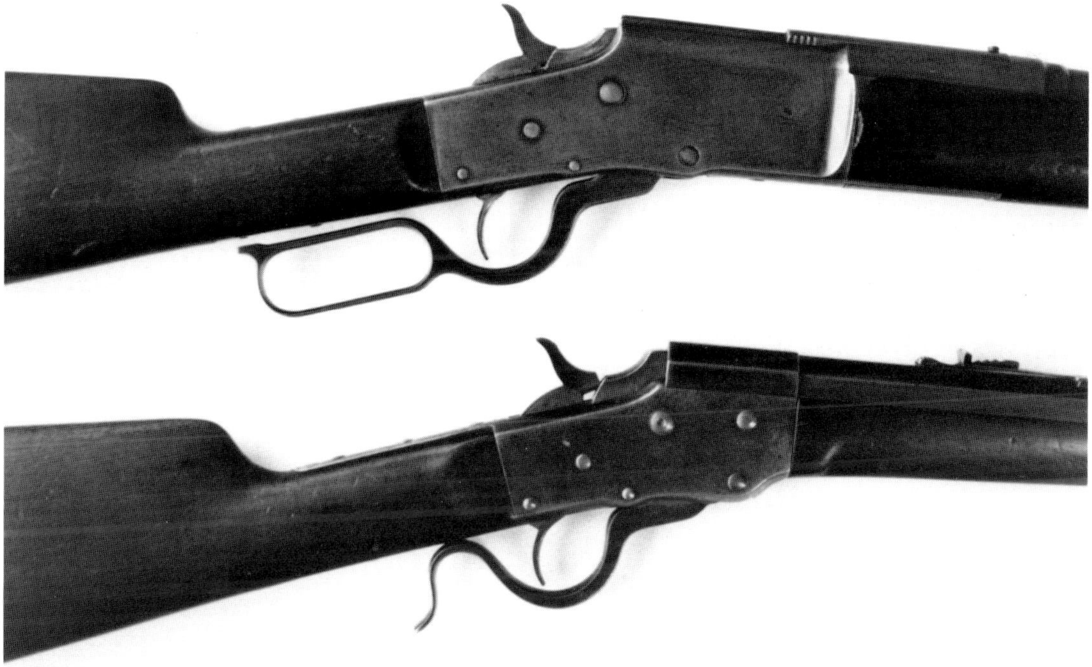

B.

Plate 3-74B. Top: (SN 203 cal .45-75). Bottom: (SN 3856 cal .32-40) single-shot detachable-interchangeable barrel model. *Courtesy Richard A. Failes*

Plate 3-75. (SN 203) Detail of rear sight on large-frame repeater. Homemade but interesting. *Courtesy of Richard A. Failes*

B.

A.

Plate 3-76. (Top: SN 904 cal .45; Bottom: SN 3832 cal. 32-40) Top rifle has 24-inch part round, part octagonal barrel. Bottom is single-shot detachable-interchangeable barrel model. Stock has cheekpiece. *Courtesy Harold R. Jonah*

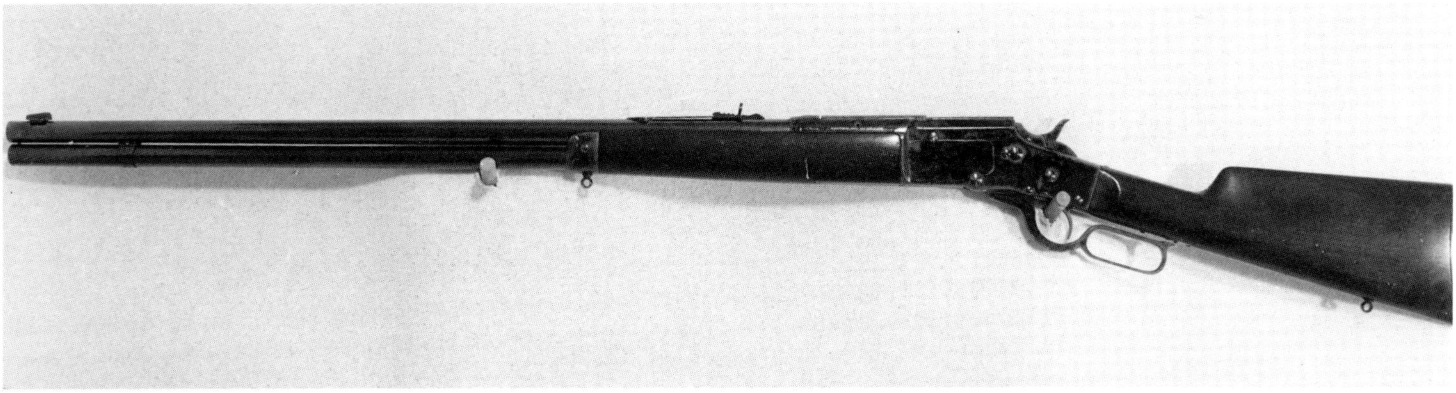

A.

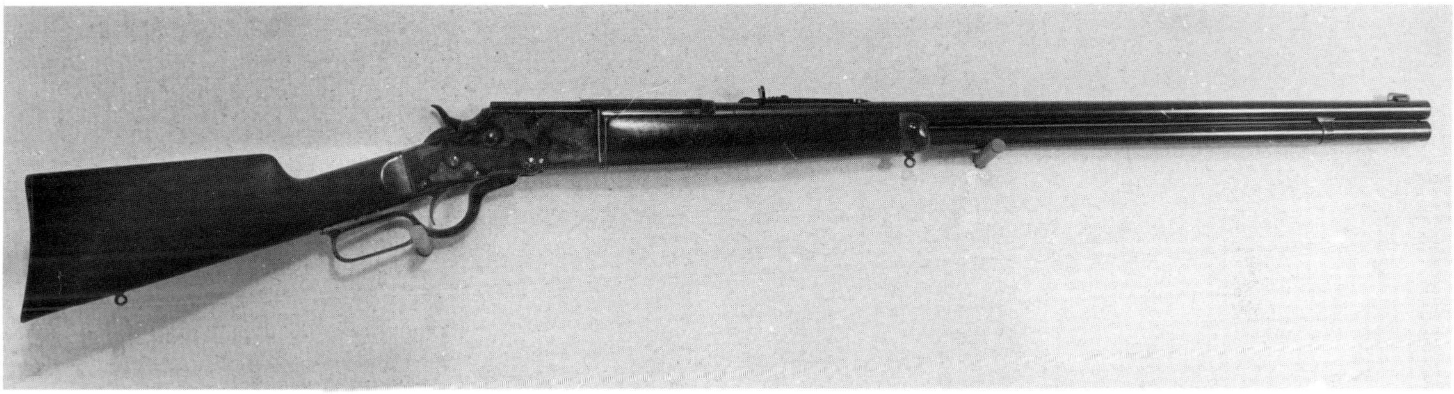

B.

Plate 3-77. (SN 2590 cal .45-85) *Courtesy A.J. Newberry*

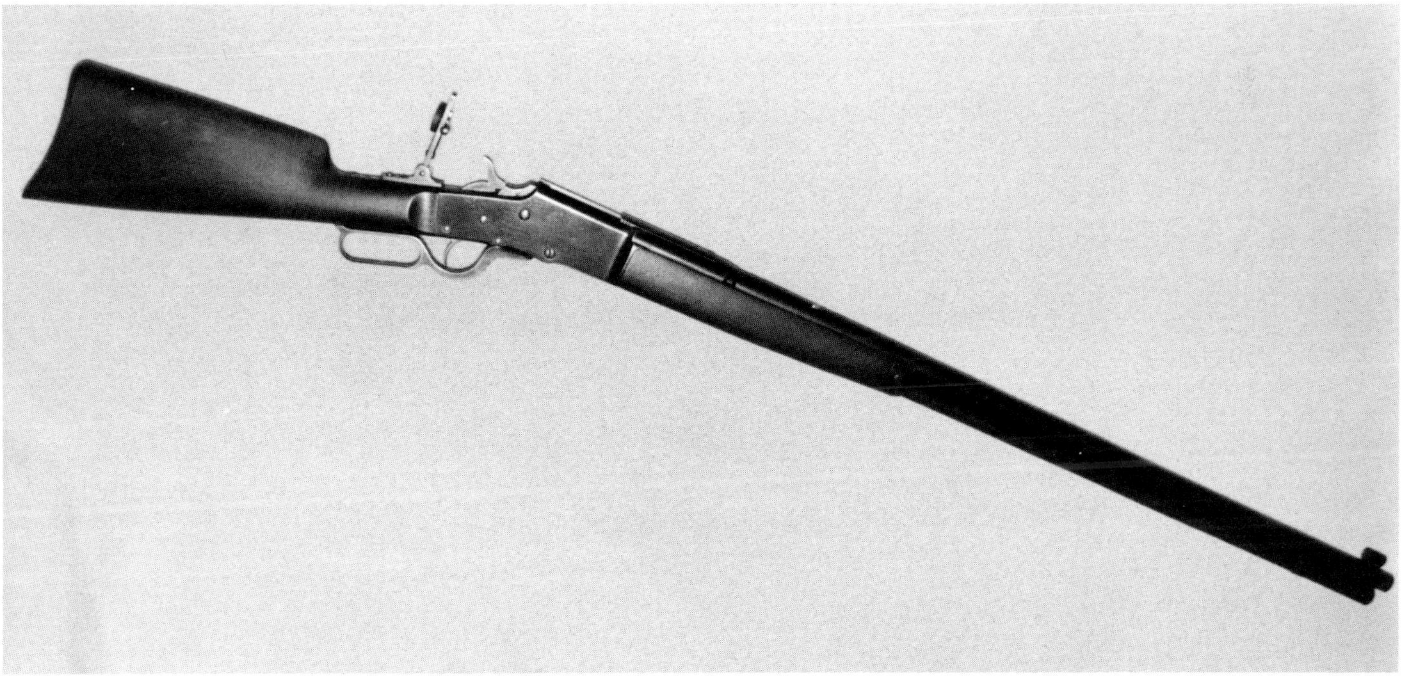

Plate 3-78. (SN 989 cal .45) Large-frame repeater, 7-screw receiver, 28-inch round barrel. Rear sight: Stevens windgauge added at later date. Front: standard Lyman hooded. *Courtesy Allen Smith*

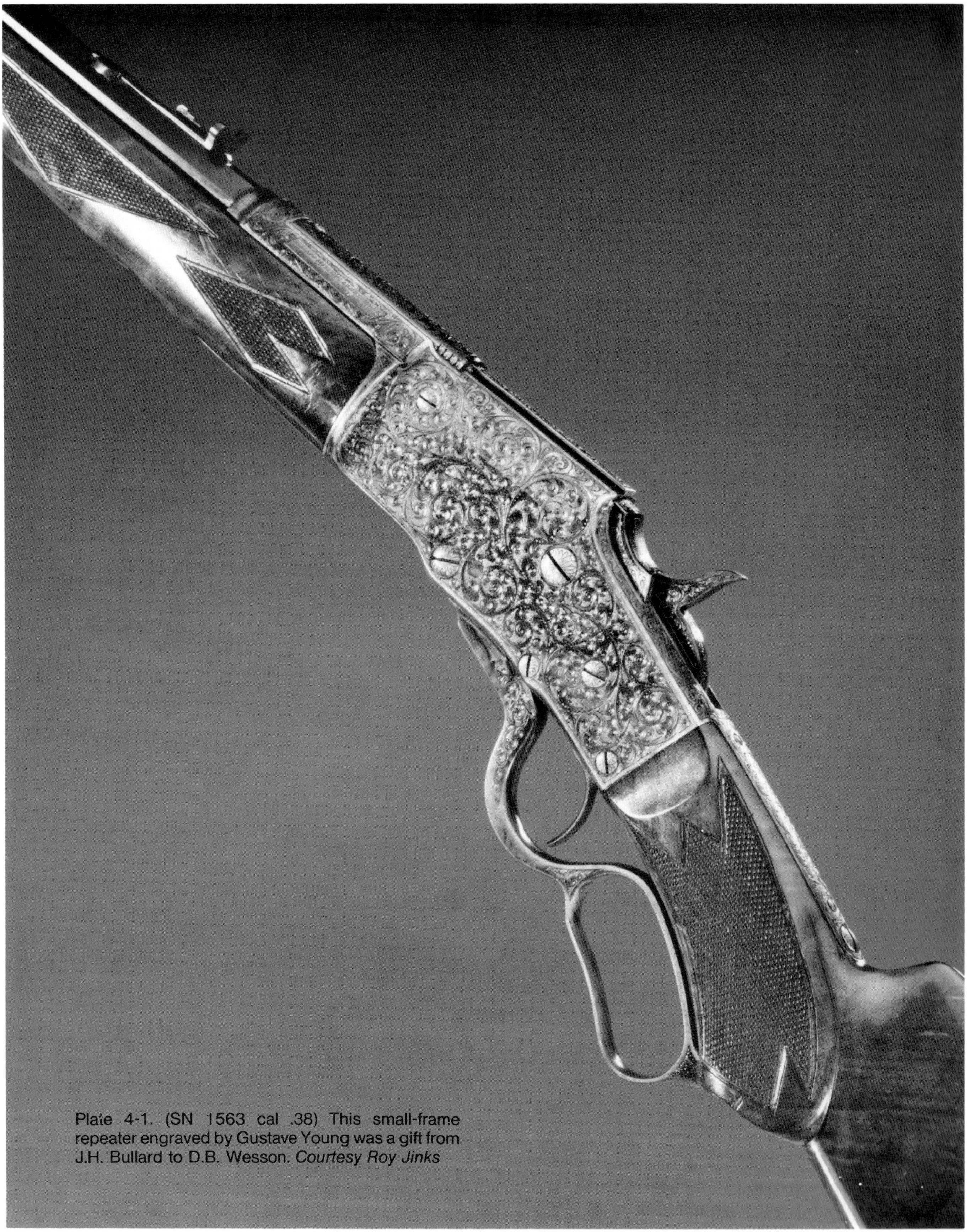

Plate 4-1. (SN 1563 cal .38) This small-frame repeater engraved by Gustave Young was a gift from J.H. Bullard to D.B. Wesson. *Courtesy Roy Jinks*

4

SMALL-FRAME REPEATERS

As stated previously, the same patent covering the large-frame repeater also covered the small-frame repeater. It came in calibers up to but *not* including .40. The .44 Winchester was the only exception to this rule. It was offered in the small-frame version in one of the Bullard catalogs, but rifles of this caliber have never been encountered by the author.

SERIAL NUMBER NOMENCLATURE

The small frames occupy the block of numbers 1500 through to 2000 for a possible maximum production figure of 500. No small-frames have appeared with serial numbers outside this range and no second batch of small-frames was ever produced, to the author's knowledge. To further underscore the rarity of the small-frames, only one rifle has appeared in the 1900-2000 range of serial numbers. The tables at the back of the chapter will show this quite graphically. It is, however, within this small group of rifles that the finest known Bullard exists. It is serial number 1563, in a .38 caliber, fully engraved by Gustave Young. The receiver, lever, hammer, etc. are case-hardened and the wood is of fancy walnut burl, with pistol grip, shotgun butt and checking. It was a gift from James H. Bullard to his old employer and friend Daniel B. Wesson (Plates 4-1, 4-2, 4-3, 4-4). Little else is known about this rifle.

CALIBER MARKINGS AND LOCATIONS

Of the few small-frame repeaters logged in the tables thus far, all have had their calibers marked in the "C" position. (Plates 4- 5C, 4-6). The only documented calibers are as follows: .32, .38, and .32 special.

COMPANY ADDRESSES

The only *documented* address used on the small-frames is the number 2 one from the tables:

BULLARD REPEATING ARMS COMPANY
SPRINGFIELD, MASS. U.S.A. PAT AUG, 16, 1881.

(See Plates 4-6 and 4-15) There may be others, but they have not been documented.

VARIATIONS IN DESIGN

The small-frames are much more standardized on their receiver design than are the large-frames. Instead of only one inspection sideplate as the large-frames have, the small-frames have inspection sideplates on both sides of the frame. The left side of the receiver always has six slotted screw heads and the right side always has three slot heads showing — one at the cover; the breechblock pivot pin and the guard lever pivot pin. This is because the hollow, internally threaded and the externally threaded male screws seen on the 7-screw large-frames are used exclusively in the small-frames. Of course the receiver's own proportions are much smaller than the large-frame receiver and both the forearm and buttstock are correspondingly reduced in size to fit accordingly (Plates 4-9, 4-10). The A style of dustcover serration design seems to predominate in the small-frames with four grooves and four protrusions. The use of extractor access holes is haphazard, although the use of them is prevalent in the 1800 serial number ranges (Plates 4-5, 4-6). See the tables for a complete breakdown. Again two trigger styles are used, but as noted here on serial number 1563, while it has a pistol grip it also uses the trigger style with less curvature, a style generally reserved for straight-gripped stocks (Plates 4-1, 4-4, 4-10). Most small-frame Bullard rifles are also marked Model 1886, indicating that the greatest production or sales of these rifles took place after 1886 (Plates 4-5).

Where shotgun butts and the gutta-percha plates are used, the turkey motif is used predominantly on the company logo. Please see the tables at the back of the book.

Barrel lengths for the small-frames are more or less standardized at 26 inches, but this can and does vary widely, as shown in the tables.

A further variation, noticed only very recently, involves the lever and its spring-pawl. The author has not seen or gathered enough photographs to draw any conclusions as to which style is rarest, but compare the pawl and lever design in Plate 4-4 to the pawl and lever design in Plate 4-5. The area supporting the pawl in Plate 4-4 is much heavier and beefier in appearance than that in Plate 4-5. Which design was used more is unknown. The rifle in Plate 4-4 is not marked Model 1886, but in Plate 4-5 the rifle is marked Model 1886. Whether two types of pinion links were used, as they were on the large-frames, is unknown at this time.

Rifling .32-40 Bullard had one twist in 16 inches
.38-45 Bullard had one twist in 16-18 inches

The .32-40 Bullard rifle bore was .306 inches in diameter and had a groove diameter of .311 inches.

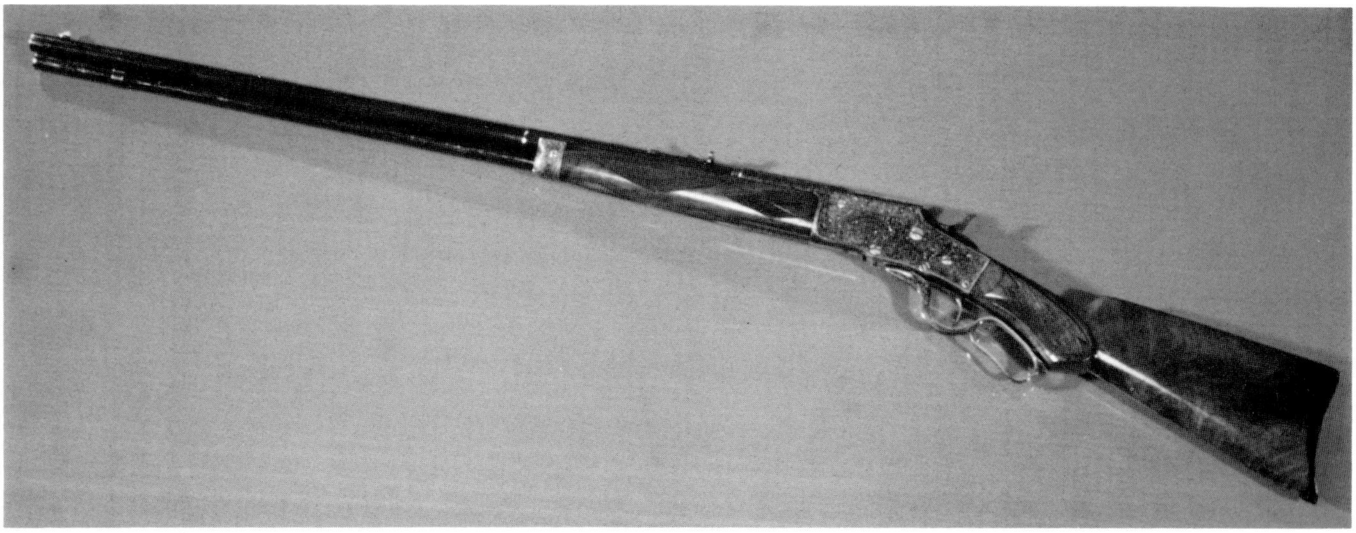

Plate 4-2. (SN 1563 cal .38) *Courtesy Roy Jinks*

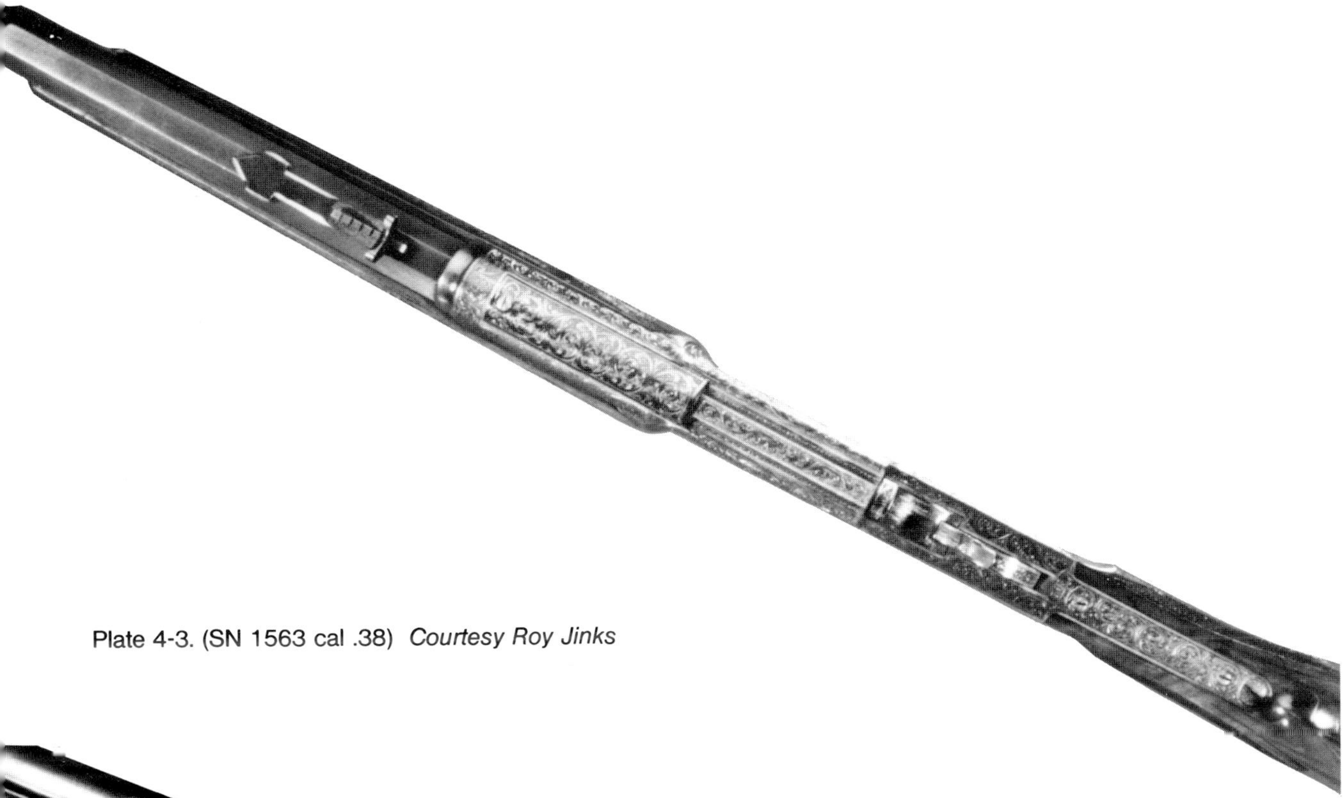

Plate 4-3. (SN 1563 cal .38) *Courtesy Roy Jinks*

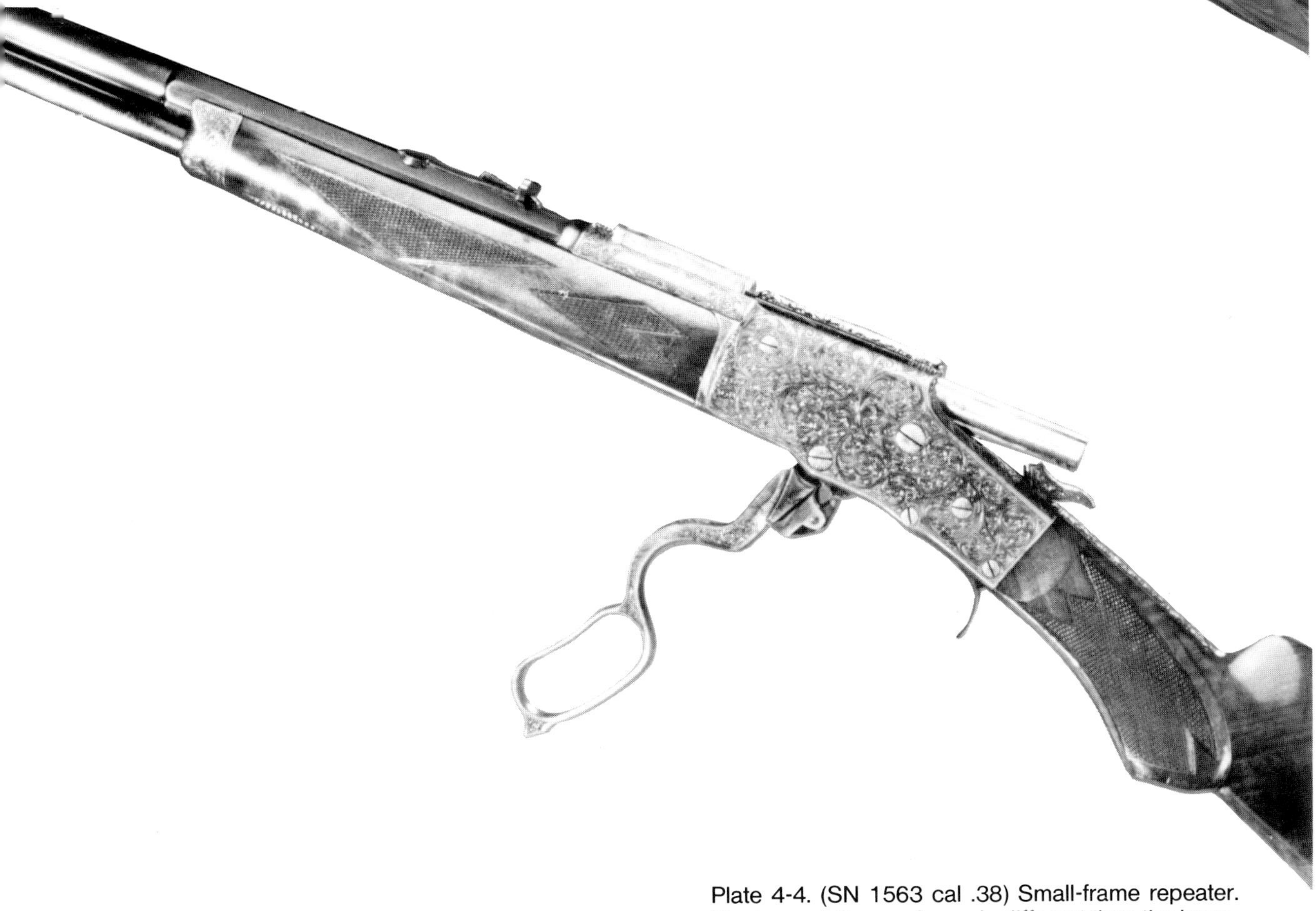

Plate 4-4. (SN 1563 cal .38) Small-frame repeater. The lever at the pawl area is different than the large-frames and the small-frame in Plate 4-18. *Courtesy Roy Jinks*

Plate 4- 5. (A, SN 1853 cal .38) The spring-pawl supports are less substantial on this rifle than the ones on SN 1563, Plate 4-4. *Courtesy William T. Bundy, Jr.*

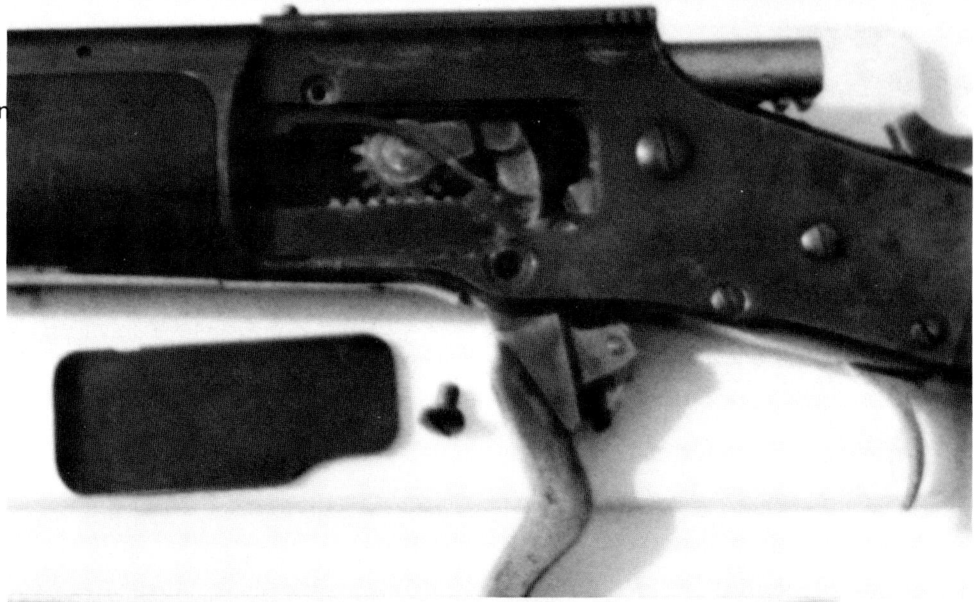

A.

B.

Plate 4-5. (B, C SN 1853 cal .38) Picture B shows clearly the style and location of the words 'model 1886' and A-style finger serrations on dustcover. Picture C has C-location caliber markings and shows extractor hole and both script styles and their locations. (Upper part of A in USA is faint.) *Courtesy William T. Bundy, Jr.*

C.

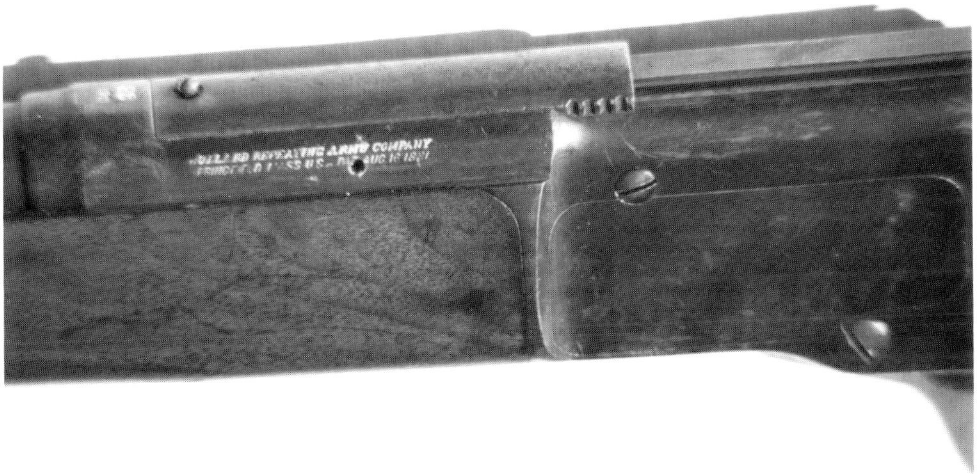

Plate 4-6. (SN 1853 cal .38) Small-frame repeater showing 'A' style dustcover finger serrations, caliber markings, breech address and drilled extractor access hole. This rifle also marked model 1886. Detail of mechanism can be noted in top photograph. *Courtesy William T. Bundy, Jr.*

A.

B.

Plate 4-7. (SN 1853) Top photograph: small-frame repeater with rule to show size of receiver. Bottom photograph: detail of right side of same rifle. Note inspection plate on right side of receiver as well as left. all small-frames appear to have been manufactured with inspection plates on both sides of receiver. Large-frames only have inspection plates on left side of receiver. *Courtesy William T. Bundy, Jr.*

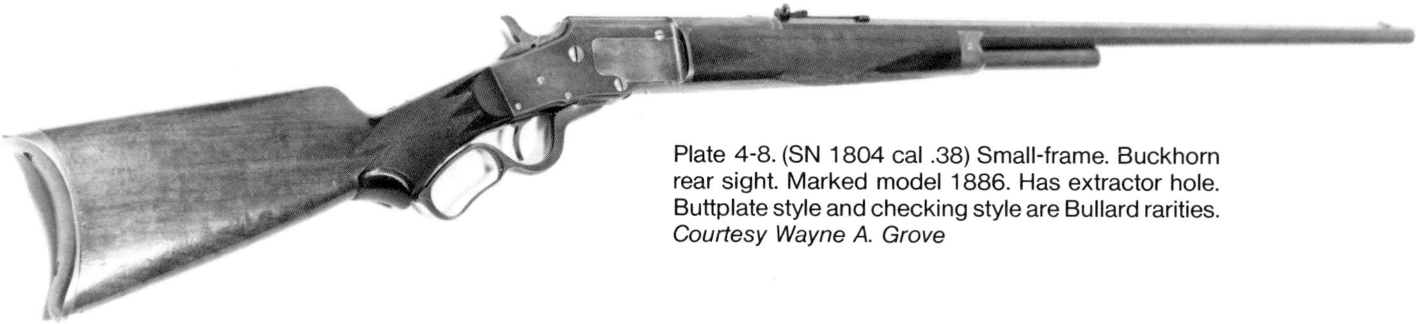

Plate 4-8. (SN 1804 cal .38) Small-frame. Buckhorn rear sight. Marked model 1886. Has extractor hole. Buttplate style and checking style are Bullard rarities. *Courtesy Wayne A. Grove*

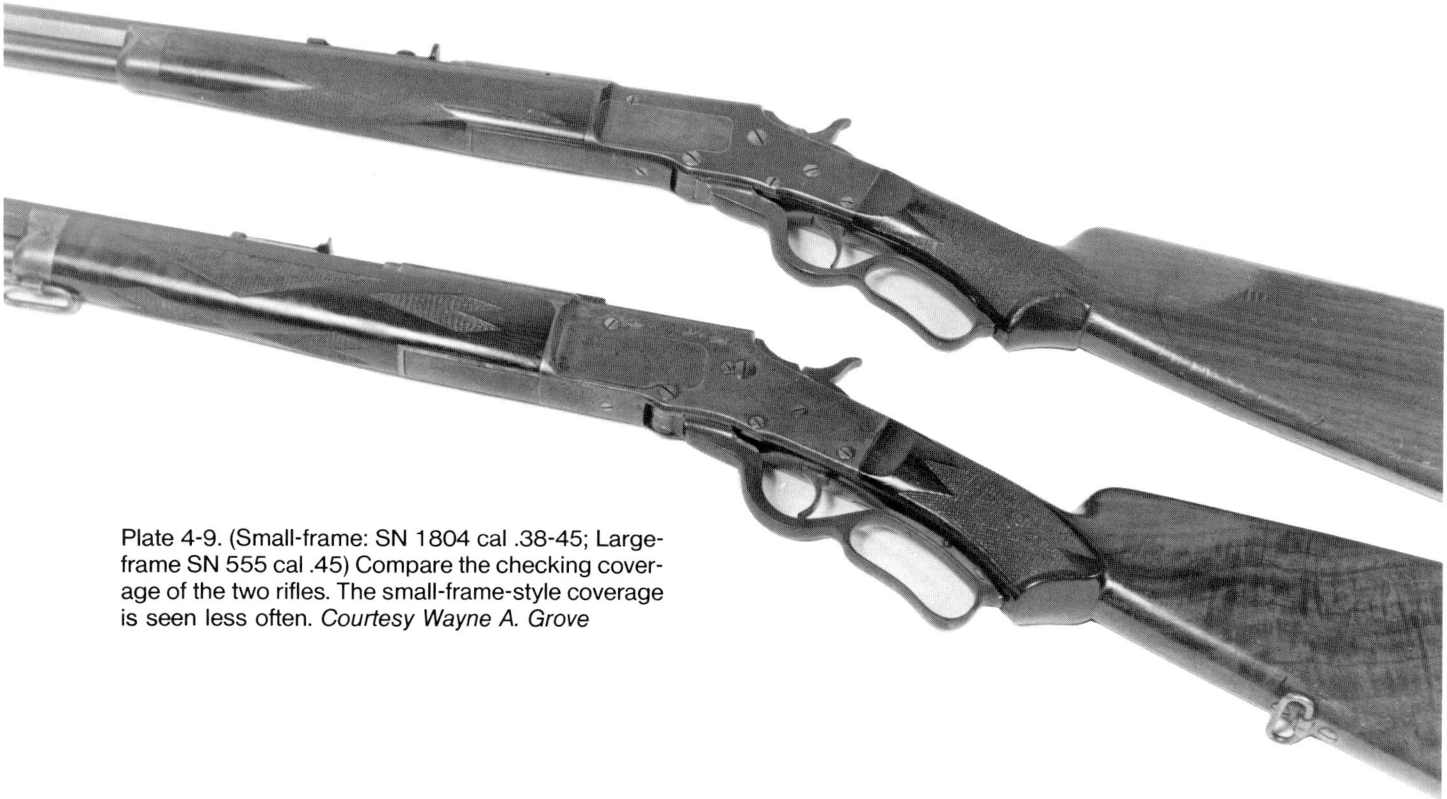

Plate 4-9. (Small-frame: SN 1804 cal .38-45; Large-frame SN 555 cal .45) Compare the checking coverage of the two rifles. The small-frame-style coverage is seen less often. *Courtesy Wayne A. Grove*

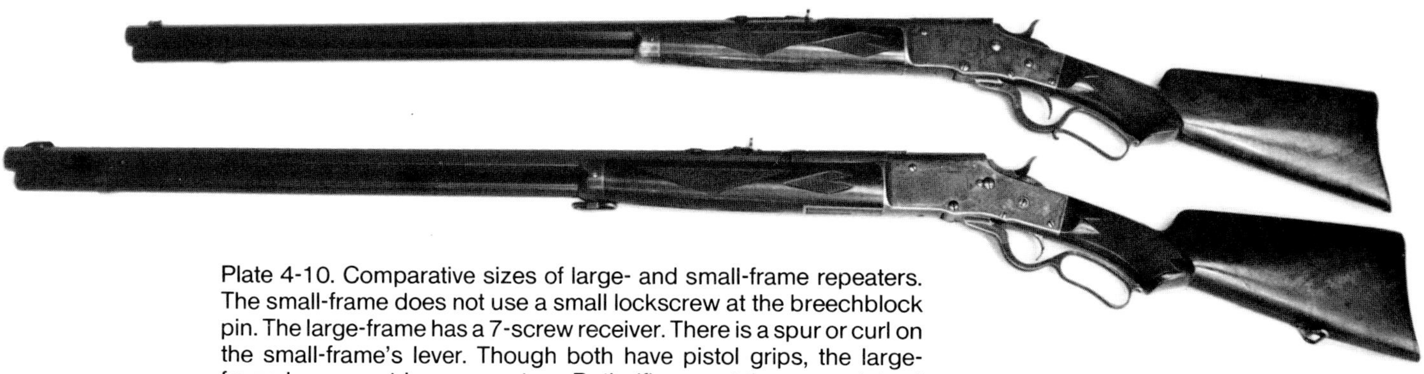

Plate 4-10. Comparative sizes of large- and small-frame repeaters. The small-frame does not use a small lockscrew at the breechblock pin. The large-frame has a 7-screw receiver. There is a spur or curl on the small-frame's lever. Though both have pistol grips, the large-frame has more trigger curvature. Both rifles sport the same style of checking. *Courtesy O.L. Peacock, Jr.*

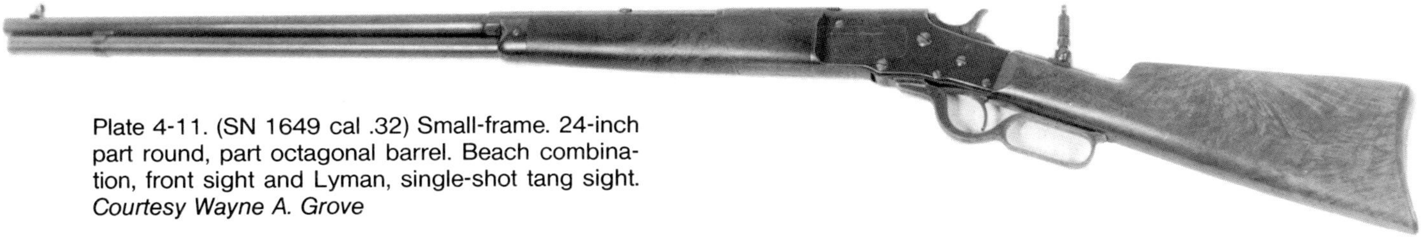

Plate 4-11. (SN 1649 cal .32) Small-frame. 24-inch part round, part octagonal barrel. Beach combination, front sight and Lyman, single-shot tang sight. *Courtesy Wayne A. Grove*

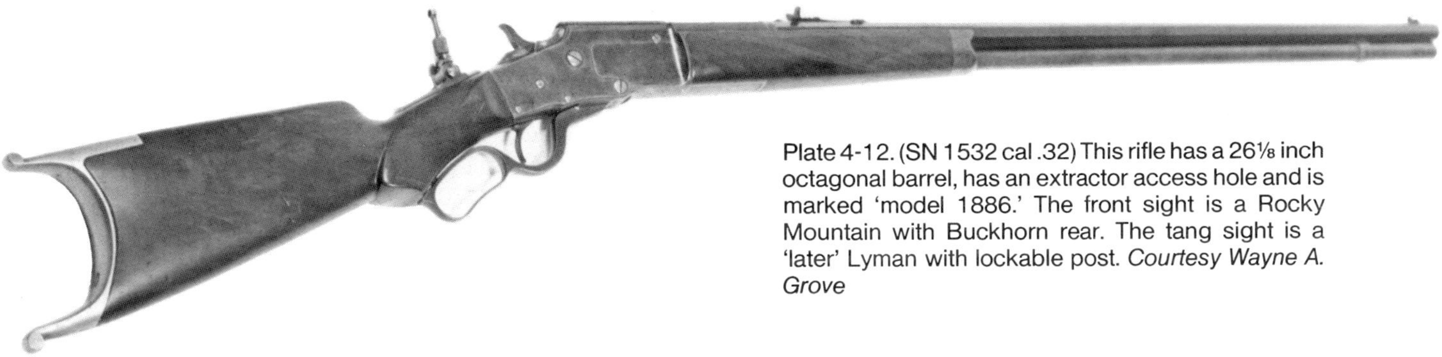

Plate 4-12. (SN 1532 cal .32) This rifle has a 26⅛ inch octagonal barrel, has an extractor access hole and is marked 'model 1886.' The front sight is a Rocky Mountain with Buckhorn rear. The tang sight is a 'later' Lyman with lockable post. *Courtesy Wayne A. Grove*

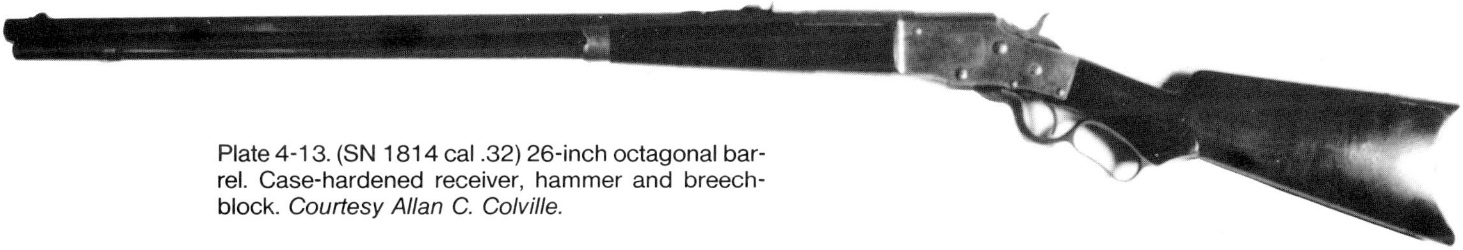

Plate 4-13. (SN 1814 cal .32) 26-inch octagonal barrel. Case-hardened receiver, hammer and breechblock. *Courtesy Allan C. Colville.*

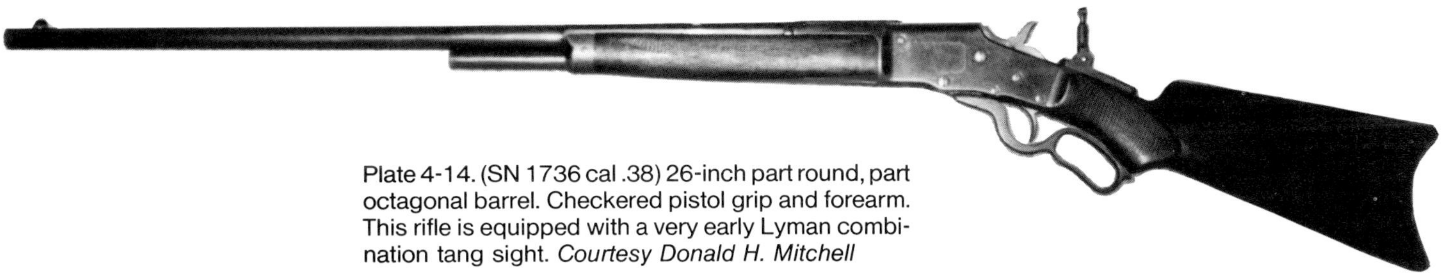

Plate 4-14. (SN 1736 cal .38) 26-inch part round, part octagonal barrel. Checkered pistol grip and forearm. This rifle is equipped with a very early Lyman combination tang sight. *Courtesy Donald H. Mitchell*

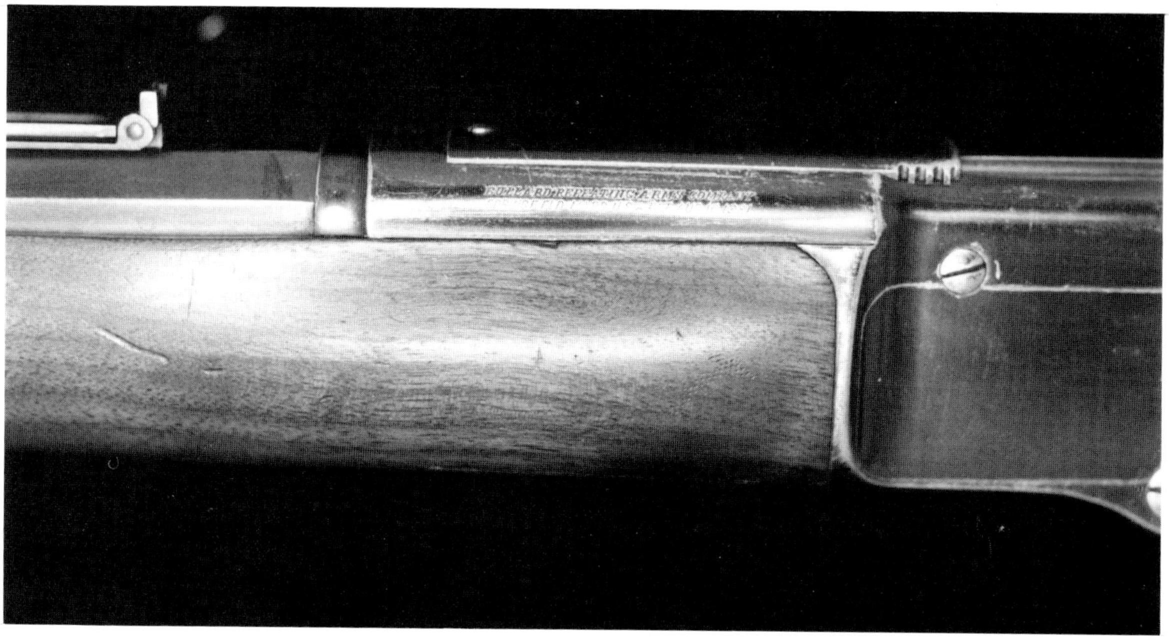

Plate 4-15. Close-up of number 2 style of address and 'B' style of dustcover serrations. Compare the address stamping with Plate 4-6 and you will see that the exact same die must have been used on these rifles, for on both addresses the upper portion of the A in USA is lightly stamped, compared to all the other letters. *Courtesy Verle W. Hunt*

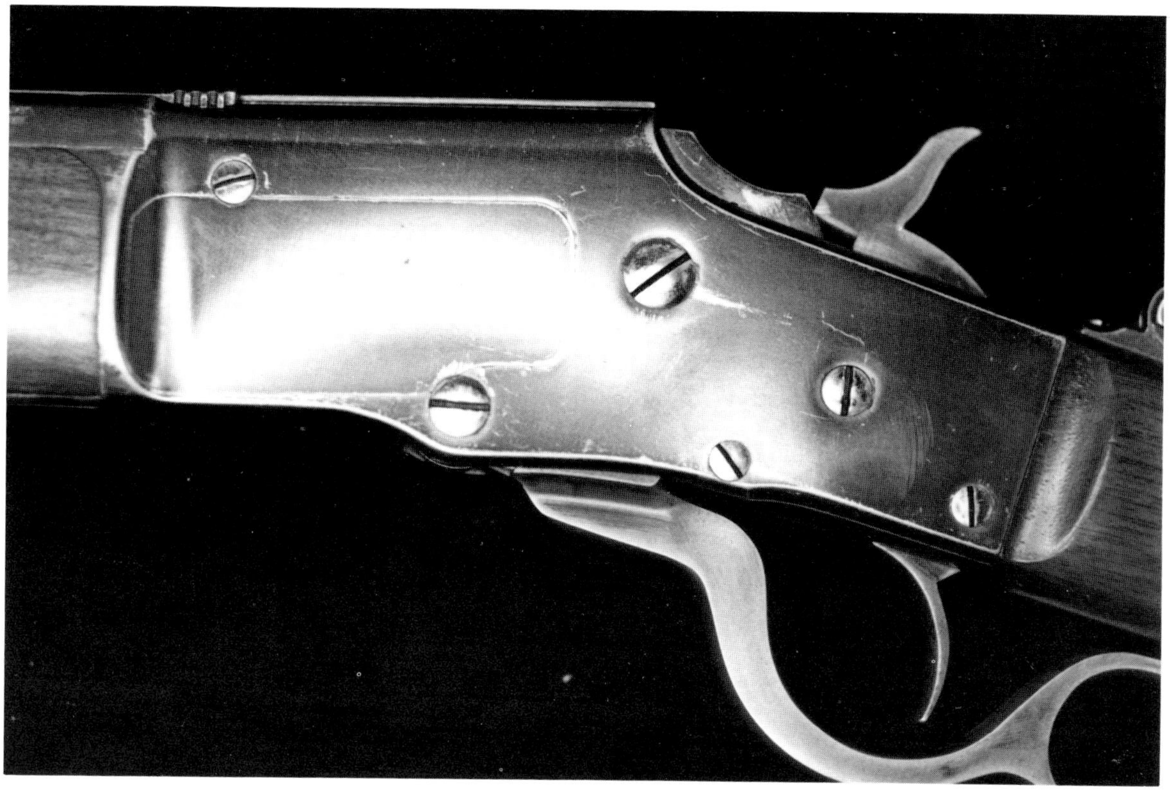

Plate 4-16. B-style finger serrations and case-hardened receiver, lever, hammer and breechblock. Examine the screw pattern closely. This is the only documented screw pattern and number (6) for the small-frame repeaters. Compare to the screw patterns in Chapter 3 where all known receivers have a small-lockscrew at the breechblock pin. *Courtesy Verle W. Hunt*

5

SINGLE-SHOT SOLID-FRAME MODELS

The single-shot Bullard rifle in its solid-frame version must have appeared towards the latter part of 1883 or very early 1884, as an article on the single-shot action appeared in the *Forest and Stream* issue of March 20, 1884.[1] A newspaper clipping sent to the author by J.P. Rehling indicates that Mr. W. Milton Farrow shot a .40-70 Bullard rifle on the Milwaukee range May 6, 1884. At some point in 1884 Mr. W. Milton Farrow was hired as assistant manager of the Bullard Arms Company to promote their new rifles in competition matches. Since his fame and expertise were well known in the world of single-shot target shooting, it stands to reason that the Bullard Company had a single-shot rifle already on hand for him to promote.

To the writer's knowledge, all solid-frame single-shot Bullards use the August 16, 1881, patent date. (See Plate 5-2, Number 2 style in Tables.) This poses some problems which will be dealt with later in the chapter on detachable-interchangeable barrel single-shot models. Almost all single-shot rifles appear with serial numbers starting in the 3500 range and going as high as 4100. I have been told by reliable sources that some single-shots appear in the 2XXX range, but have so far been unable to confirm this. Whether these rifles in the 2XXX range are solid-frame or not also remains a mystery at this time. The single-shot was based upon the mechanism of the repeater, with the exception of the rack and pinion and cartridge-carrier assembly, which was obviously not needed in a single-shot design. As the catalogs issued by the Bullard Company indicate, many of the parts between the repeater and the single-shot were interchangeable. These parts included levers, braces and brace links. This action was quite strong and very smooth, and those who own one today can attest to the fine workmanship of the rifle. Over the course of its production life, the single-shot rifle was probably available in almost any caliber imaginable. Documentation done by the author shows solid-frame models available in calibers from .40-90 Sharps to .38-90, and .38-55 down to .22-15. The tables at the back of the book will verify this information. The barrel bore could be designed to handle either patched or lubricated bullets, whichever the customer deemed most preferable.

All types of sights then popular on any single-shot target or hunting rifle were available and could be fitted at the factory. These were primarily made by Winchester and Lyman. The following was written by Mr. Edward Iwanchuk, a noted shooting enthusiast with a particular love of the single-shot rifle: "Various Bullard S.S. target rifles are seen with the Winchester windgauge front sight, some with spirit level inserts (on special order). The rear tang sight varies, as most seen are equipped with the mid-range vernier peep sight at a cost of 4.00 dollars and also a mid-range vernier peep sight manufactured by the Marlin-Ballard Company. (The S.S. target model is depicted in the 1887-1888 catalog.) In the last case a Marlin-Ballard windgauge front sight accompanies the mating tang sight, however this combination is seldom seen, and could be presumed scarce."

On January 28, 1886, Solomon K. Hindley and Edwin S. Field decided to file a patent application. That patent was granted July 6, 1886, and covered the solid-frame single-shot Bullard rifle. Why the delay in patenting a mechanism that they had used for at least two years? Research indicates no solid-frame single-shot with any patent date stamped on it except for the August 16, 1881 date (the same one used for the repeater in Plate 5-2).They were granted patent number 345058 for a "breech loading firearm" (Plate 5-1) on July 6, 1886.

More research will be required before it can be known conclusively whether the small-frame repeater or the solid-frame single-shot had the lesser production rate. One can be sure that both types had a low production volume.

Judging from the available photographs, it can be seen that the solid frame single-shots were often fitted with pistol grips, both checked and unchecked, in combination with Swiss butt plates.

Documented serial numbers from 3532 to 3660 show only solid-frame models, but after that the detachable-interchangeable models begin to predominate (see Tables). When the detachable-interchangeable model appears, it seems to supersede the solid-frame almost entirely, although a few still appear in the higher single-shot serial number ranges.

The standard Bullard practice of building arms only as the trade demanded seems to have been followed here too, as the number of single-shot Bullards uncovered has never been significant.

Of all the Bullard single-shot rifles chambered in .32-40, most all were chambered for the Winchester round, not the Bullard one.

In this writer's opinion, the only circumstances that could alter the number of single-shot rifles made would be the discovery of a military order of single-shot muskets or carbines. But even if such an order was found, the numbers would not change drastically because the large orders needed to sustain the company economically obviously never materialized.

The options available on the solid-frame rifles were too numerous to list here, for as can be seen by the tables and the literature in Chapter Nine, almost any combination of barrel length caliber, stock or wood was available to the discerning buyer. It is best to study the tables to see what really was produced, rather than relying on catalog descriptions.

Again it goes without saying that Bullards were produced for individuals on an as-ordered basis, therefore the hand-fitting of parts no doubt contributed to the higher costs of these rifles.

By all accounts these single-shots were well made and very accurate. They were heavily promoted but never became a commercial success. The 1880s saw many economic storms, but these alone should not have been able to sink the factory.

Also interesting, but undocumentable because of the few solid-frame single-shots seen, is the theory that there may be two frame sizes available in the solid-frames. The catalogs always list the .22-caliber single-shots separately from all the other larger calibers. The author has so far been unable to confirm this, but believes that the .22 receiver may be of smaller overall proportions or may just be thinner in its frame thickness. Someone who owns two solid-frame single-shots could photograph them together to provide photographic proof of the size difference between a .22 and its larger brothers. That the detachable-interchangeable barrel model of the Bullard single-shot came in two frame sizes is expressed very clearly in a letter from the Bullard Company to a Mr. Jackson (see Chapter 9).

One other undocumentable difference on the single-shots — apparently existing on one or both styles of single-shots — is rounded top receivers instead of the common or standard faceted receivers. Anyone who has one of these rifles and would like to add to our information on Bullards is more than welcome to contact the author. These rounded receivers could have been prototype models and even rarer than the standard single-shots.

FOOTNOTES

1 Philip B. Sharpe, "Early American Cartridge Rifle," *The Rifle in America* (New York: Funk and Wagnall Co., 1938), p. 53.

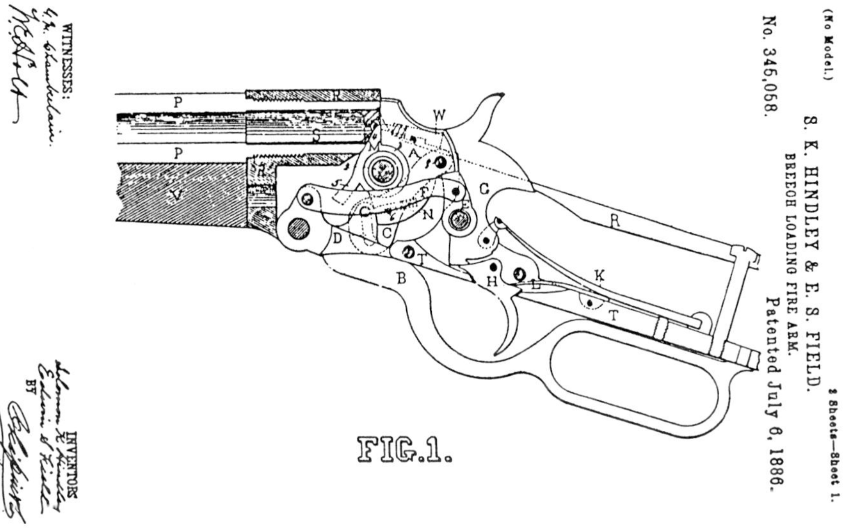

(No Model.) 2 Sheets—Sheet 2.

S. K. HINDLEY & E. S. FIELD.

BREECH LOADING FIRE ARM.

No. 345,058. Patented July 6, 1886.

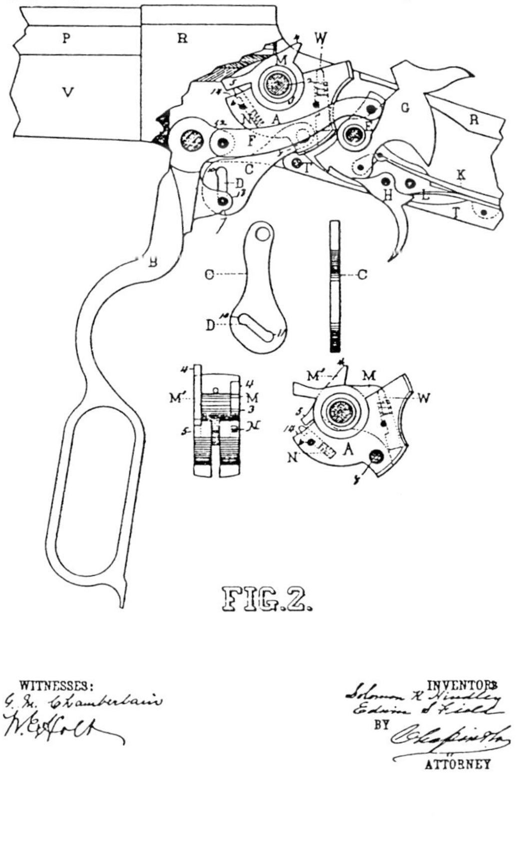

FIG.2.

WITNESSES:

INVENTORS

BY

ATTORNEY

UNITED STATES PATENT OFFICE.

SOLOMON K. HINDLEY AND EDWIN S. FIELD, OF SPRINGFIELD, MASS.

BREECH-LOADING FIRE-ARM.

SPECIFICATION forming part of Letters Patent No. 345,058, dated July 6, 1886.

Application filed January 28, 1886. Serial No. 190,043. (No model.)

To all whom it may concern:

Be it known that we, SOLOMON K. HINDLEY and EDWIN S. FIELD, citizens of the United States, residing at Springfield, in the county of Hampden and State of Massachusetts, have invented new and useful Improvements in Breech-Loading Fire-Arms, of which the following is a specification.

This invention relates to improvements in breech-loading fire-arms, the object being to provide an improved swinging breech-block, and mechanism for operating and locking the same, and an improved extractor and extractor-spring.

In the drawings forming part of this specification, Figure 1 is a longitudinal vertical section of a portion of the stock, frame, and barrel of a breech-loading gun, and a side elevation of the breech mechanism thereof, all constructed according to our invention, said figure showing a cartridge-shell in section in the barrel, and the breech and extractor mechanism in the positions they occupy after firing. Fig. 2 is a side elevation of the rear part of the barrel and stock, and of the frame immediately adjoining the latter, of the breech mechanism shown in the position which the parts thereof occupy after the extraction of the shell, and when they are in a position to permit of reloading the gun, and edge and side elevations of the breech-block with the extractors thereon, and of the breech-block link.

In the drawings, R is the frame of the gun, P is the barrel, and V is the stock.

A is the breech block, pivoted on the pin 2 in the frame, and adapted to have its upper end, in which is located the firing-pin W, swing against and close the rear end of the barrel.

M M' are the extractors, two being shown in the detached views of the breech-block in Fig. 2; but one may be omitted, if desired. Each extractor consists of a ring, 3, having the arms 4 and 5 thereon, the first for engagement with the cartridge-head, and the second for engagement with the spring-actuated extractor-pin N, which is located in the lower end of the breech-block. Said ring 3 of the extractor is let into the side of the breech-block, so that its outer side and that of said arms are in the same plane as the side of the latter, and it has a free swinging motion on the said block to the extent shown in Fig. 2, where its two positions are shown. The side of the rear end of the barrel P is cut away sufficiently to receive the said arm 4 of the extractor, and when the breech-block is against the end of the barrel, as in Fig. 1, to allow said arm 4 to take the position there shown. The said extractor-pin N is, as aforesaid, located in the breech-block behind the extractor-arm 5, and is thrown outward by a coil-spring shown in dotted lines thereon. One side of said pin is notched, as shown in dotted lines, and a pin, 6, passes through the breech-block under pin N, and within said notch therein, whereby the latter is secured in the block, but is free to have a limited endwise movement therein. The outer end of pin N is V-shaped to adapt it to enter a notch, 7, in a projection on the guard-lever B, when the parts are in the position shown in Fig. 1, and serve to hold said lever in the position there shown. The breech-block has a central slot about midway between its sides, as shown in Fig. 2, in which is pivoted one end of the link C at 8, and near the opposite end of said link is formed the slot D, having the curved recesses 10 11 at opposite ends thereof. The edge of the breech-block at 14, Fig. 2, is slightly curved for a purpose below described. The guard-lever B is pivoted near the lower side of the frame R, and is provided with the usual finger-loop, and has thereon the projection 12 opposite its pivot, and the projection 13 below the latter. Two brace-links, F, (only one of which is shown in the drawings,) have one end pivotally connected to said projection 12 on the guard-lever, and they extend each side of the breech-block to the rear of the latter, and are there pivotally connected to the swinging brace E, which is hung on the hammer-pivot. The said projection 13 on the guard-lever B is slotted to receive the lower end of the breech block link C, and the latter is connected to said projection by a pin passing through the latter and the slot D therein.

G is the hammer, pivoted in the usual manner in the frame, H is the sear, K is the main spring, and L is the sear-spring.

The operation of our improvements is as follows: The parts of the breech mechanism

Plate 5-1.

Plate 5-2. (SN 3532 Sharps .40-90) Solid-frame model. Close-up of receiver address. Patent date used was same as for repeaters, Aug. 16, 1881. *Courtesy Steven Petris*

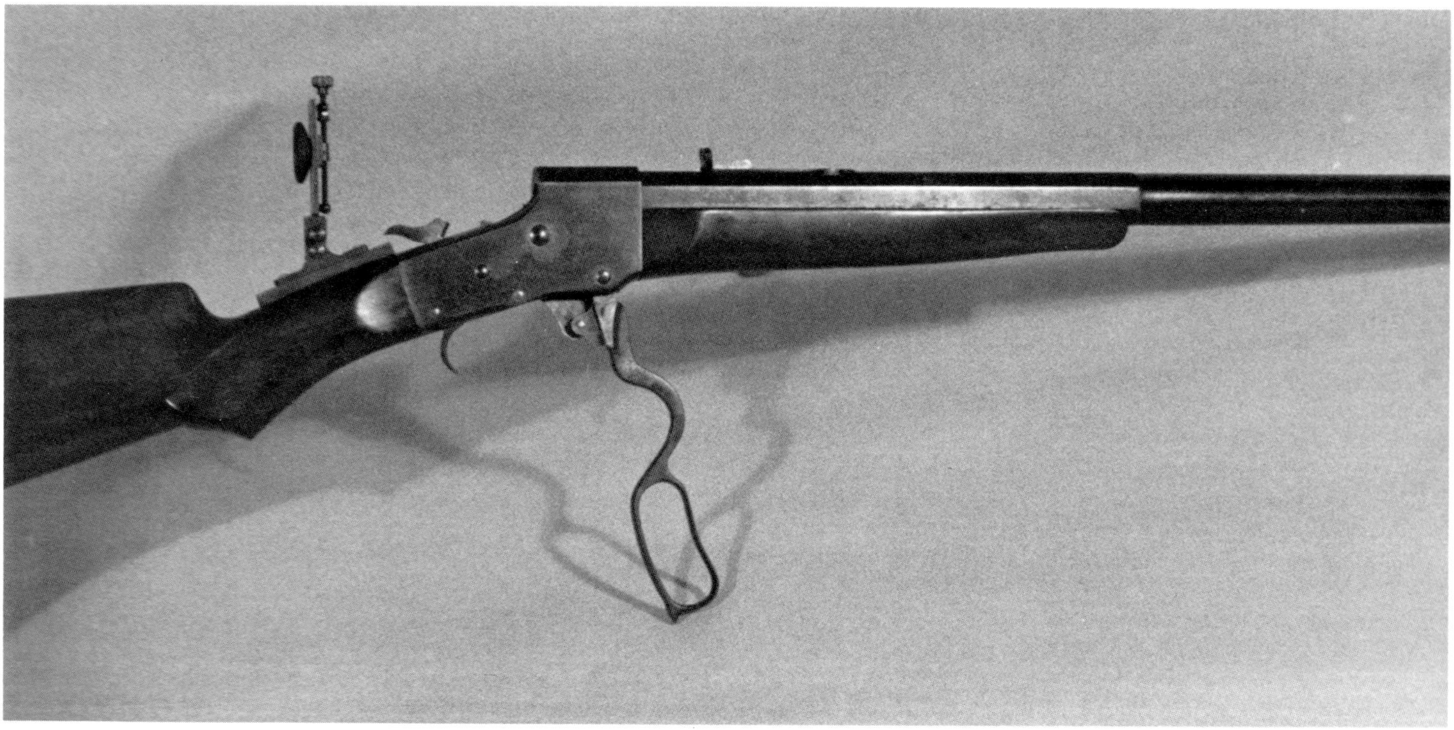

Plate 5-3. (SN 3532 Sharps .40-90) Solid-frame, single-shot. Action open. *Courtesy Steven Petris*

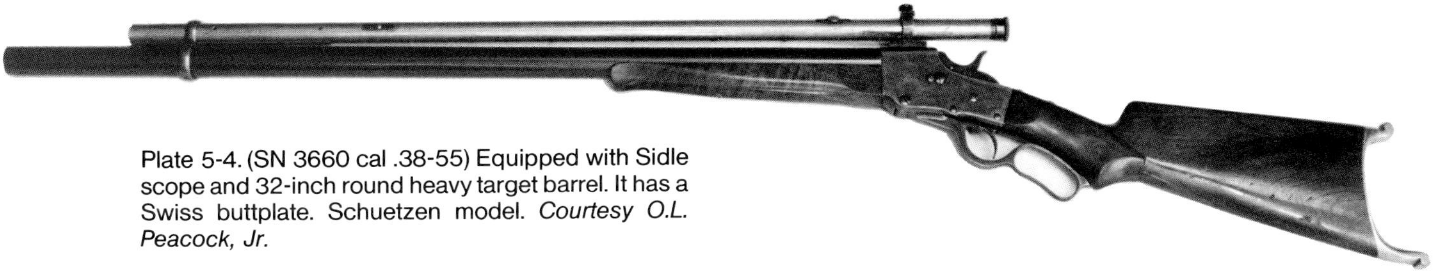

Plate 5-4. (SN 3660 cal .38-55) Equipped with Sidle scope and 32-inch round heavy target barrel. It has a Swiss buttplate. Schuetzen model. *Courtesy O.L. Peacock, Jr.*

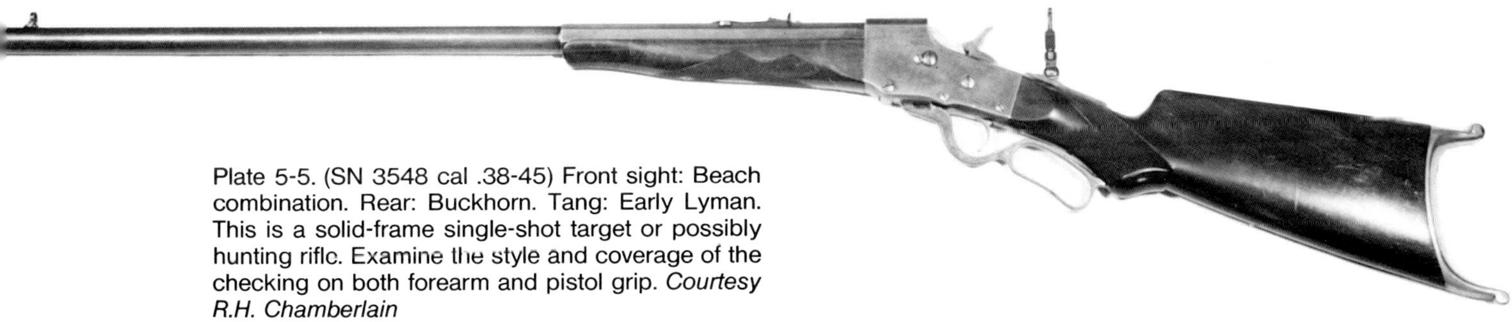

Plate 5-5. (SN 3548 cal .38-45) Front sight: Beach combination. Rear: Buckhorn. Tang: Early Lyman. This is a solid-frame single-shot target or possibly hunting rifle. Examine the style and coverage of the checking on both forearm and pistol grip. *Courtesy R.H. Chamberlain*

Plate 5-6. (SN 3657 cal .38-55) Solid-frame, single-shot. 30-inch part round, part octagonal barrel. Nickel-plated receiver with still another style of checking coverage, especially noticeable on fore-arm. This rifle is stamped with July 6, 1886 Patent date, should have used the Aug. 16, 1881 date. Front sight: Beach combination. Tang: Ballard mid-range. *Courtesy Matthew A. Bailey*

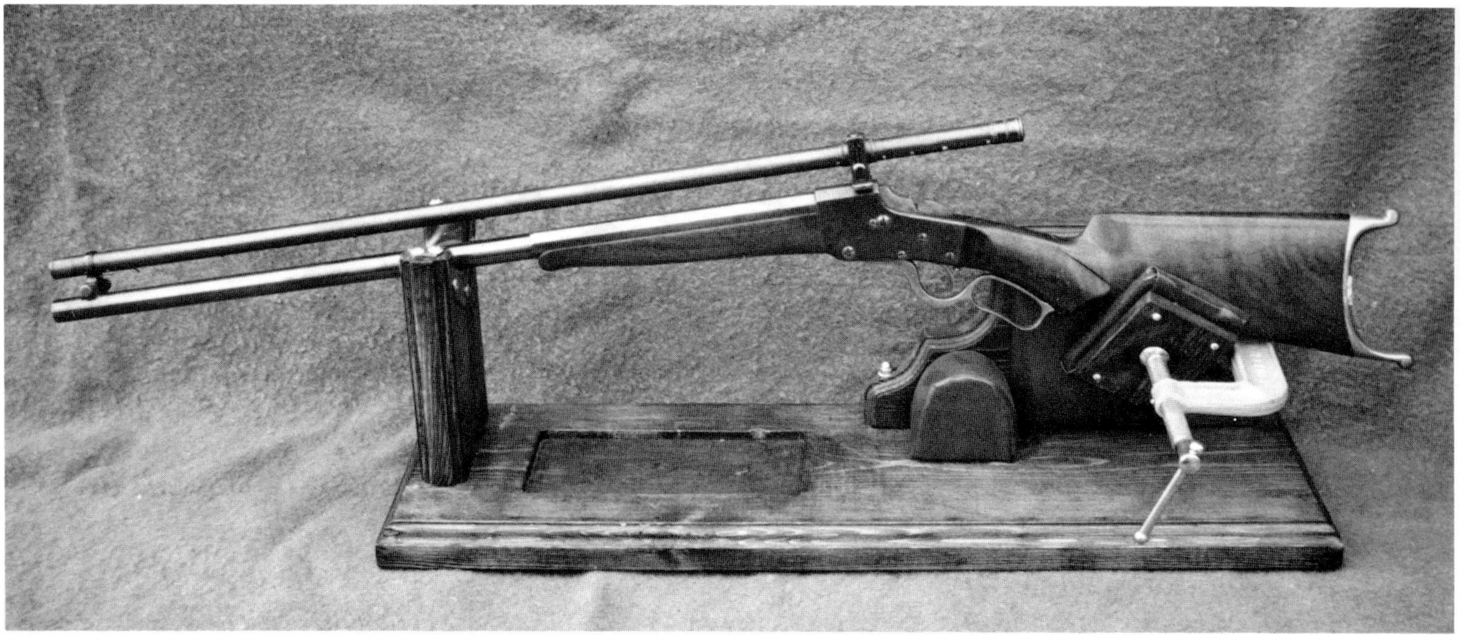

Plate 5-7. (SN 3559 cal .32-40) It has a 25-inch barrel mounted with Malcolm scope. *Courtesy R.E. Keeler*

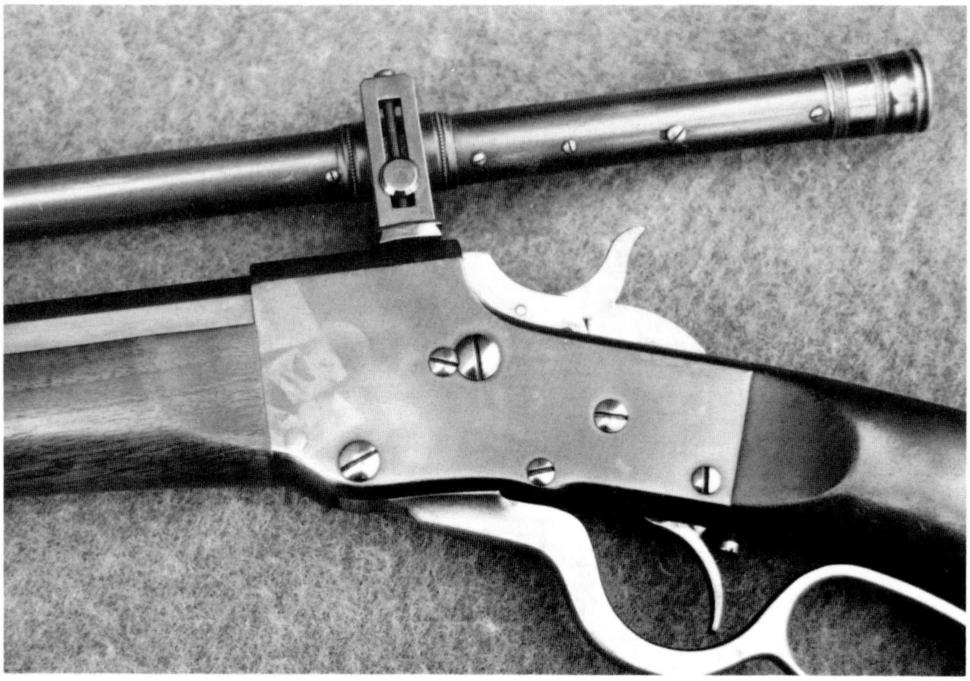

Plate 5-8. (SN 3559) Solid-frame. Close-up of Malcolm globe sight mounts. *Courtesy R.E. Keeler*

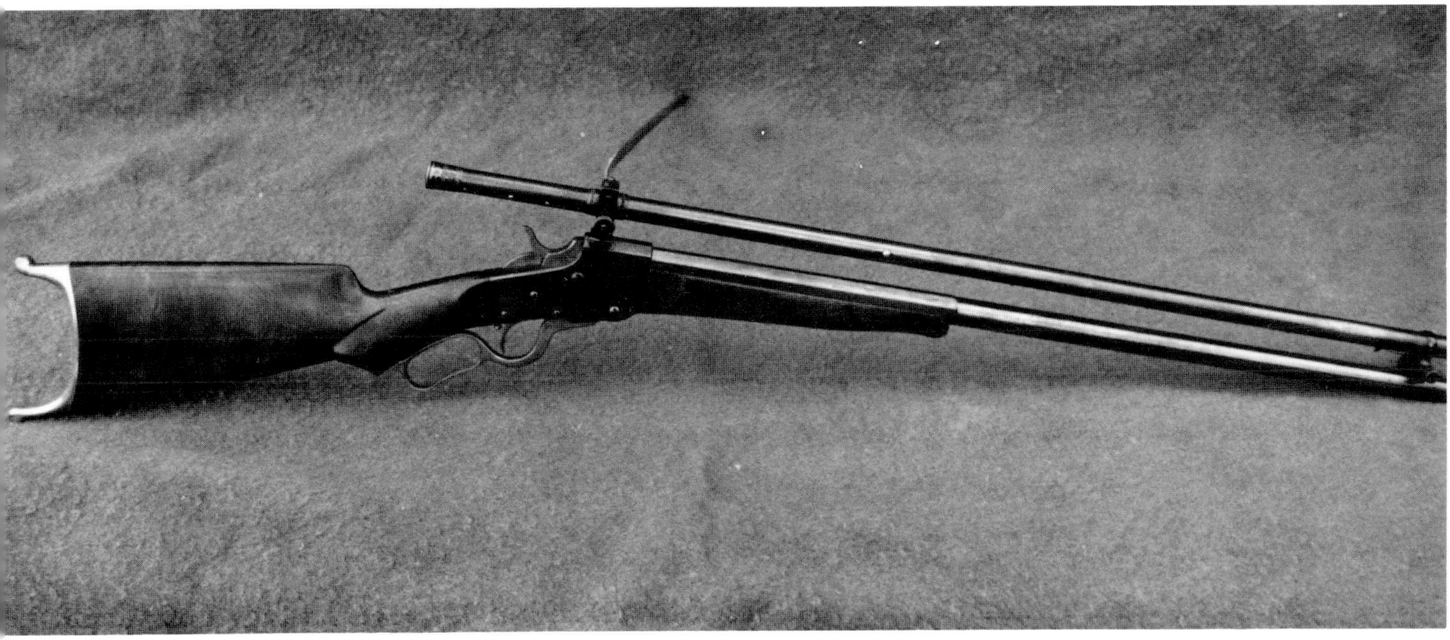

Plate 5-9. (SN 3559 cal .32-40) Swiss buttplate and right-side overall view. *Courtesy R.E. Keeler*

Plate 5-10. (SN 3559) Overhead detail view of rear sight mount. *Courtesy R.E. Keeler*

1

2

3

Plate 6-1. 1. Solid-frame model with later Lyman tang sight with lockable post. 2. (SN 3882 Winchester cal .50-95) Large-frame. Detachable-interchangeable barrel model. This rifle has two barrel retaining screws. 3. Small-frame. Detachable-interchangeable barrel model. This rifle has a Stevens 'tip-up' rear sight and an early Lyman tang sight. *Courtesy O.L. Peacock, Jr.*

6

SINGLE SHOT DETACHABLE-INTERCHANGEABLE BARREL MODELS

This mechanism was patented on October 30, 1888. The patent number was 391953 (Plate 6-5). To the author's knowledge, this date was never used on any Bullard rifle as part of the address or patent information. The detachable-interchangeable models carried the patent date of July 6, 1886, already detailed in the previous chapter (Plate 5-1). The factory's use of irrelevant patent dates on their single-shot rifles and the reasons behind their use are still a mystery.

The patent 391953 was filed for on May 9, 1887, and the first known advertising was dated January 15, 1887 (Plate 9-7). An advertisement carried in the February 1887 issue of *The Rifle* states, "Bullard rifles with detachable barrels — now manufacturing in .22, .32, .38 with .40 and .50 to come later." These rifles are also mentioned in the 1887 catalog.

There were two frame sizes for these rifles, one for calibers .38 and smaller, and a larger width frame for the .40 and larger calibers. That is why the above mentioned advertisement announced that the .40 and .50 calibers were to come later. Had there been no difference in the frame, they could have supplied large calibers right from the start. Mr. Field's letter to Mr. Jackson (Chapter 9) makes this abundantly clear and even dates the beginning of production of these rifles. Judging by the photographs, the large- and small-frames do not appear to be vastly different in overall dimensions, in the manner of the repeaters. The differences in size seem more apparent in the frame width and breech flats or facets which are wider on large-frames. Examining the photographs, you will see a slightly different lever design between the large-frame and small-frame rifles (Plate 6-1). Where the lever merges with the frame at its pivot point on the large-frames, the leading edge of the lever appears to be truncated and less flowing than its small-frame counterpart. A second small distinction between them is that on the small-frames the forearm seems to meet the frame exactly level or in line with the line cutting through the receiver where the barrel and breech separate from the frame. On the large-frames the forearm line meets the frame at a slightly higher point than the line through the receiver (Plates 6-28, 6-1).

The method used for holding the barrel in the receiver was simple, solid and strong. This model of rifle appears to have almost supplanted in popularity the solid-frame model when it appeared in 1887. This popularity was probably due to its versatility in allowing the owner the luxury of two or more different caliber barrels on one frame, with no sacrifice in accuracy! Note that if you had a .40-caliber or large-frame rifle, you could only buy barrels of larger or the same caliber, but not smaller. The same was true for the small-frames, in that you could not fit a barrel of larger caliber than .38 to a small-frame, but you could go to a smaller caliber. The inherent strength and ingenuity of the design can be seen in Plates 6-6, 6-7, 6-9.

The barrel, breech and forearm were removed from a specially machined dovetail in the receiver with the simple removal of one or in some instances two screws on the left side of the receiver (Plates 6-1, 6-9).

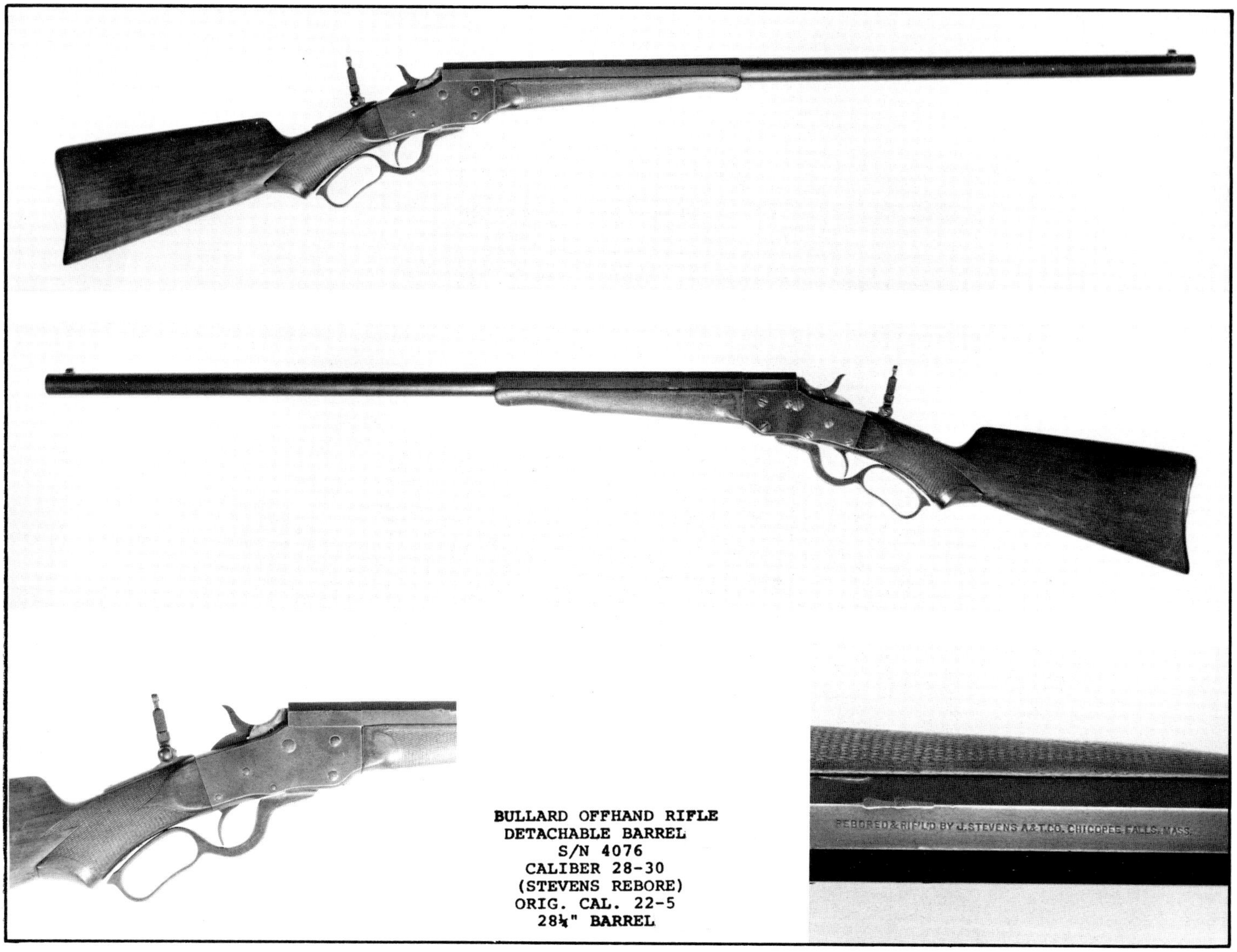

Plate 6-2. (SN 4076 cal .28-30) Highest Bullard serial number yet discovered by the author. *Courtesy Fred H. Cates*

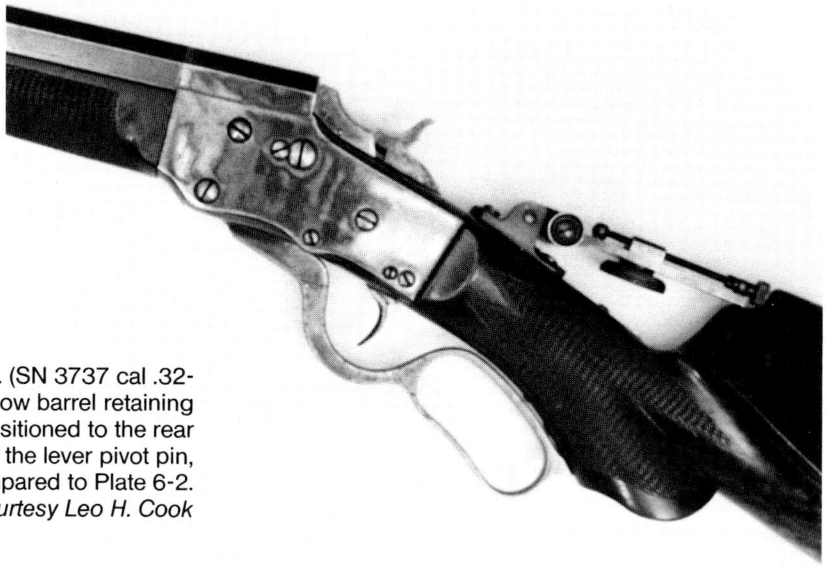

Plate 6-3. (SN 3737 cal .32-40) Note how barrel retaining screw is positioned to the rear of the lever pivot pin, compared to Plate 6-2. *Courtesy Leo H. Cook*

A curious situation arises when viewing Plates 6-1, 6-2, and 6-3 in regards to the positioning of the retaining screw. It will be noted that seldom is the retaining screw drilled through the receiver in the same location. This means that only the original barrels supplied with a particular receiver would fit it. Barrels from another detachable model would not likely mate to a receiver with a different serial number. This anomaly further points to the fact that the Bullard factory never really got far from expensive hand labour. Had these arms been mass-produced, the holes surely would have been drilled in a uniform location.

With the exceptions mentioned in Chapter 5, all detachable-interchangeable models appear in the 3500 to 4100 serial number range. None have been accounted for by the author outside this range of numbers (see tables).

One specialized version of the detachable-barrel model is mentioned in the May 1887 issue of *The Rifle*, page 8. It advertises that a sample has been made of this model with three different barrels *and* detachable triggers (see Plates 6-15, 9-18). They were as follows: (1) 14-pound, 30-inch, .32-caliber, heavy Sheutzen barrel with set triggers; (2) 10-pound, .32-caliber with regulation 3-pound pull trigger; (3) .50-caliber smoothbore. This rifle was also shown in some advertising of the period. How many were manufactured remains a mystery, but I believe that serial number 3780 is one of these rarities or possibly the actual one advertised. It is of .32 caliber with a 30-inch long barrel and double-set triggers. Note the specially designed lever to accommodate the larger space necessary for the double triggers. The triggers themselves appear to be externally screwed to the bottom of the receiver to facilitate their easy removal and replacement. Unfortunately, at the time of press this rifle was unavailable for closer scrutiny. One other difference in the receiver occurs at the barrel lock and screw. There appears to be one lock screw and one pin holding the barrel in the frame. This differs from the one- and two-screw methods of locking the barrel to the receiver shown in Plate 6-1 Without doubt this is one of the finest and rarest of all Bullard single-shot rifles.

The Bullard Company heavily stressed its desire to be a force to be reckoned with in the world of target shooting by hiring such notable shooters as Mr. W. Milton Farrow and by sponsoring shooting events such as the Bullard matches. At least four of these events were held. As well, the factory grounds were used by a local shooting club for their events. More will be said of these matches later, in the chapter on advertising, as these events were held with one main purpose, the promotion of Bullard arms.

The utter simplicity and inherent strength of the Bullard detachable-barrel system is further underscored by the superb drawings rendered by Craig R. Hostetter for this book. These drawings, though not to scale, give all the dimensions of the barrel and sleeve (Plate 6-17, 6-19) for a small-frame single-shot.

Plates 6-4, 6-13 and 6-20 show the locations of the serial number and caliber stampings on the single-shot rifle. No doubt there are several other serial number stampings on the internal receiver parts.

Having stated earlier, in Chapter 5, that the solid-frame single-shots used the 1881 style of address or the number 2 address in the tables, it can now be seen from Plates 6-12, 6-16 and 6-22 that the number 4 address type with the 1886 date was used for the detachable-barrel models. The author has not seen any deviation from this pattern, but it would not come as a surprise if a rifle or two were stamped incorrectly, as was the case with one small-frame repeater.

Close examination of the many photographs in this chapter, such as Plate 6-24, shows the myriad small differences encountered on the receivers of these rifles. Some have one barrel retaining screw, some have two (possibly standard on the large-frames, but not enough of them have been seen to substantiate this), while others have one screw and one smaller diameter pin. The pins may have been the result of experimentation on the factory's part to cut out the cost of threading holes, but the cost of machining a pin to close tolerances would also be high, so what advantage they hoped to gain is still not known. These pins show up too frequently to be an outside modification, and again the skill required to machine the pins to close enough tolerances to allow them to fit tightly would probably be beyond the scope of most gunsmiths. The true reasons for the variety of barrel pinning methods may never be known unless factory correspondence is found relating to the manufacture of the rifles.

Similar to the solid-frame single-shots, these rifles came with a full assortment of options, including barrel lengths and weights. But unlike Winchester no catalog yet uncovered assigns a number to a barrel weight and no grading system like Winchester's appears for differentiating between wood grains, save what Bullard refers to as a "fancy wood." The fancy woods are just that, burl or crotch walnut with lots of figure, but it is beyond the author's expertise to say where they would fit in on Winchester's wood-rating system. The readers can decide for themselves where and how the wood should be graded (See back cover).

SOME SERIAL NUMBER LOCATIONS ON THE BULLARD DETACHABLE-INTERCHANGEABLE BARREL MODEL SINGLE-SHOT RIFLE.

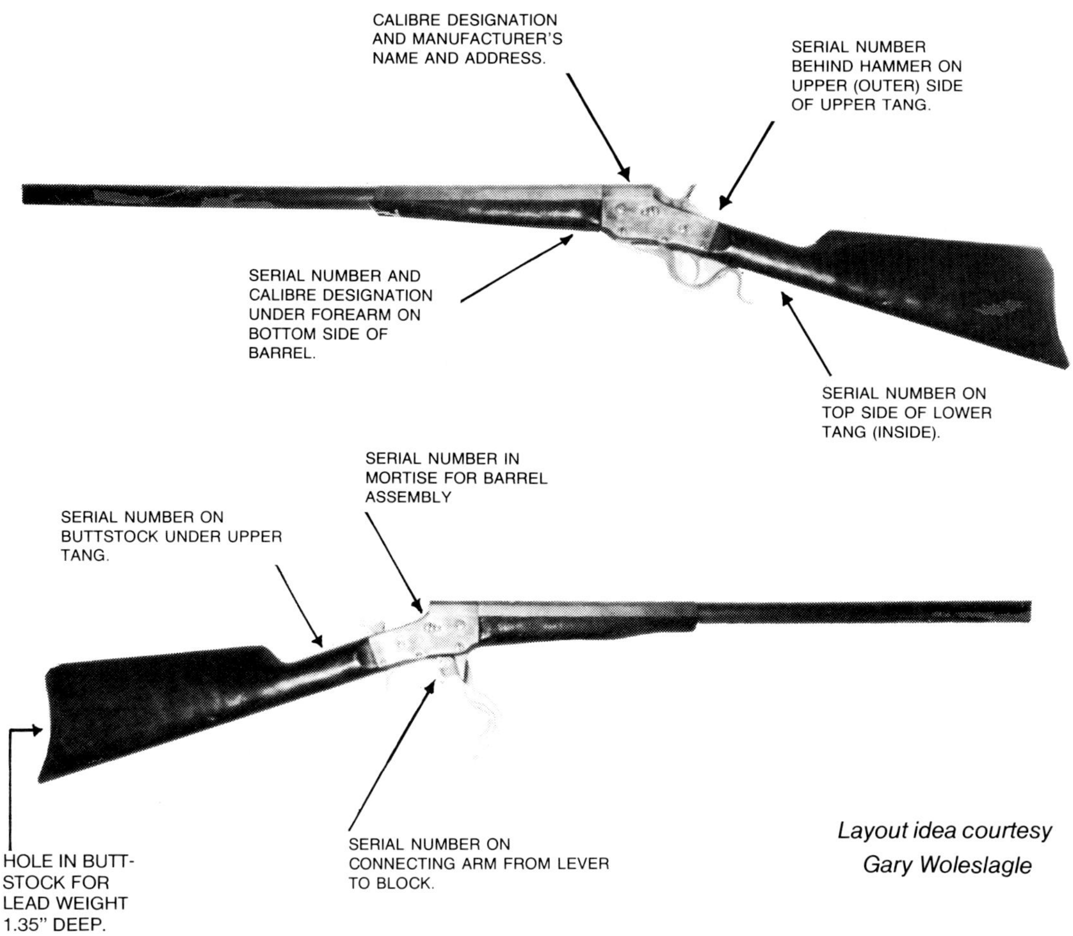

Layout idea courtesy Gary Woleslagle

Plate 6-4. (SN 4038 cal .32-40).

UNITED STATES PATENT OFFICE.

EDWIN S. FIELD AND SOLOMON K. HINDLEY, OF SPRINGFIELD, MASSACHUSETTS.

FIRE-ARM.

SPECIFICATION forming part of Letters Patent No. 391,953, dated October 30, 1888.

Application filed May 9, 1887. Serial No. 237,523. (No model.)

To all whom it may concern:

Be it known that we, EDWIN S. FIELD and SOLOMON K. HINDLEY, citizens of the United States, residing at Springfield, in the county of Hampden and State of Massachusetts, have invented new and useful Improvements in Detachable Barrel-Connections for Fire-Arms, of which the following is a specification.

This invention relates to breech-loading fire-arms, and pertains to improvements in detachable barrel-connections therefor, the object being to provide a receiver or frame for a fire-arm adapted to receive and hold a detachable breech secured to the end of a barrel whereby several barrels may be used in succession with one receiver; and the invention consists in the peculiar construction and arrangement of the receiver, the barrel-breech, and means for securing the latter in the receiver, all as hereinafter fully described, and pointed out in the claim.

In the drawings forming part of this specification, Figure 1 is a perspective view of a gun-receiver or frame having a detachable barrel-breech applied thereto, said receiver and breech being constructed according to our invention, said figure showing the rear end and a part of the side of the receiver broken off and a part of a barrel and its fore-arm, a breech-block, and an extractor. Fig. 2 is a side elevation of the front end of the receiver or frame (the rear end being broken off) with the detachable barrel-breech therein, said figure showing a part of a barrel connected with said breech and a portion of the fore-arm thereunder, and a portion of the breech-block in position against the rear end of said breech. Fig. 3 is a front end elevation of the receiver with the detachable breech secured therein. Fig. 4 is a perspective view of the extractor, and showing in dotted lines the position of a securing-screw. Fig. 5 is a perspective view of one of the screws employed to attach the detachable breech to the receiver.

In the drawings, 3 is the metallic receiver or frame of the arm, to which are attached the stock, the breech mechanism, and the barrel; but instead of making said receiver as is ordinarily done to adapt it to have the barrel screwed directly to the end thereof, it is constructed with the sides or "cheeks" *a a* thereof extending parallel with each other toward the forward end of the receiver and left separated at that point, as shown in Fig. 3, and having their inner opposite sides properly formed and finished, as below described, to receive between them the barrel-breech 4. At the rear of the barrel-breech 4 the usual breech-block 5 is hung between the sides of the receiver, and any suitable lock mechanism may be used in connection with said breech-block. Fig. 1 shows the barrel-breech 4 not quite back to its place in the receiver 3, and it is left in that position to more clearly show the manner of applying the shell-extractor 6 to the rear end of the barrel and the manner of engaging the extractor (which is shown in said figure partly drawn out) with the breech-block 5. The said detachable barrel-breech 4 is a metallic block bored longitudinally and internally screw-threaded from its front end to provide for screwing the rear end of the barrel 7 into it, the rear end of the barrel being screwed against an annular collar, *c*, at the rear end of the breech 4, the border of said collar being in a plane with the bore of the rear end of the barrel, as shown. The letter *v*, Fig. 3, indicates the plane of the screw-thread in breech 4.

In practice each of the several barrels which it may be desired to use in the gun, whether shotgun or rifle barrels, has one of said breeches 4 firmly screwed onto its rear end, and the several barrel-breeches are interchangeably fitted into the receiver 3 between its cheek-pieces *a*. The breech 4 has a longitudinal groove, *n*, in each side thereof, in which engage two longitudinal ribs, *w*, on the inner opposite sides of the cheek-pieces *a* of the receiver, and the end of said breech, being entered between the ends of said cheeks, is easily pushed back to its position in the receiver, as shown in Fig. 2, and is then secured in the receiver by a screw or pin, 8, passing transversely through the receiver and the under side projecting part of the breech. As an additional precaution against any possible motion of the barrel-breech 4 in the receiver when the gun is fired, two screws, 9, are put in the end of said breech, which serve as keys to aid in preventing any upward movement of the forward end of the breech or that end nearest the muzzle of the barrel. The downhanging portion *h* under the barrel-breech serves no purpose but to fill the space between the ends of the receiver, and its inner side is made of a form to adapt it to the contents of the latter.

An extractor, 6, having a shank, J, with a notch in one side, as shown, and a head, K, thereon, is fitted to slide in a cylindrical perforation between the outside of the barrel and the inner side of the breech 4, Fig. 1 showing said extractor in that position and partly drawn out. A screw (indicated by dotted lines in Fig. 4) passes through the side of the breech 4 and has its point engaging in said notch in the side of the extractor-shank, whereby the extractor, while allowed to move longitudinally, is retained between the barrel and its breech.

In practice the head K of the extractor is let into the side of the barrel so that it comes to a proper position behind the rim of the cartridge when the latter is put in the barrel, and one end of the head K (the lower end shown in Fig. 4) engages with the end of a recess, *e*, (see Fig. 1,) on one side of the pivoted breech-block 5, so that when said block is swung to the position shown in said figure after the arm has been fired the extractor is drawn out and the shell is extracted from the barrel. Each barrel and breech used in the receiver is provided with an extractor adapted to the bore of the barrel.

The perforation *f* in the pending part of the breech 4 is made to provide convenient means for attaching the fore-arm 10 (portions of which are shown in Figs. 1 and 2) under the barrel.

What we claim as our invention is—

The frame of a breech-loading fire-arm, having its parallel sides *a a* extending toward the barrel and separated at their forward termination, and each having a longitudinal rib, *w*, on its inner side, combined with a barrel-breech, 4, substantially as described, to enter between said sides of the frame, having longitudinal grooves *n* in its outer sides, in which said ribs engage, and a fastening pin passing transversely through said frame and breech, substantially as set forth.

EDWIN S. FIELD.
SOLOMON K. HINDLEY.

Witnesses:
G. M. CHAMBERLAIN,
WM. H. CHAPIN.

(No Model.)

E. S. FIELD & S. K. HINDLEY.

FIRE ARM.

No. 391,953. Patented Oct. 30, 1888.

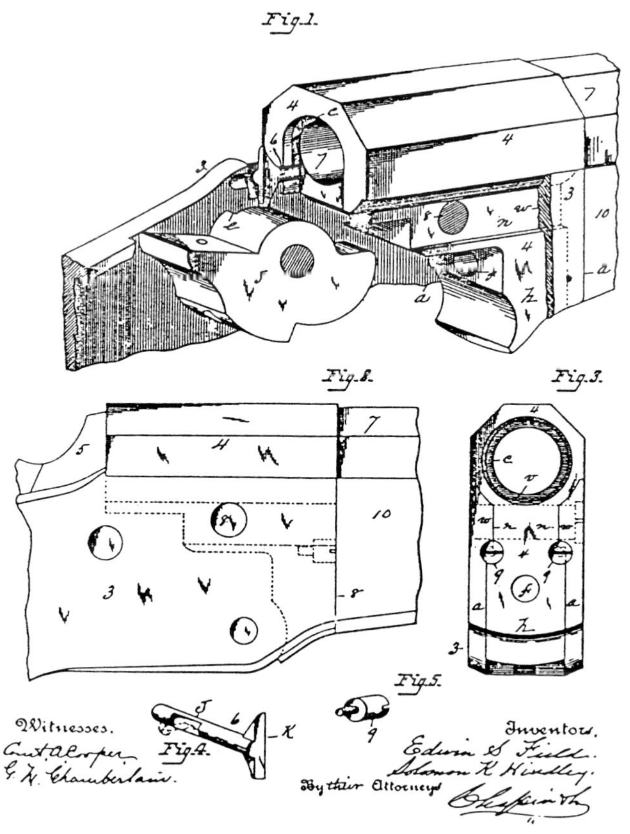

Plate 6-5.

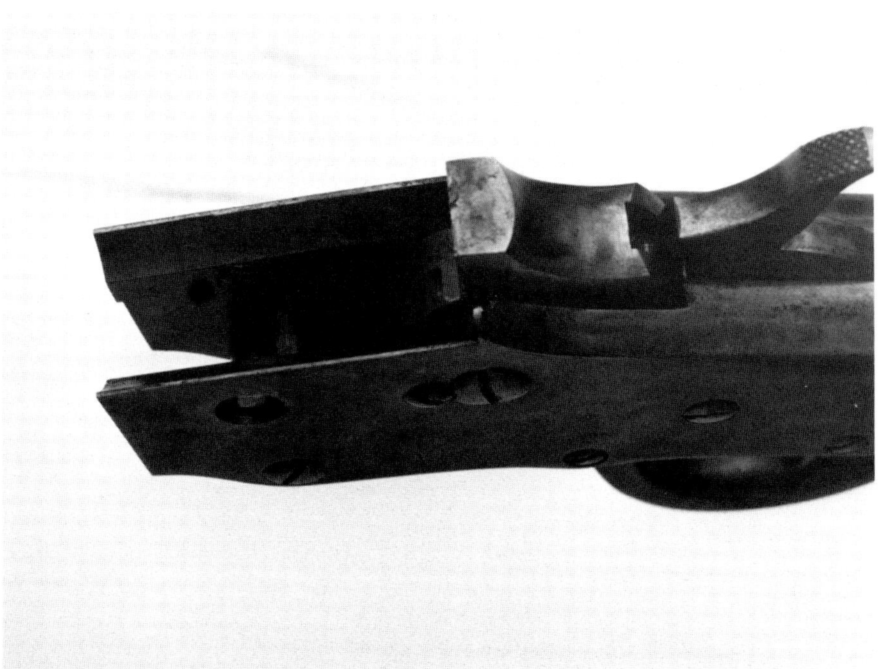

Plate 6-6. Close-up of receiver dovetail which also shows up well in patent drawings. *Courtesy B.A. Boucher*

Plate 6-7. Upper view of receiver disassembled. Each barrel came with its own extractor. *Courtesy B.A. Boucher*

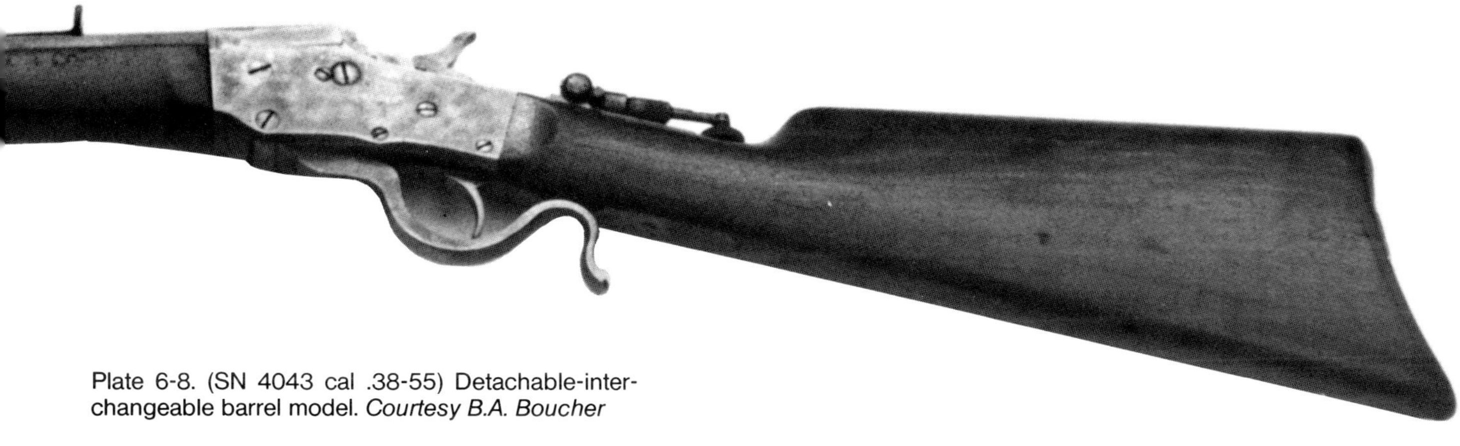

Plate 6-8. (SN 4043 cal .38-55) Detachable-interchangeable barrel model. *Courtesy B.A. Boucher*

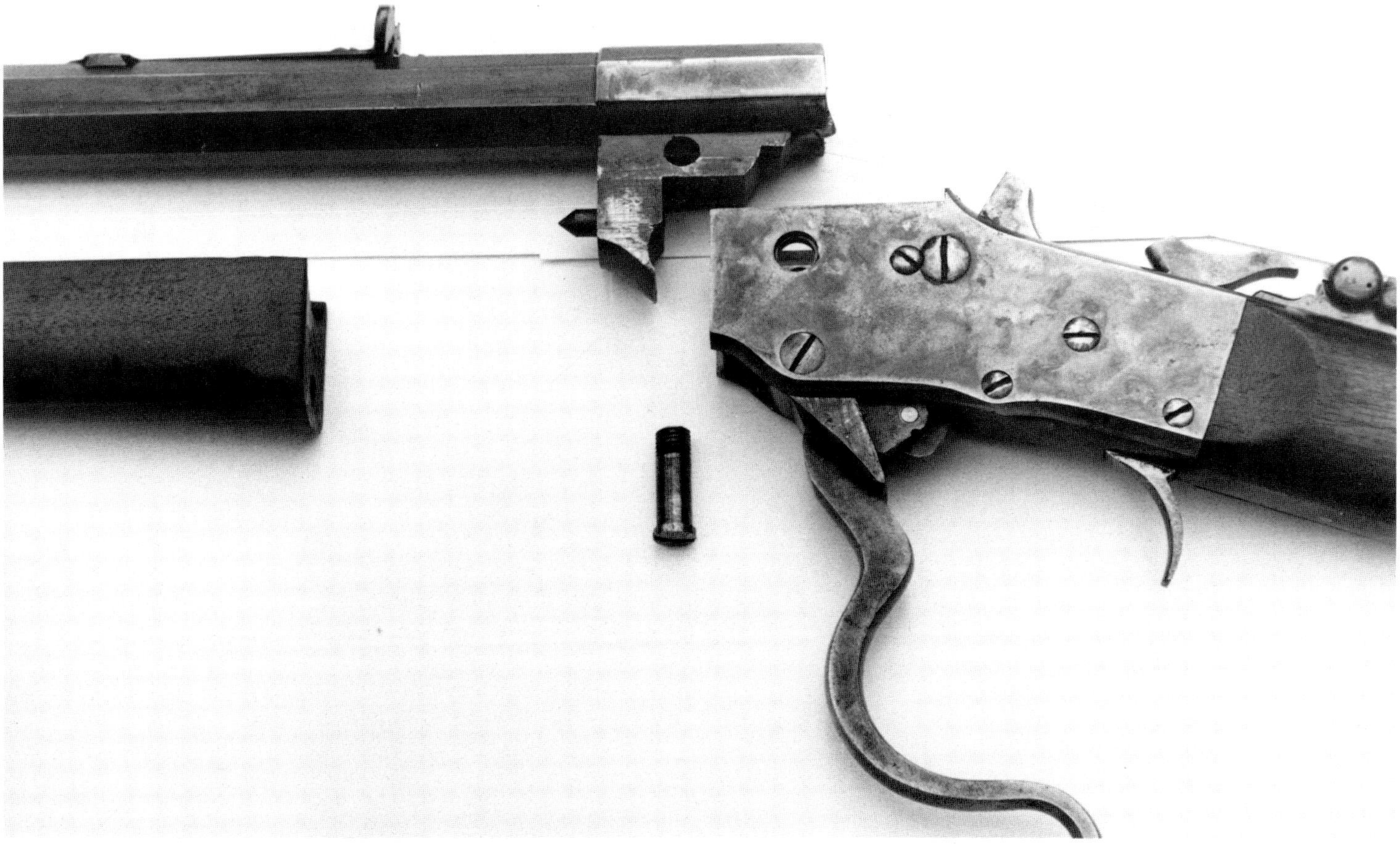

Plate 6-9. (SN 4043 cal .38-55) One screw holds the barrel assembly to the action — simplicity itself. *Courtesy B.A. Boucher*

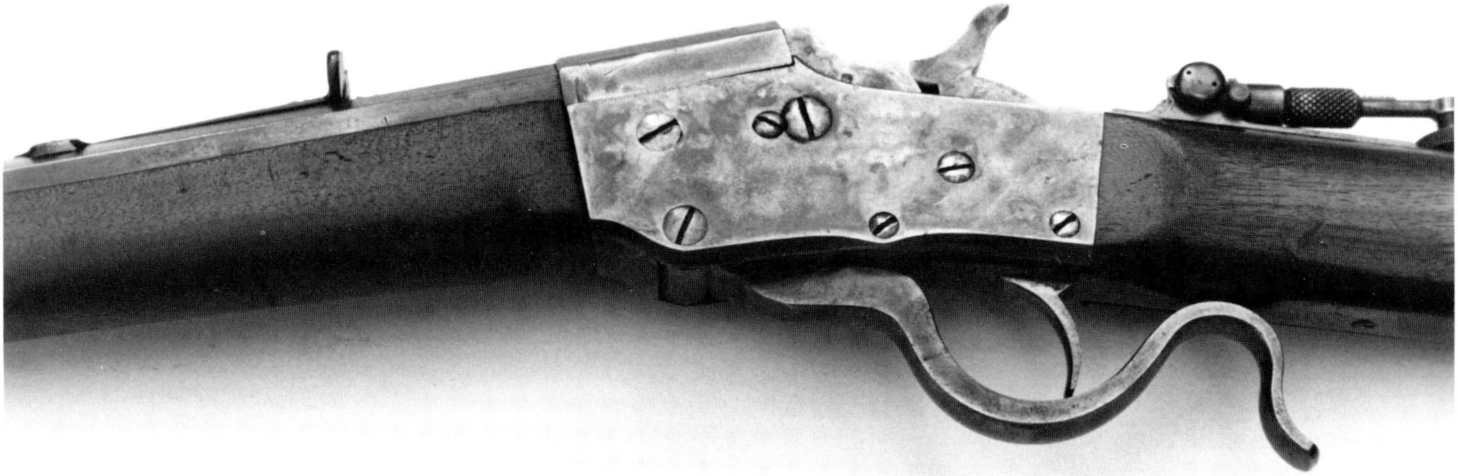

Plate 6-10. (SN 4043) Note the barrel lockscrew is located almost vertically above the lever pivot pin. The serpentine or 'S' lever is frequently seen on these rifles. *Courtesy B.A. Boucher*

Plate 6-11. (SN 4043) Right-hand close-up receiver and serpentine finger lever design. *Courtesy B.A. Boucher*

Plate 6-12. (SN 4043) Close-up of number 4-style address and serial number. Examine the faint stamping of the serial number behind the hammer. *Courtesy B.A. Boucher*

Plate 6-14. (SN 4043) Detail of Bullard gutta-percha buttplate design. Turkey motif used on many single-shots, small-frames and some large-frames. It should be emphasized that there is a substantial difference in buttplate size between small-frame and large-frame rifles of all types owing to the dimensional differences in their stocks. *Courtesy B.A. Boucher*

Plate 6-13. (SN 4043) Above: serial number stamp on forearm. Hole in forearm fits peg shown right. Right: detail showing locations and styles of caliber and serial number locations under barrel. *Courtesy B.A. Boucher*

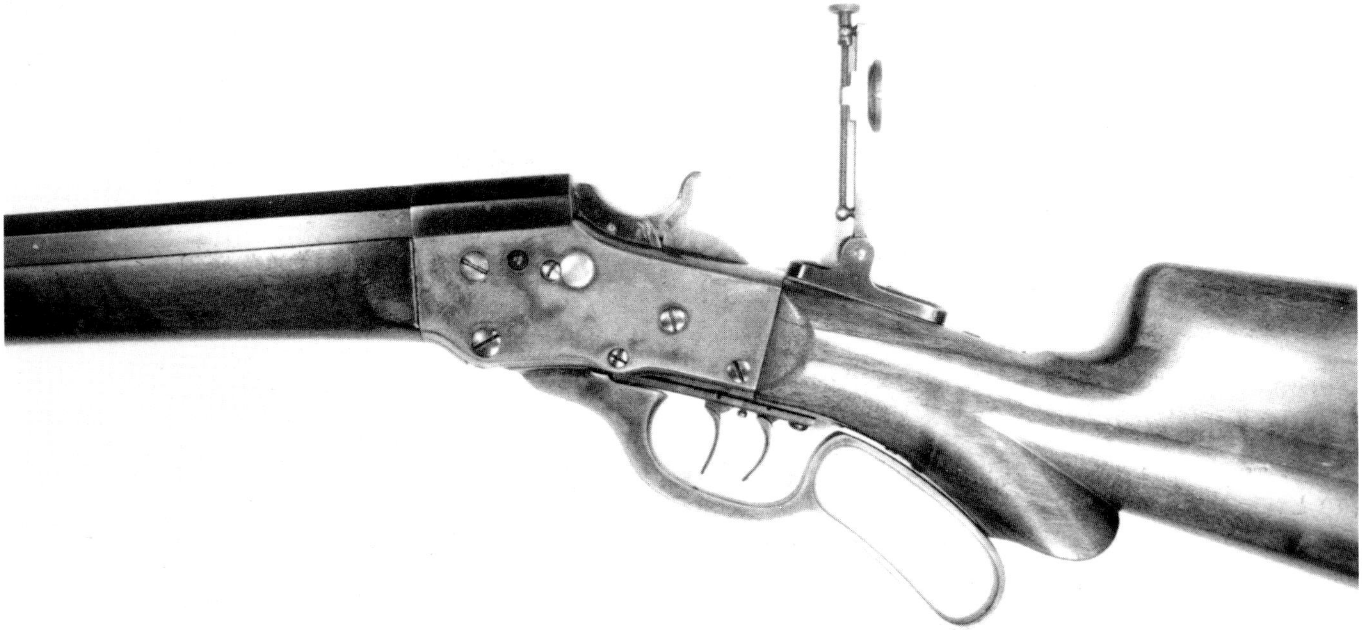

A.

B.

Plate 6-15. (SN 3780 cal .32-40) A. Note extra pin in receiver just behind barrel lock pin. Also that breechblock pin head has no slot on it. This type of slotless head appears infrequently. B. Equipped with a Ballard mid-range tang sight. Note retaining plate and holding screw on trigger assembly which allows triggers to be changed quickly. The author believes this rifle is one of the few Bullards with quickly detachable triggers. See chapter nine, for full details. *Courtesy Herb Glass*

A.

Plate 6-16. Top: left view of receiver. Note position of barrel locking pin in relation to the lever retaining pin. Compare other plates to see the difference. Middle: right-side view. Note serpentine lever. The line that separates the breech from the receiver is readily apparent in this photograph. Bottom: standard number 4 on tables, two-line receiver address and patent stamp. Caliber markings were generally heavily stamped compared to the same markings on the repeaters, which had to be rolled on. *Courtesy Richard A. Failes/Photos by Ted Jez*

B.

C.

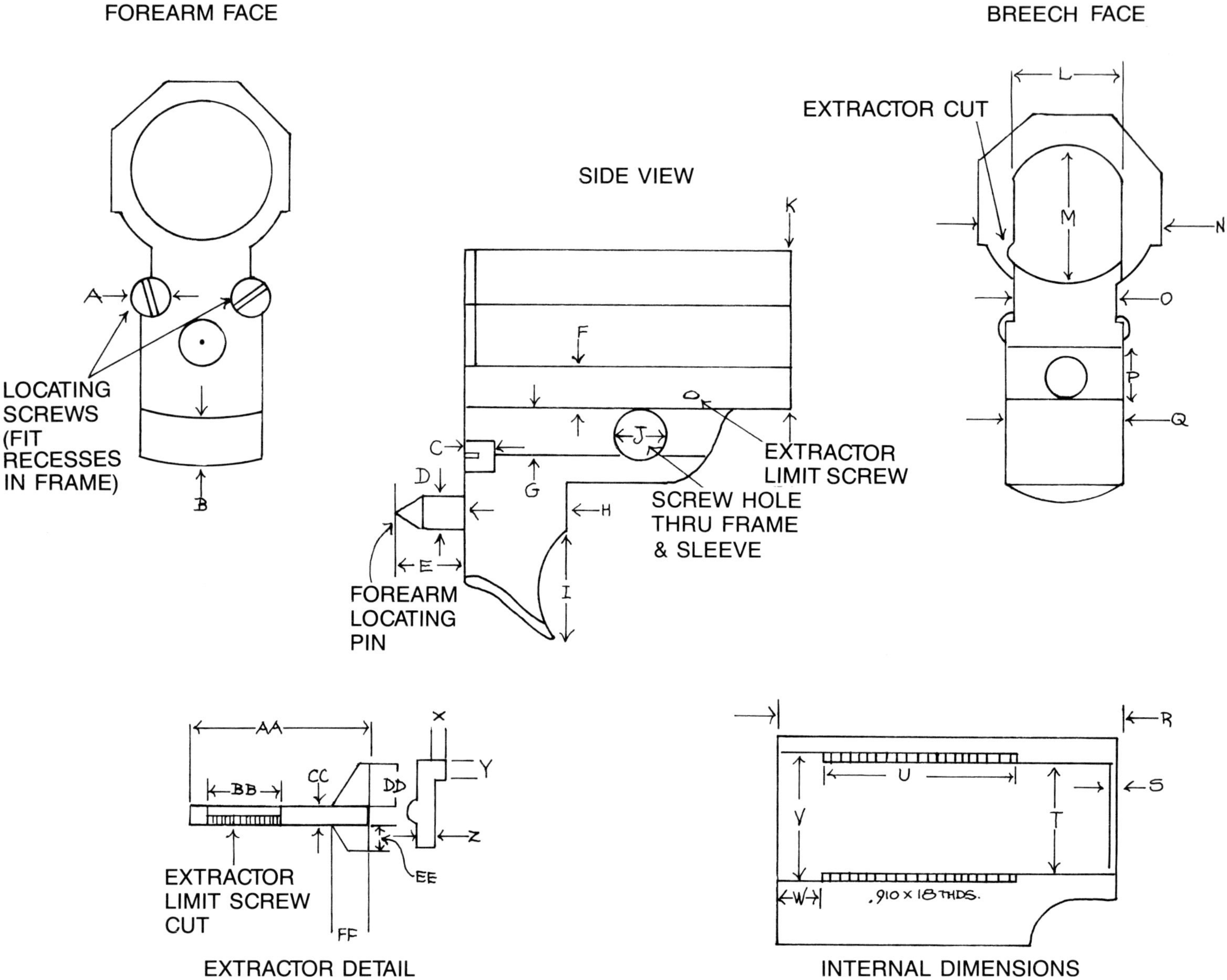

TABLE OF DIMENSIONS
(EXPRESSED IN INCHES, TO THE NEAREST .001″)

A	.270	I	.650	Q	.740	Y	.100
B	.285	J	.312	R	2.00	Z	.093
C	.160	K	1.000	S	.040	AA	1.100
D	.250	L	.650	T	.840	BB	.415
E	.375	M	.750	U	1.100	CC	.150
F	.230	N	1.100	V	.920	DD	.250
G	.255	O	.610	W	.250	EE	.150
H	.600	P	.312	X	.062	FF	.220

ORIGINAL CONCEPT — CRAIG R. HOSTETTER
ILLUSTRATION (NEW RENDERING) — PAULA CAMERON

Plate 6-17. (SN 3953 original caliber .22-10CF rebored to .25-21) Complete dimensions for detachable-interchangeable barrel model, small-frame.

A.

Plate 6-18. (SN 3953) Top: detail of breech, barrel lug and cartridge extractor, which is operated by a finger on lever cam. See Plate 6-5 for detail. Note rails or grooves that mate with their opposites in receiver to make for a very solid and accurate rifle. Middle: headspace is built in when barrel is threaded into breechblock. See Plate 6-17. Bottom: lever open. Cartridge being extracted. Patent drawings show how cam operated by the lever moves the extractor. *Courtesy Craig R. Hostetter*

B.

C.

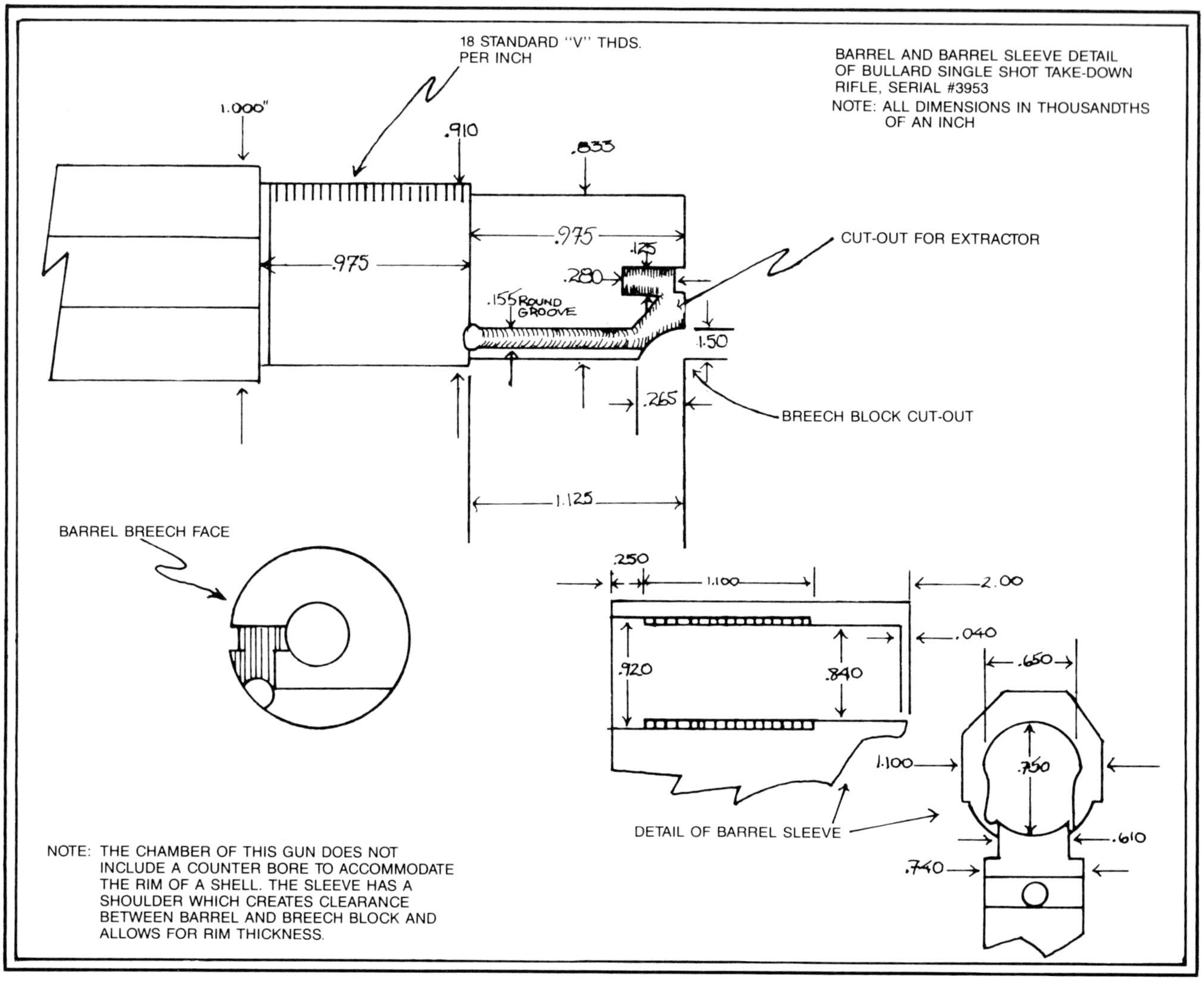

Plate 6-19. ORIGINAL CONCEPT — CRAIG R. HOSTETTER
ILLUSTRATION (NEW RENDERING) — PAULA CAMERON

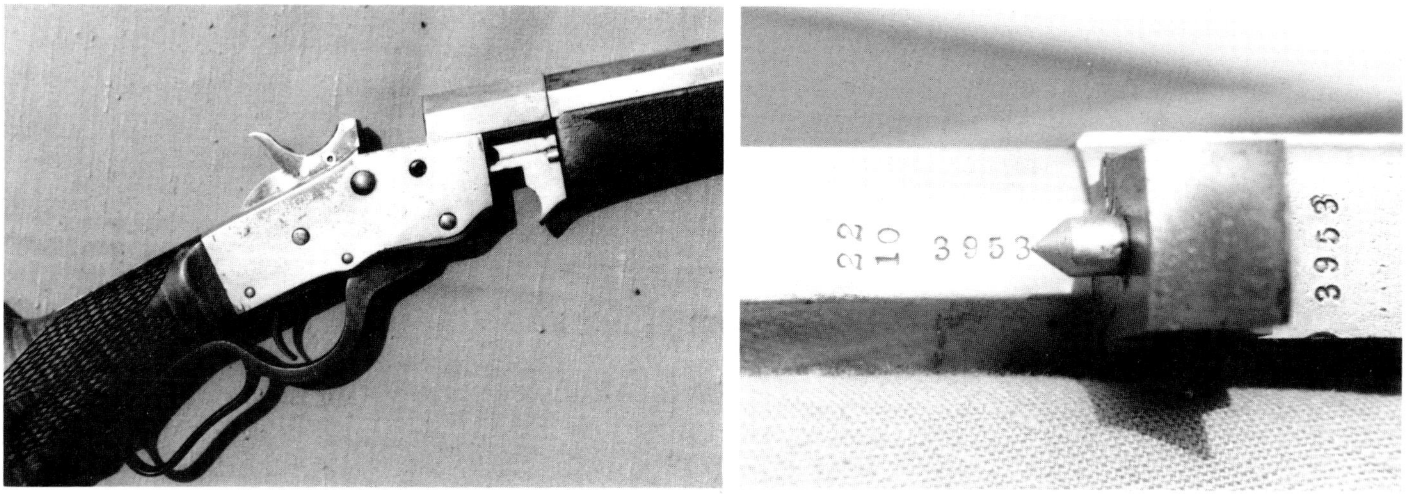

Plate 6-20. (SN 3953) Above: detail of barrel and breech separation. Hole of barrel lug can just be seen in front of receiver. The lock pin fits through this hole and threads into right side of receiver to secure barrel to receiver. Examine the coarseness of the checking on this rifle. Right: close-up of caliber and serial markings on barrel and mounting lug. *Courtesy Craig R. Hostetter*

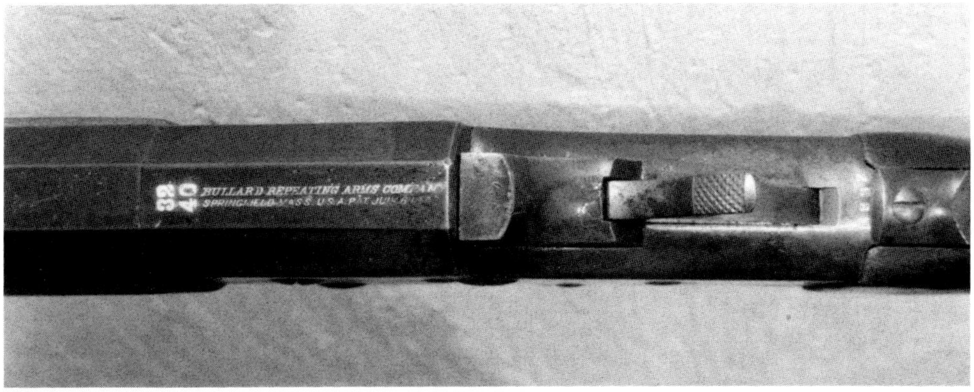

A.

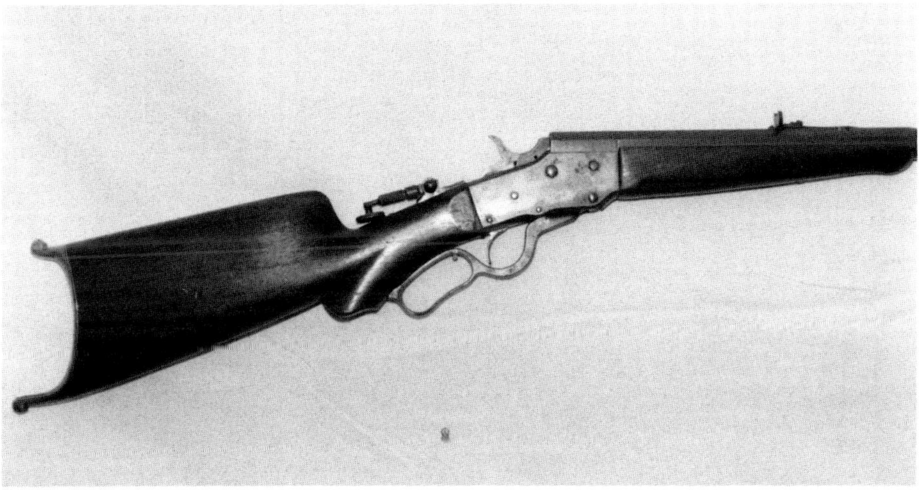

B.

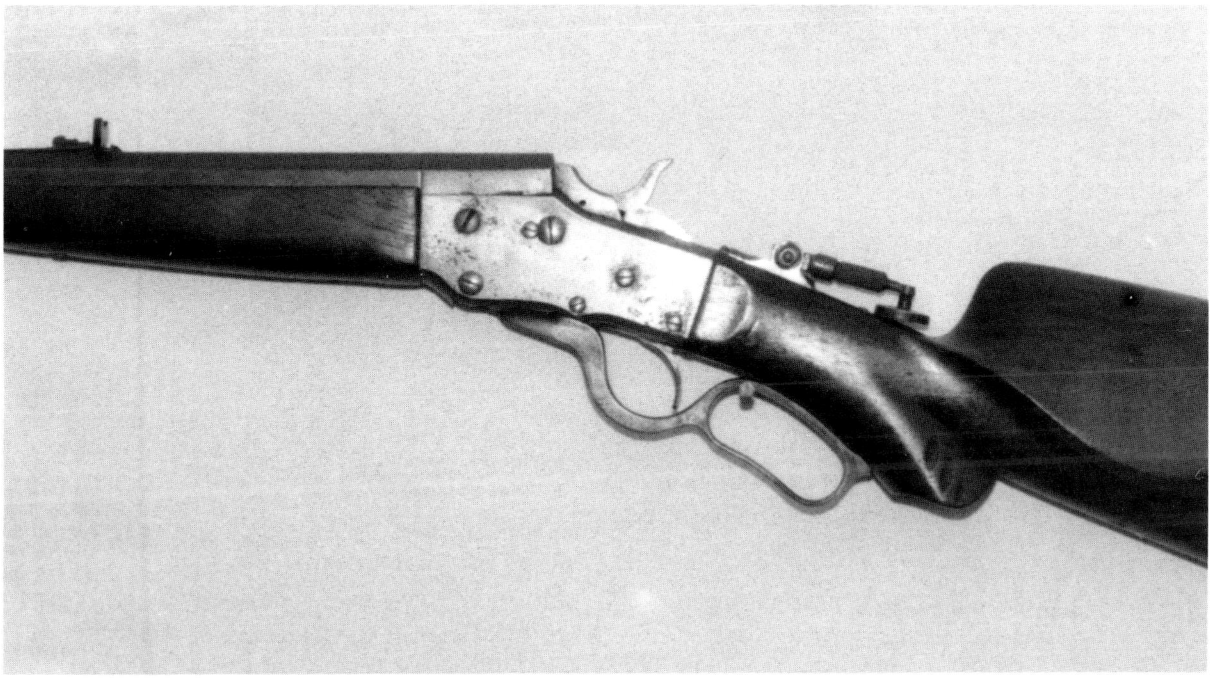

C.

Plate 6-21. (SN 3832) Detachable-interchangeable barrel model with Swiss buttplate. Note location of serial number behind hammer and standard style of address. *Courtesy Harold R. Jonah*

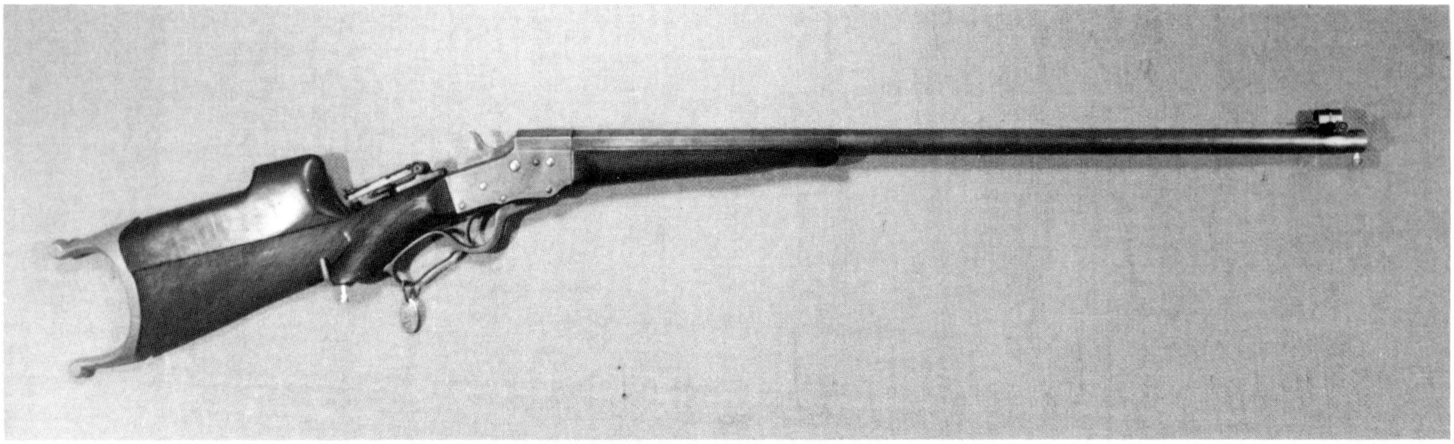

Plate 6-22. (SN 3897 cal .22 rimfire) Top: caliber markings and standard two-line address in style number 4 from the tables. Right: overall view of rifle. *Courtesy The J.M. Davis Gun Museum (Accession No. 9399)*

A.

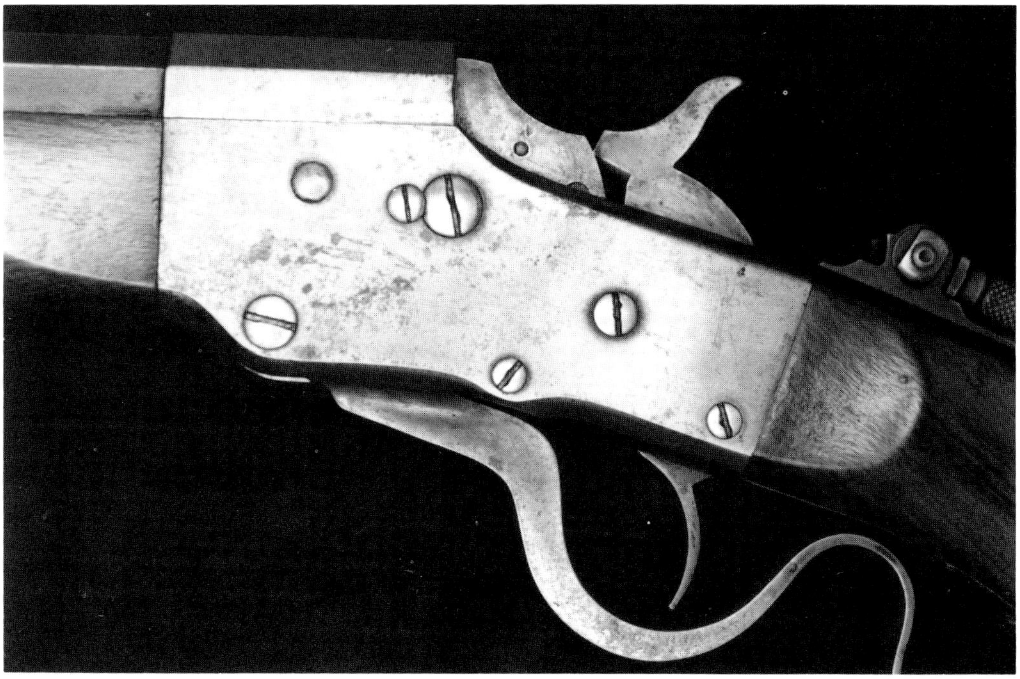

Plate 6-23. Detachable-interchangeable barrel model, again with serpentine finger lever. Another difference is noticeable on the barrel lockscrew. It must be screwed in from the right side of the receiver, not the left side, which was the more normal method, or it is a pin only with no slot cut in the head. See chapter on experimental arms. *Courtesy Verle W. Hunt*

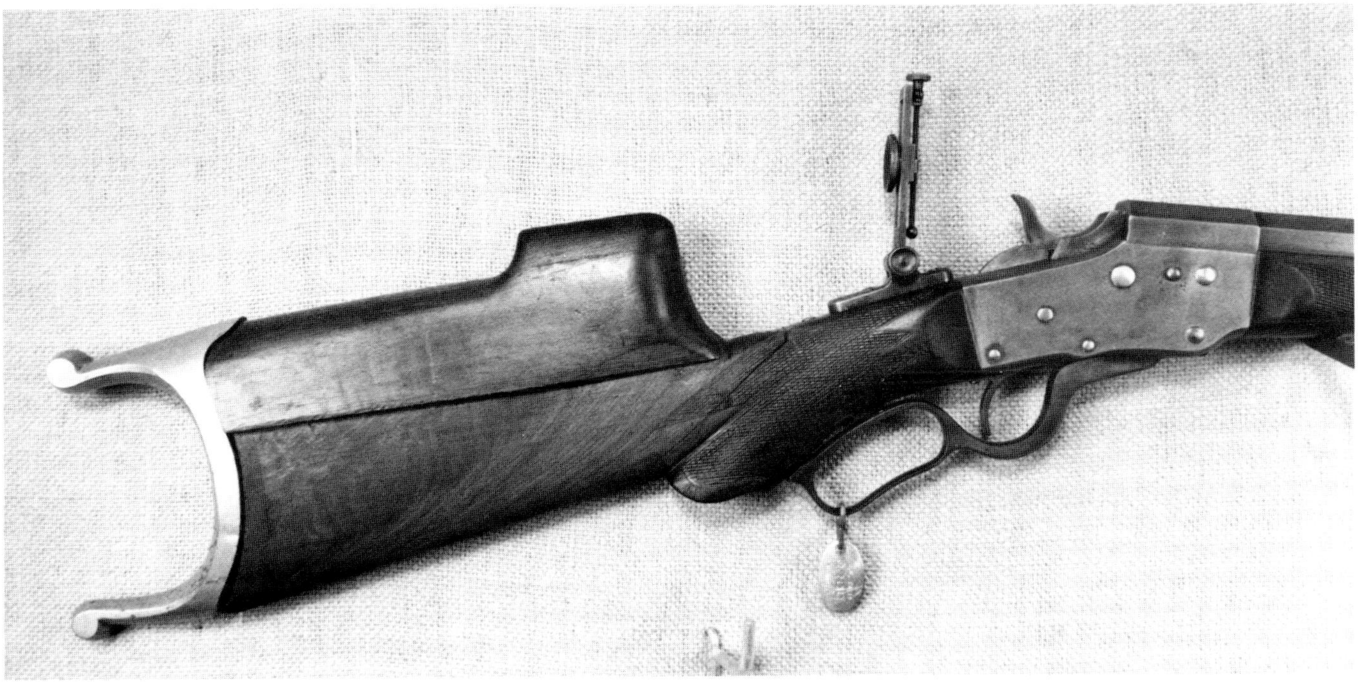

A.

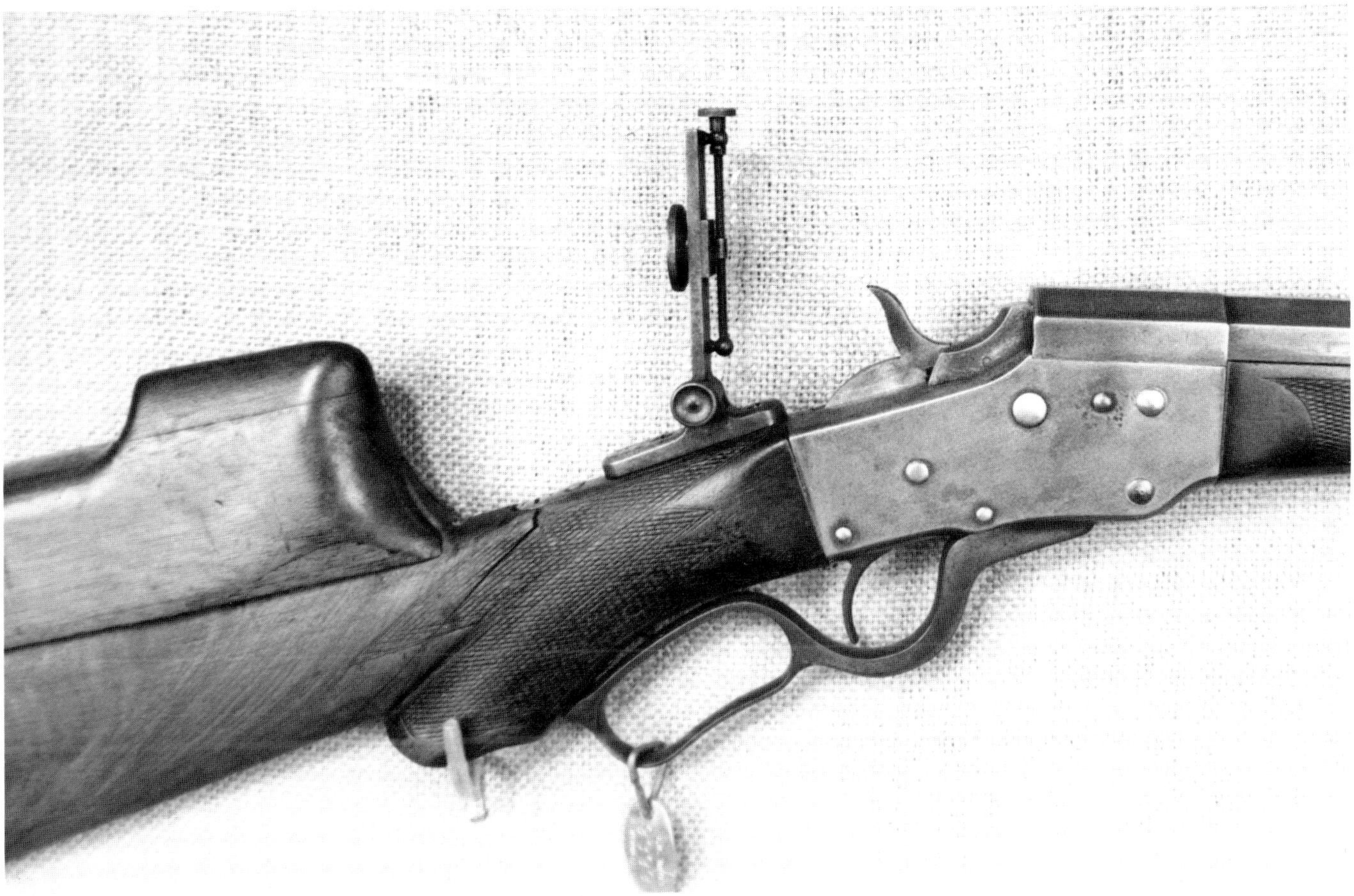

B.

Plate 6-24. Top: view of heavily customized buttstock, obviously a homemade affair to accommodate the owner's shooting style. Bottom: close-up view of rear tang sight. Note style of checking and extra lock pin directly behind main barrel lockscrew. *Courtesy The J.M. Davis Gun Museum*

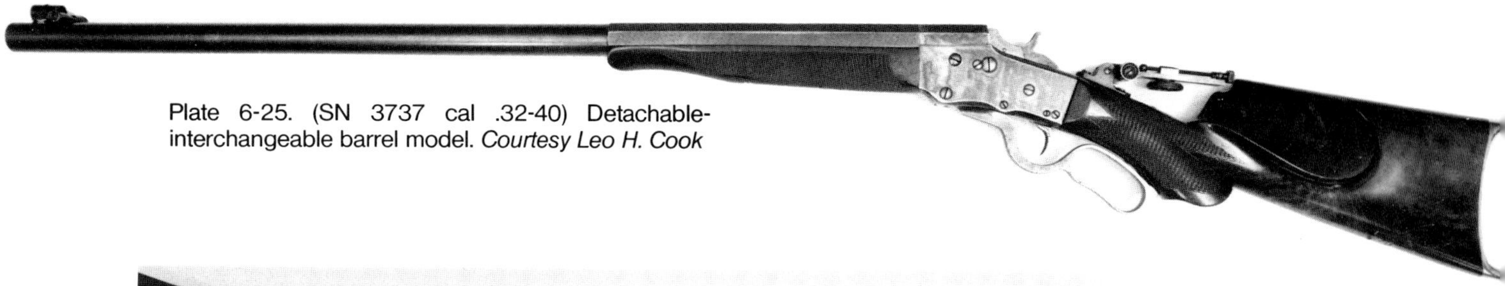

Plate 6-25. (SN 3737 cal .32-40) Detachable-interchangeable barrel model. *Courtesy Leo H. Cook*

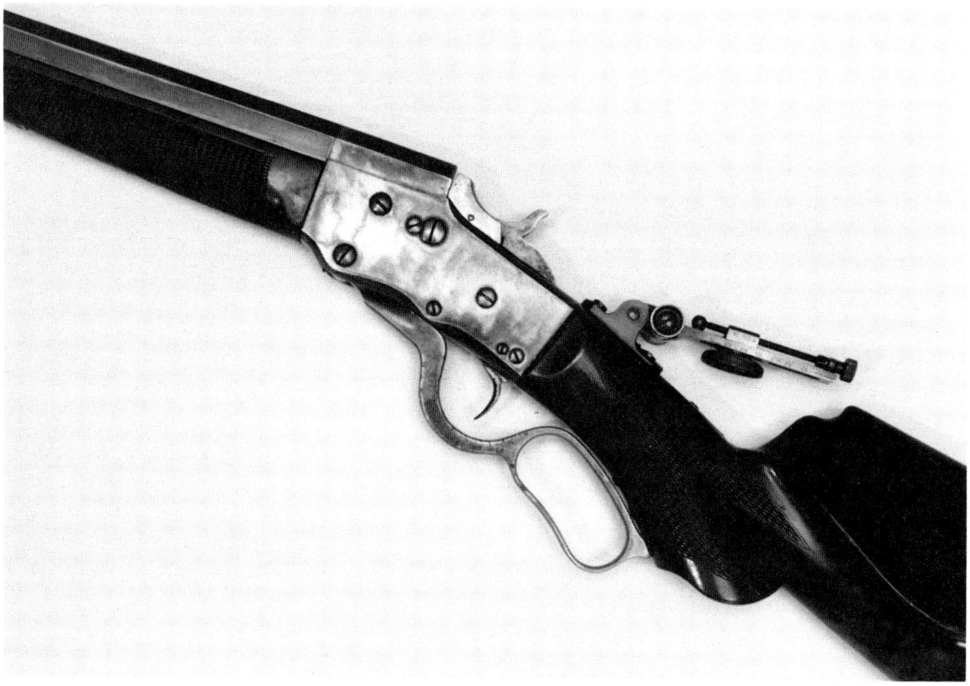

Plate 6-26. (SN 3737 cal .32-40) Locking screw is located behind the lever pivot pin. *Courtesy Leo H. Cook*

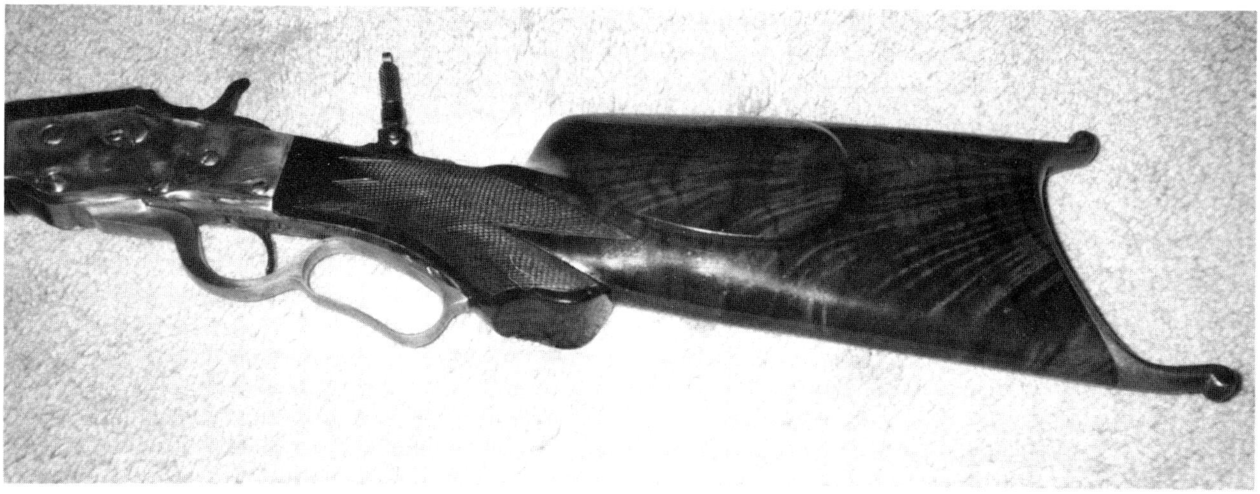

Plate 6-27. (SN 3915 cal .32-40) To the author's knowledge, Bullard wood was only available in standard and fancy grades. No elaborate system such as Winchester's for grading wood and its grain was ever used by Bullard. At least no proof of any complex system has ever turned up. The author has heard of a catalog purporting to contain engraving, checking and wood grades used by the Bullard factory. However, as this catalog has not been seen by the author, the original statement made above must stand. *Courtesy Robert Green*

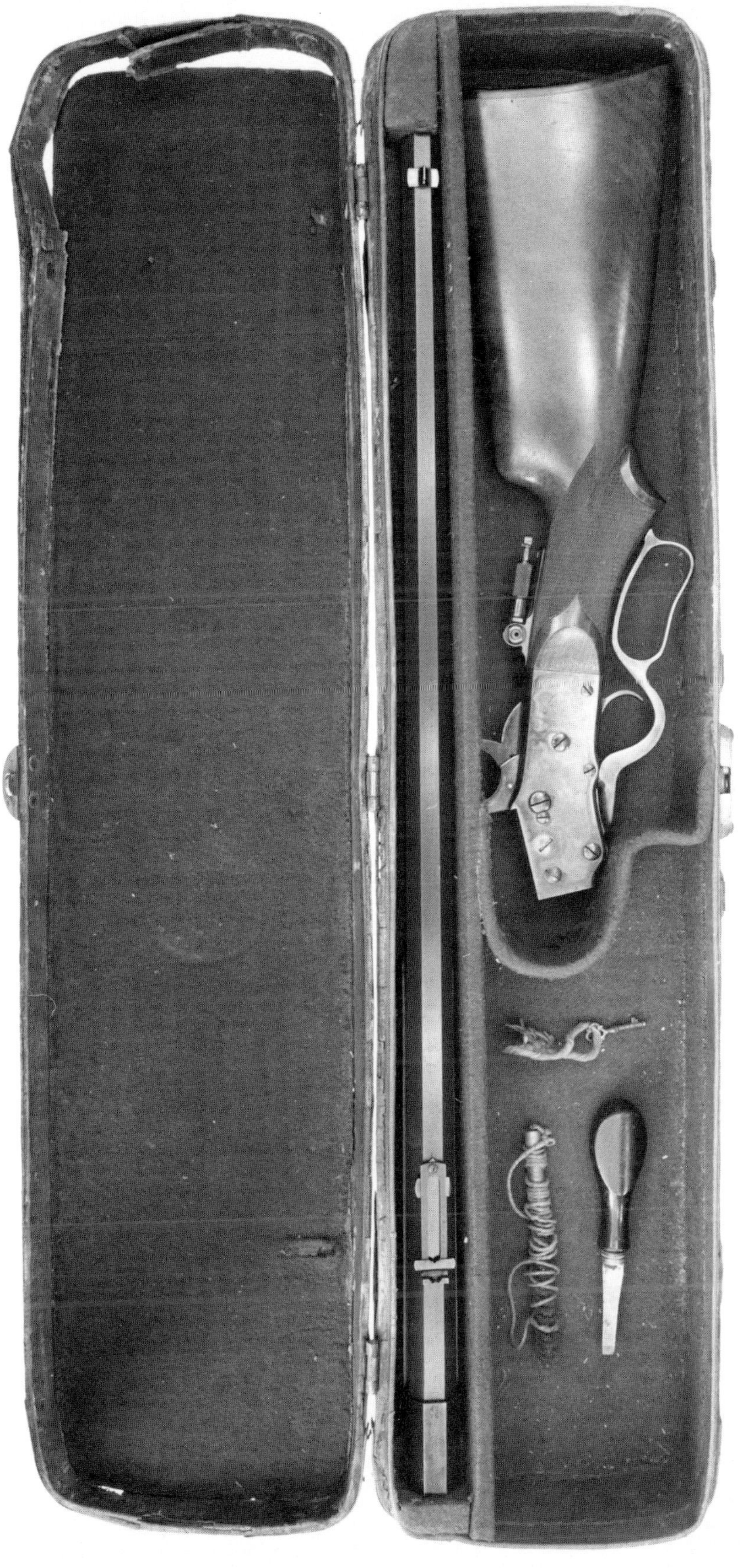

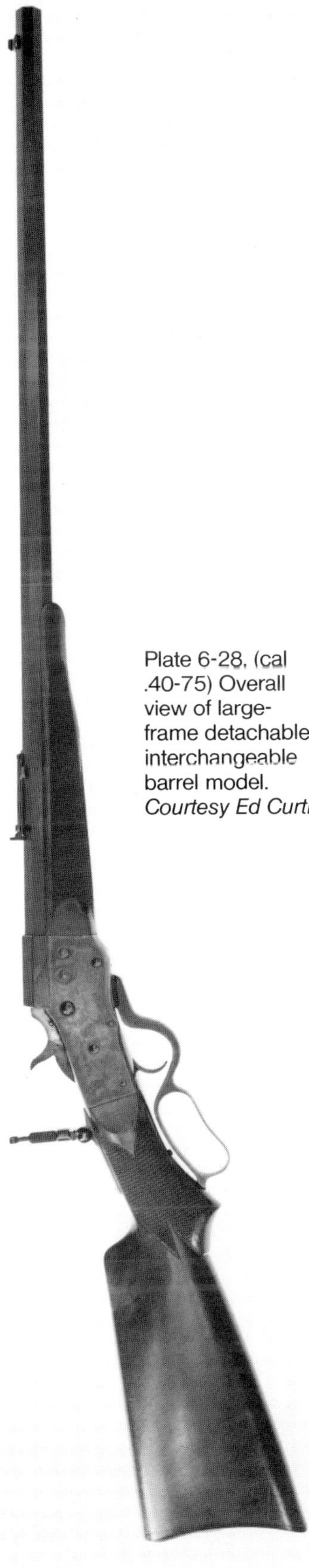

Plate 6-28, (cal .40-75) Overall view of large-frame detachable-interchangeable barrel model. *Courtesy Ed Curtis*

· BULLARD ·

REPEATING ARMS COMPANY

SPRINGFIELD, MASS.

May 16 1885

Ordnance Dept

Washington D.C.

Gentlemen

The undersigned manufacturers of magazine repeating rifles. are finishing two military rifles. and wishing to obtain the Govt standard bayonet and butt plate. respectfully request that we may obtain the same from the U.S. Armory in this city: two angular bayonets and butt plates for the same.

An early reply will greatly oblige

Very Respectfully

Bullard Repeating Arms Co

By S K Hindley

Secty

Plate 7-1. Courtesy the National Archives, record group 156, file #2972

7

MILITARY AND EXPERIMENTAL MODELS, ALL TYPES

Bullard was no different than any other firearms company of its time in that it desired to stablilize its work force and profits through military contracts. With that end in mind, the Bullard Company was to offer to any government who would purchase them the following: single-shot military muskets and carbines, and repeating rifles in both carbine and musket style with magazine cutoffs. Another variation of the Bullard repeating action was the removal of the tubular magazine and the substitution of a receiver-mounted box-type magazine — quite an accomplishment for a company in business for only about eight years.

The first known documentable evidence of the Bullard Company's desire to manufacture military arms was a letter dated May 16, 1885, from the Bullard Company to the Ordnance Department, asking to be supplied with two angular bayonets and two buttplates for two military rifles they were finishing (Plate 7-1). This letter was received by the Ordnance Department on May 18. On May 19 the Ordnance Department sent a memo to the National Armory in Springfield, Massachusetts, giving the Armory authority to "make the sale, provided the rifles are for *experimental* purposes." This memo was received at the National Armory on the May 23. A letter was sent from the Armory to the Ordnance Department indicating that the triangular bayonet and buttplates were wanted for experimental arms. The Armory also asked the Bullard Company for written confirmation of the intended use of these bayonets and buttplates. This was written and sent out May 27, 1885. On May 29 the sale to the factory was completed.

Unfortunately no other documentation occurs until March 9, 1888, when the Bullard Company asks to submit a rifle to the Ordnance Department for test purposes. This rifle was to be known as the Hindley rifle and it included a detachable magazine. Whether or not any other Bullard arms, with the exception of the Hindley, were submitted for government tests is unknown at this time. Plate 9-20 indicates that there were. It would appear logically that a repeater or single-shot would have been tested sometime in the period between 1885 and 1888, but until this is proven it is only supposition.

The 1887 Bullard catalog lists all the military models as being available, with the exception of the Hindley variation (see Chapter 9).

On May 29, 1888, Solomon K. Hindley received a patent for his magazine gun which was pure Bullard except for the magazine placement and type. On June 5, 1888, he received a second patent for a magazine for firearms. Readers will see that the magazine design varies slightly from that shown in the test documents. As pointed out in the test, there were very obviously serious shortcomings in having an open-topped, gravity-fed cartridge magazine. These shortcomings were partially addressed by the patents Hindley secured, but they were still only gravity-fed magazines with all the pitfalls associated with a non-positive feed system. Their bulkiness and awkward placement on the rifle would also have been detrimental to their having been chosen

(No Model.)

S. K. HINDLEY.

MAGAZINE GUN.

No. 383,641. Patented May 29, 1888.

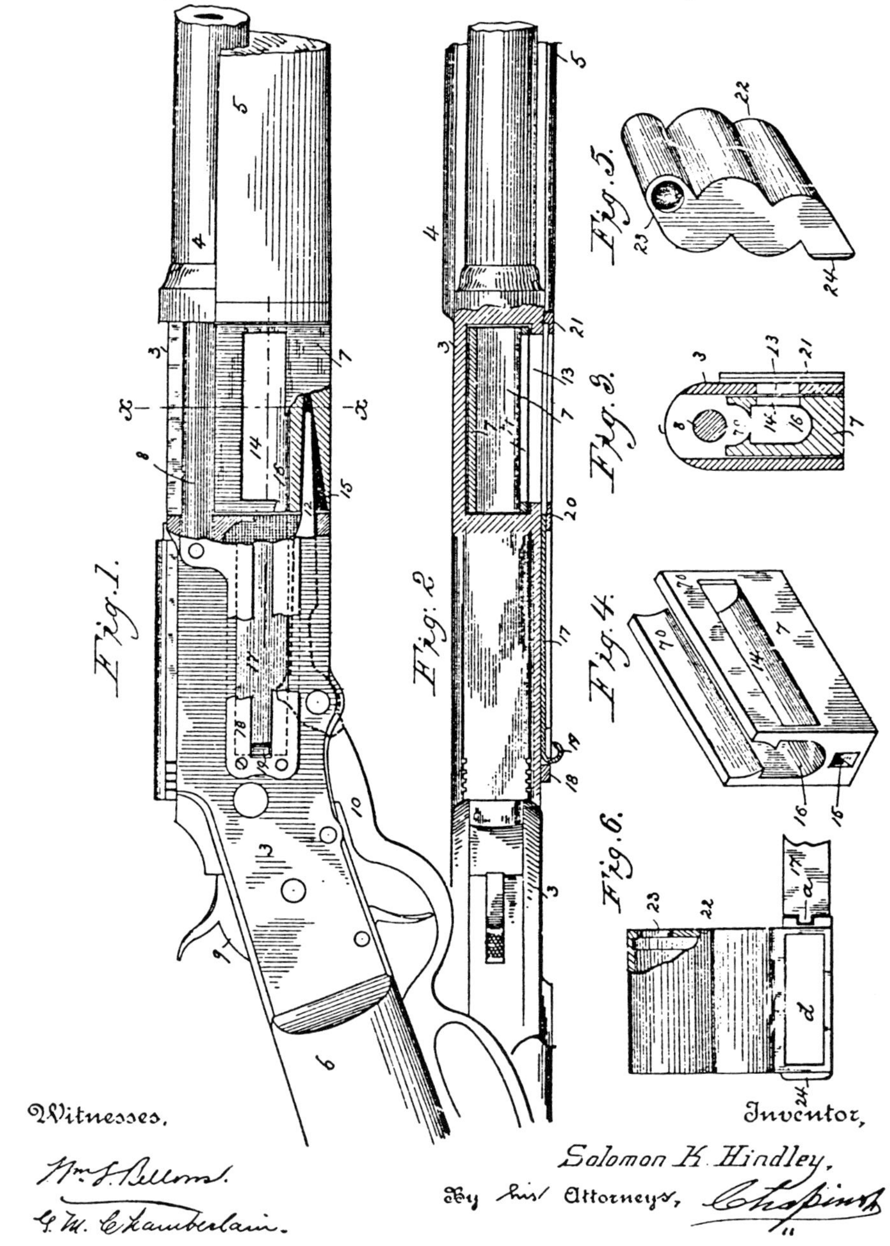

Plate 7-2.

for production over other systems such as the Mauser and Lee-Metford (Plates 7-2, 7-3). No models of these magazines were submitted to the Patent Office.

Only two Hindley variations have been documented by the author and these vary slightly from each other. The sideplate on the magazine side of each rifle is different; the levers and hammers are different. The test indicates the rifle submitted gave problems in cartridge feeding. Hindley was given a small amount of time to rectify this problem (Plate 7-4). This was accomplished by slightly enlarging the area in both the carrier and slot in receiver side to prevent jamming of shells when feeding them to the carrier. The difference in lever and hammer design again points to the fact that the company was still far from a mass-producer. The author does not know if one or both of these known muskets were submitted for the tests. Obviously these muskets were hurriedly put together, with little time taken by the factory to ensure their proper functioning under test conditions. The serial numbers of these muskets were 2524 and 2529. Both had round barrels of 28 inches and serial number 2524 had a five-groove rifling pattern. But musket serial number 2529 had a six-groove pattern of rifling, like the one used in the government test. The government test contains these statistics. From the photographs, it appears that serial number 2529 has had modifications in the magazine port that concur with the paragraph described in Plate 7-4 making serial number 2529 the probable test musket.

Between these two muskets, a tubular magazine carbine with a magazine cutoff device was produced with a serial number of 2521. The barrel is 22 inches long and round, with six grooves and lands in its rifling pattern. The upper tang carries the Bullard address and patent dates, a departure from the normal location (Plates 7-5 through 7-9). The magazine cutoff device also shows up on several other documented rifles. All information and photographs of this device in Plates 7-10 through 7-15 were supplied by John Witzel and apply to serial number 2538. It is also chambered for a .45-70 government cartridge, has a 26-inch round barrel and is stocked with bird's-eye maple. It also has a half magazine (Plate 7-14). Serial number 1272 also has this cutoff device and is equipped with a 28-inch round barrel.

According to Lee R. Boddy, a man who by the early 1960s had accumulated a collection of some 87 Bullard rifles of all descriptions, "Bullard made four muskets: 2 in .45-70 single-shot; 1 in .35-55 (he owned the gun); 1 in .45-70 two-shot experimental; 1 in .45-70 saddlering carbine (he owned the gun). The musket serial numbers run from 2495 to 2500, including carbine." If Mr. Boddy's serial numbers are correct, then there may be several other military rifles out there, but the author has been unable so far to corroborate this information, in so far as the serial numbers are concerned.

Jack Behn, in his book entitled *".45-70" Rifles*, mentions in his Bullard chapter an advertising circular in which the company purports to show military single-shots being made, although Behn has never seen nor heard of one elsewhere.[1] Does this mean that a military order of single-shot muskets was sold, or is it a posed picture produced in the hope of securing some contracts? This author cannot answer that question and letters to museums around the world have failed to provide an answer. Surely a number of examples were made to interest possible buyers.

If no military orders were sold, then it appears that Bullards of all types are very rare pieces, and if an order were sold, it must have been a small one, which would still leave Bullard rifles as rare items.

Bullard's military rifles provide the historian-collector with some of the most interesting and intriguing questions ever. The author hopes to do more research in the future to further illuminate this subject. (Refer to charts at back for full details of these rifles.)

FOOTNOTES

1 Jack Behn, *".45-70" Rifles* (Highland Park, New Jersey: Gun Room Press, 1956).

(No Model.) 2 Sheets—Sheet 1.

S. K. HINDLEY.

MAGAZINE FOR FIRE ARMS.

No. 384,161. Patented June 5, 1888.

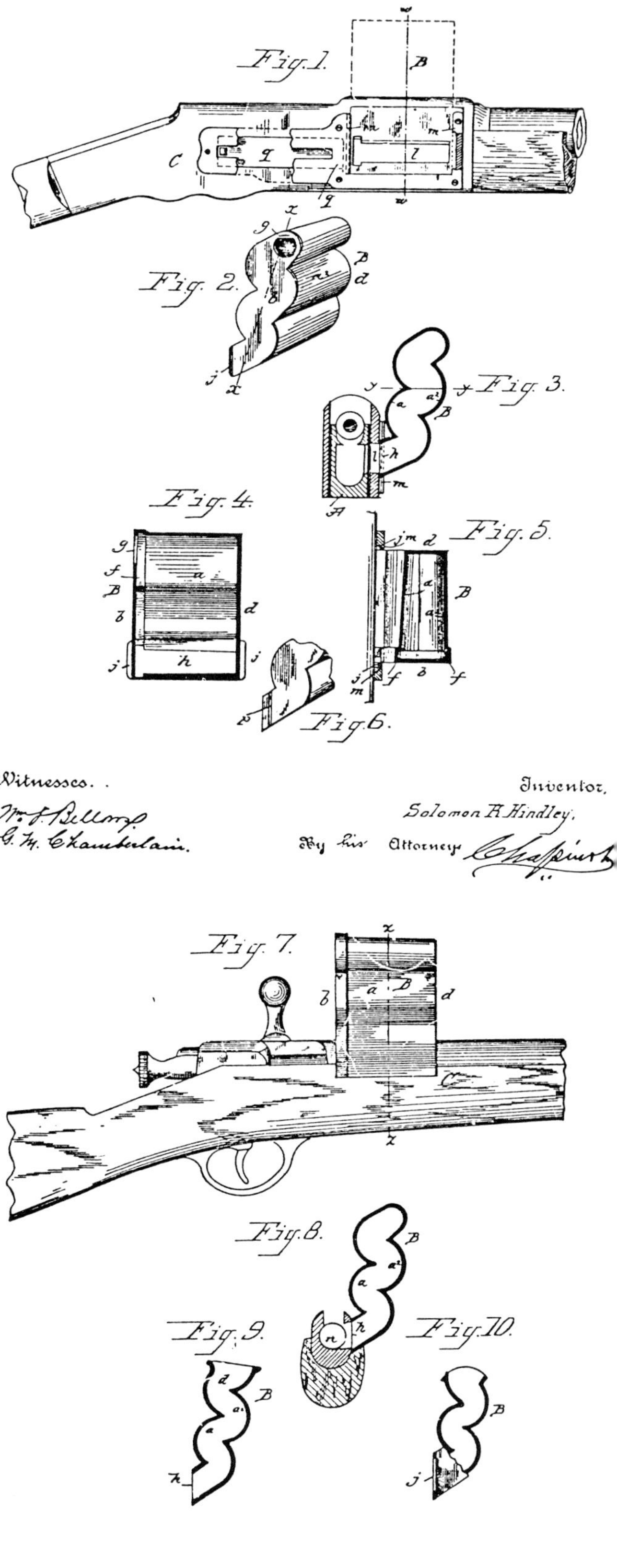

UNITED STATES PATENT OFFICE.

SOLOMON K. HINDLEY, OF SPRINGFIELD, MASSACHUSETTS.

MAGAZINE FOR FIRE-ARMS.

SPECIFICATION forming part of Letters Patent No. 384,161, dated June 5, 1888.

Application filed February 23, 1888. Serial No. 265,046. (No model.)

To all whom it may concern:

Be it known that I, SOLOMON K. HINDLEY, a citizen of the United States, residing at Springfield, in the county of Hampden and State of Massachusetts, have invented new and useful Improvements in Magazines for Fire-Arms, of which the following is a specification.

This invention particularly relates to a cartridge-magazine applicable to that class of fire-arms in which the frame or stock of the arm at the rear of the barrel is provided with an aperture capable of the reception of a cartridge, and from which aperture the said cartridge may be transferred or projected into the said barrel-bore, and is most particularly designed for use in connection with breech-loading fire-arms in which a breech piece or pin is arranged at the rear of the barrel and in an axial line therewith, and so as to be moved backward to open and forward to close the breech, said breech-pin moving into and through the said cartridge-aperture, carrying the cartridge lying therein into the barrel in the latter instance and back and away from the said cartridge-space in the former instance, then leaving said cartridge space free for the reception therein of another cartridge to be conveyed thereto, either directly from the magazine or indirectly therefrom, under the action of an interposed cartridge carrier between said magazine and said axially-disposed cartridge-space; and the said invention consists in certain features of construction and combination of parts, as will hereinafter more fully appear, and be specifically pointed out in the claims.

Reference is to be had to the accompanying drawings, in which the present invention is illustrated, and in which similar letters of reference indicate corresponding parts in all the views.

Figure 1 is a side elevation of a portion of a breech-loading fire-arm embracing the features of breech construction found in the well-known "Bullard system," and showing an adaptation thereof for the reception thereupon of the magazine of the present invention, the position of the latter upon the said fire-arm being indicated in dotted lines; and Fig. 2 is a perspective view of the magazine. Fig. 3 is a cross-section of the fire-arm shown in Fig. 1 and of the magazine applied thereupon on line *w w*, Fig. 1. Fig. 4 is a vertical longitudinal section of the magazine on line *x x*, Fig. 2. Fig. 5 is a horizontal section on line *y y*, Fig. 3. Fig. 6 is a view of a slight modification in construction, to be hereinafter referred to. Fig. 7 is a side elevation of a portion of a breech-loading fire-arm embracing the features of breech construction well known as the "Bolt system," also showing the magazine of the present invention as applied thereto. Fig. 8 is a cross-section of Fig. 7 on line *z z*, Fig. 7. Figs. 9 and 10 are cross-sectional views of magazines of the present invention when formed under slightly-modified construction.

In the Bullard system of fire-arms, as well known and as heretofore practiced, a carrier-block, as shown at A, is employed in the stock or frame at the rear of the barrel, having a body transversely concave or recessed, which portion thereof normally lies in a plane below that of the barrel and coincident with the bore of a longitudinally-arranged and forwardly-extending tubular magazine, which is disposed in the fore-arm along and beneath the barrel, and from which longitudinally-arranged magazine, cartridges are successively projected onto said carrier-block, which block is then, under the action of the operating parts, carried upwardly to present the cartridge in the axial line of the barrel, when the reciprocating breech-pin is moved forward and carries the cartridge into the barrel.

In the Bolt system of fire-arms, as well known, the breech at the rear of the barrel is provided with an aperture or cartridge-receiving space in the axial line of the barrel, through which a reciprocating bolt or plunger plays for the introduction of the cartridge into the barrel and the closing of the breech; and in the use of the magazine of the present invention, in connection with a fire-arm similar to the Bullard gun described, the magazine is applied thereto in a manner to guide and discharge cartridges therein contained to the said carrier-block when in its normal position, and when the magazine is used in connection with a Bolt gun it is applied thereupon so as to discharge cartridges therefrom to the cartridge-receiving space of said gun in line with the barrel.

As to the magazine, the same consists, essentially, of a box, B, having a width substantially corresponding with the diameter of a

Plate 7-3.

EXCERPT FROM GOVERNMENT TEST OF "HINDLEY RIFLE"

. . . On reviewing the tests it was discerned that the length of carrier and detachable magazine was such as to cause sticking with a considerable proportion of the ammunition. The board therefore decided, with the approval of the Commanding Officer, to allow the withdrawal of the guns for the changes necessary to overcome this defect. In making these changes, one side of the brass carrier was removed entirely, and a steel plate substituted.

As the tests were re-commenced, the inventor was also permitted to substitute a new extractor for the one already partially disabled. The increase in length of carrier and detachable magazine was but slight.

TEST II: RAPIDITY WITH ACCURACY

In firing as a single loader there was some delay occasioned by the slipping forward of the slide so as to interfere with the insertion of the cartridge. It was necessary to push this back into position three times. In one instance a cartridge fell out of the carrier before the lever had caused motion in the latter.

CONCLUSIONS

As regards the detachable magazine, the Board is of the opinion that in its present undeveloped condition it is open to serious objections.

It is not convenient to use the gun as a single-loader when the magazine is attached, although it is possible to do so. It would be out of the question to place the magazine in position and load it at the critical moment. In its present form, it is not possible to carry it loaded.

The inventor stated to the Board that it was his idea to make the case lighter and cheaper, so as to serve for packing and issuing the ammunition. In carrying out that idea the shape of the magazine case would have to be changed to allow of packing more advantageously.

Plate 7-4. The above is part of the government test of the Bullard military side-loading rifle musket (aka Hindley rifle). The only part that was Hindley's design was the magazine.

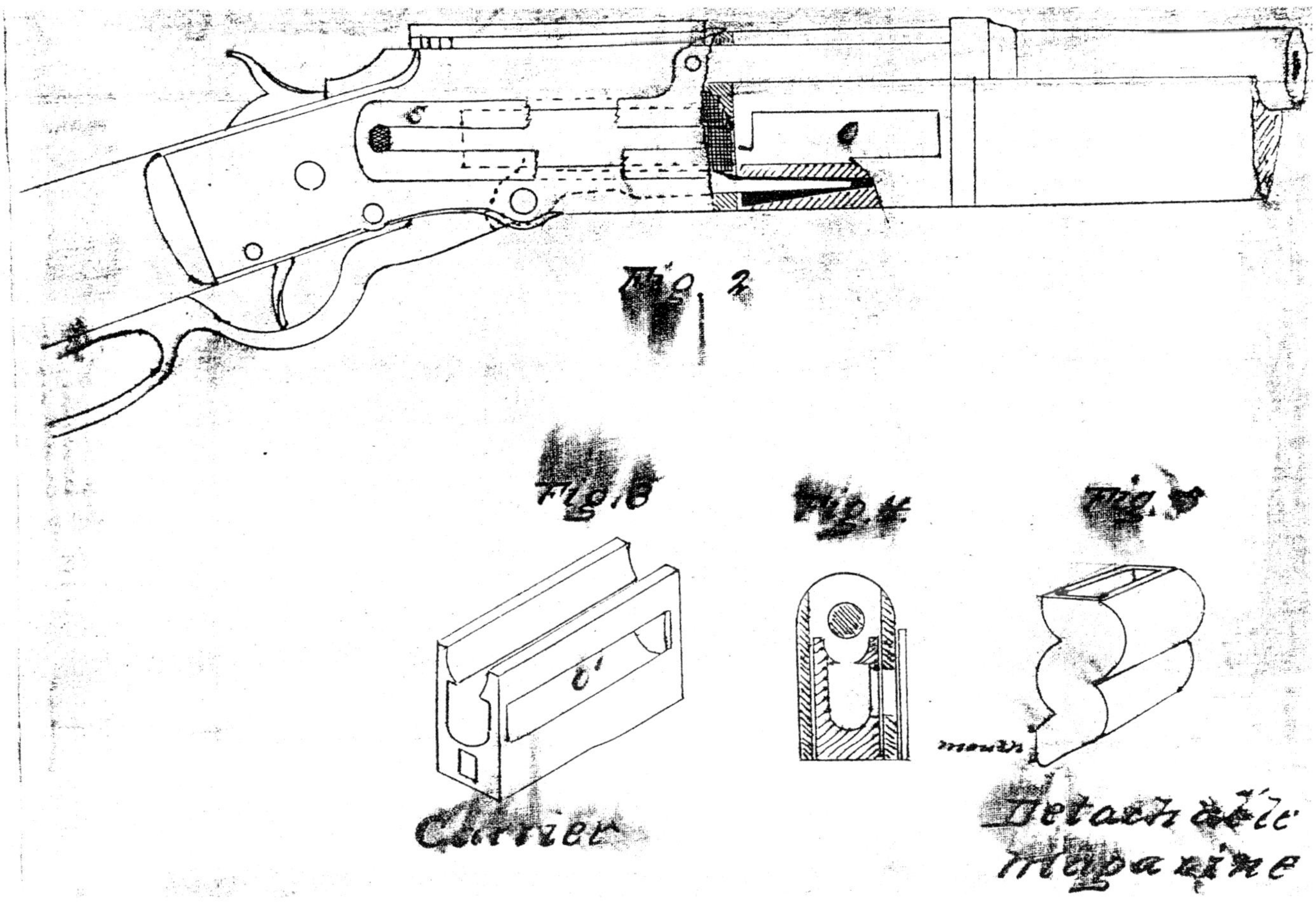

Plate 7-5. (SN 2521) The only documented Bullard saddle-ring carbine (military). Large button on receiver controls mag cutoff device. When in lower position it prevents lever from moving past bottom vertical. In a Bullard the cartridge carrier only moves on the furthest push of the lever past bottom vertical. The Bullard rifle, with its very long lever throw, is probably one of the few rifles able to accept this modification. *Courtesy O.L. Peacock, Jr.*

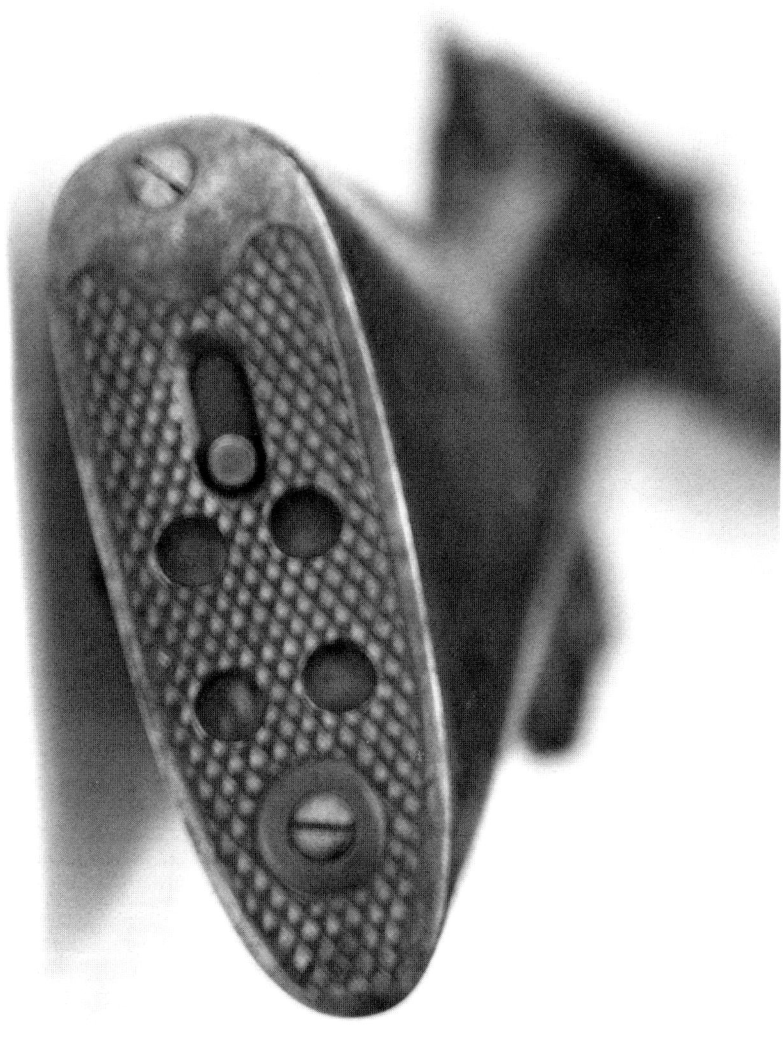

Plate 7-6. Detail of buttplate on carbine SN 2521. Moving pin upwards in slot opened the four holes that may have been for sectioned cleaning rod. *Courtesy O.L. Peacock Jr.*

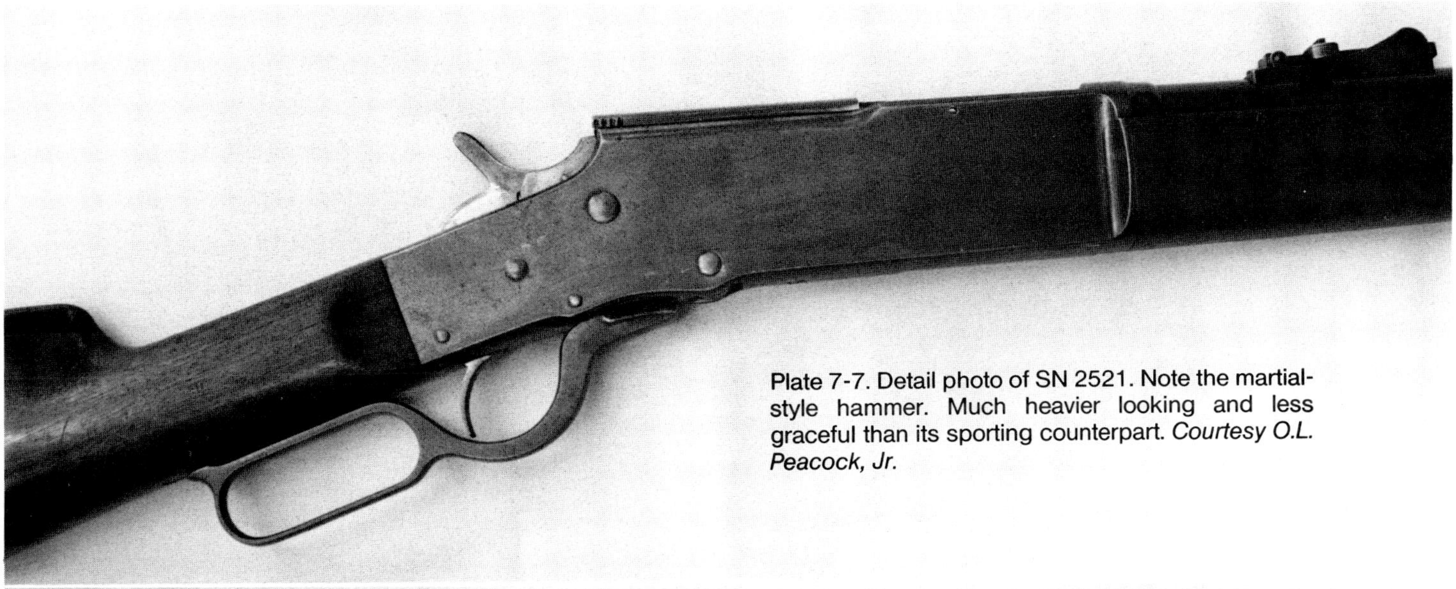

Plate 7-7. Detail photo of SN 2521. Note the martial-style hammer. Much heavier looking and less graceful than its sporting counterpart. *Courtesy O.L. Peacock, Jr.*

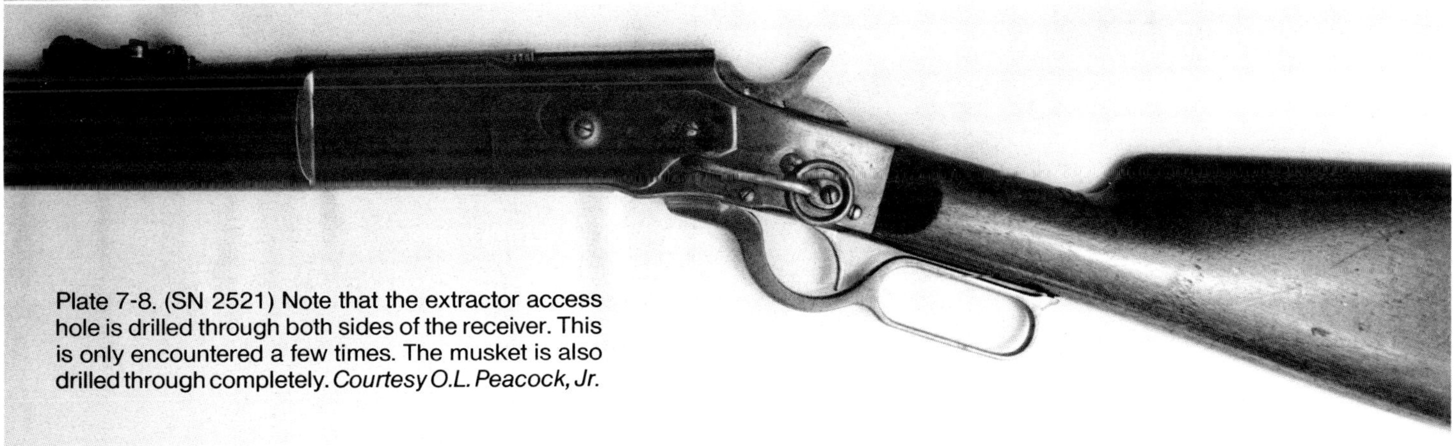

Plate 7-8. (SN 2521) Note that the extractor access hole is drilled through both sides of the receiver. This is only encountered a few times. The musket is also drilled through completely. *Courtesy O.L. Peacock, Jr.*

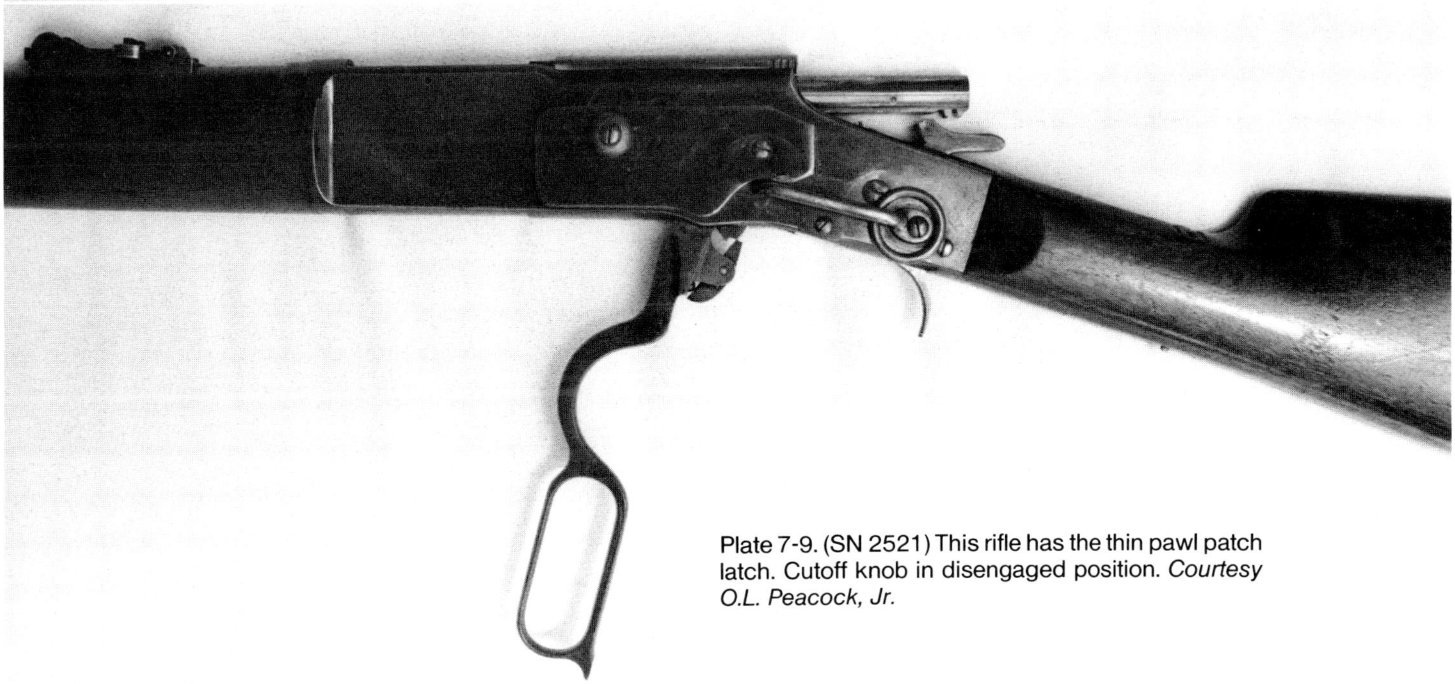

Plate 7-9. (SN 2521) This rifle has the thin pawl patch latch. Cutoff knob in disengaged position. *Courtesy O.L. Peacock, Jr.*

A.

B.

C.

Plate 7-10. (SN 2538) Top: left side of action with magazine cutoff device disengaged. Middle: cutoff device engaged. Bottom: internal view of magazine cutoff plate with mechanism in disengaged position. *Courtesy John Witzel*

A.

B.

C.

Plate 7-11. Top: engaged position. This serial number also appears in standard place behind hammer. Middle: inner mechanism of rifle. Lever in closed position. This photo shows the lever pivot pin was one that would have been used on an 8-pin receiver, for it has a small cut-away in its head to allow a small lockscrew to hold it in place. The oversized cover does that duty instead. Bottom: lever now in partially open position. These photographs show the pinion link used with the thin spring-pawl design. The pinion link arms fit outside of the lever link. *Courtesy John Witzel*

A.

B.

Plate 7-12. Top: lever in open bottom vertical position. Spent cartridge would be ejected, but if magazine cutoff were engaged further forward movement of lever would be prevented, thus the cartridge elevator could not move upward in order to load fresh cartridge, but one could manually load this rifle from the breech. Bottom: lever in open position forward of bottom vertical. Only in the lever arc distance from vertical to forward extreme does the carrier arm move thereby lifting the elevator. In other words, if the rifle was hung up level and a plumb line dropped from the lever pivot pin, only the lever movement from the plumb line forward would move the carrier arm, as shown here.

Examination of these photos shows that the projection in the inner cover of the magazine cutoff device simply blocks the carrier bar (arm) from lifting the carrier when it is in position to engage the magazine cutoff feature thereby preventing the lever's forward motion past bottom vertical. This of course would prevent the lifter bringing a cartridge up in line with the chamber, enabling the rifle to be loaded directly into the chamber by hand, yet retaining a full magazine. By engaging the cutoff device the spent shell can be ejected mechanically and a new one fed manually without emptying the tube magazine. *Courtesy John Witzel*

Plate 7-13. Left: close-up of hammer on rifle with magazine cutoff. Martial style. This type of hammer will be referred to as a martial type for the purposes of this book. This hammer type also occurs on musket SN 2524. This design may have been introduced to alleviate harsh military use, although the author has never heard of a broken hammer on a Bullard. Right: standard Bullard hammer SN 1367. Note larger diameter hammer pivot pin on SN 2538. *Courtesy John Witzel*

Plate 7-14. Left: (SN 2538) close-up of caliber markings and standard 'C' location on Bullard rifles. This rifle must have been intended for the military, for it is marked .45-70. in reality both the .45-85 Bullard and .45-70 cartridge are interchangeable. Both have 2.1-inch-long brass casings. Right: (SN 1367) this picture graphically displays a bit of rear sight silliness that occurs frequently on Bullards. The elevated ramp on the sight can only be used for a little over ⅓ its length before it butts against the first ridge of the breech, making it a useless sight for distance shooting. *Courtesy John Witzel*

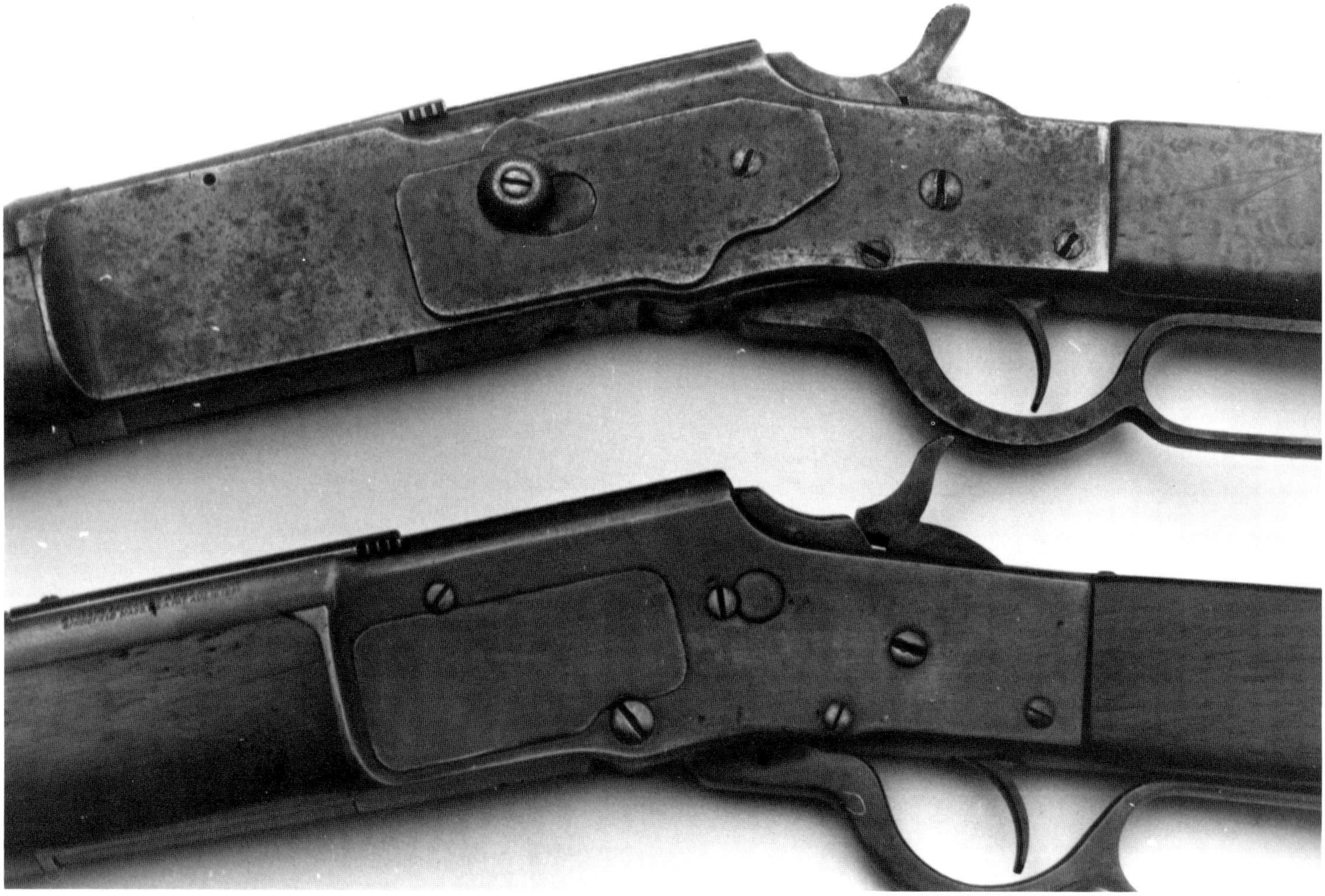

Plate 7-15. (Top: SN 2538 cal .45-70 magazine cutoff; Bottom: SN 1367 cal .45) Note breechblock pin has no slot in head and is flush with receiver surface. This is the same type of pin as used on SN 2538 but with small lockscrew to hold it in place. SN 2538 needs no such lockscrew, as the plate holds the pin in place. Another mystery! Note neither buttstocks have customary flare at wrist. *Courtesy John Witzel*

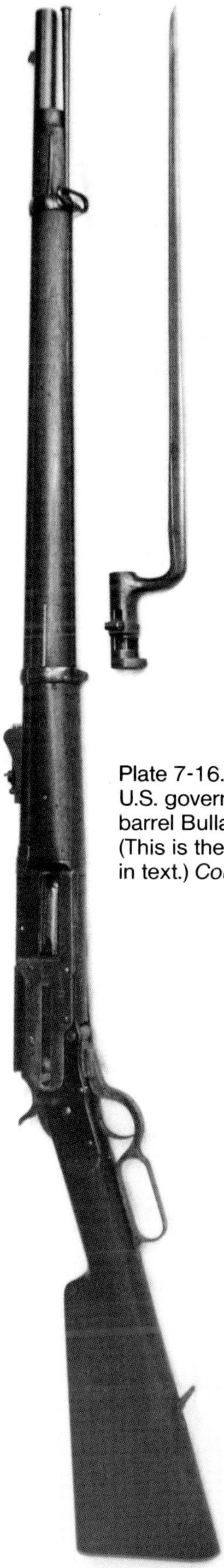

Plate 7-16. (SN 2529 cal .45-70 U.S. government) 28-inch round barrel Bullard side-loading musket. (This is the Hindley rifle referred to in text.) *Courtesy O.L. Peacock, Jr.*

Plate 7-17. (SN 2529) This rifle may have been the one tested by the Ordnance Dept. The carrier has a lip on each end. This may be all that can be seen of the steel plate mounted to the cartridge carrier to provide strength after the brass was completely cut away on the port side, as mentioned in the rifle test. This prevented shells from jamming in the port. This area certainly differs from the port area on SN 2524. This rifle also has the thin pawl latch. *Courtesy O.L. Peacock, Jr.*

Plate 7-18. (SN 2529) Action open, magazine port open. Note carrier blocks magazine port in extreme travel and top position. Magazine must have had beveled edges, fitted into sideplate from top without locking. *Courtesy O.L. Peacock, Jr.*

Plate 7-19. (SN 2529) Magazine port door open. The sliding door caused problems during the test. *Courtesy O.L. Peacock, Jr.*

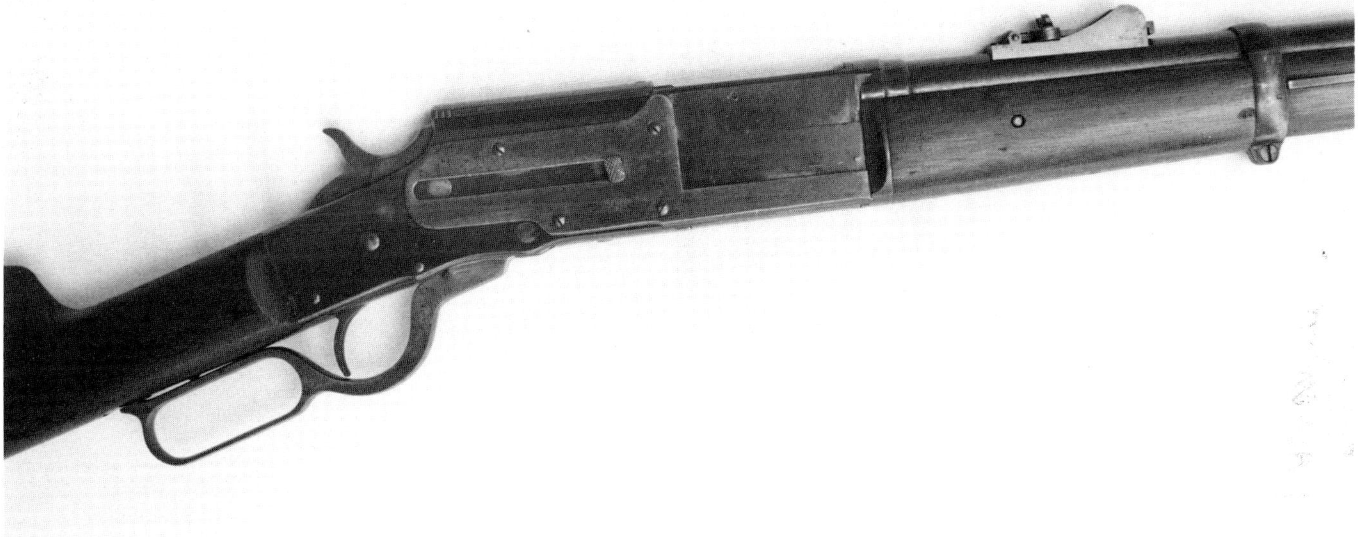

Plate 7-20. (SN 2529) Magazine port closed. The sideplate that holds the magazine differs from that used on SN 2524. Musket has Buffington rear sight. *Courtesy O.L. Peacock, Jr.*

Plate 7-21. (SN 2529) Left view of receiver. It has B-style finger serrations on dustcover. Wood forearm is left completely off receiver. *Courtesy O.L. Peacock, Jr.*

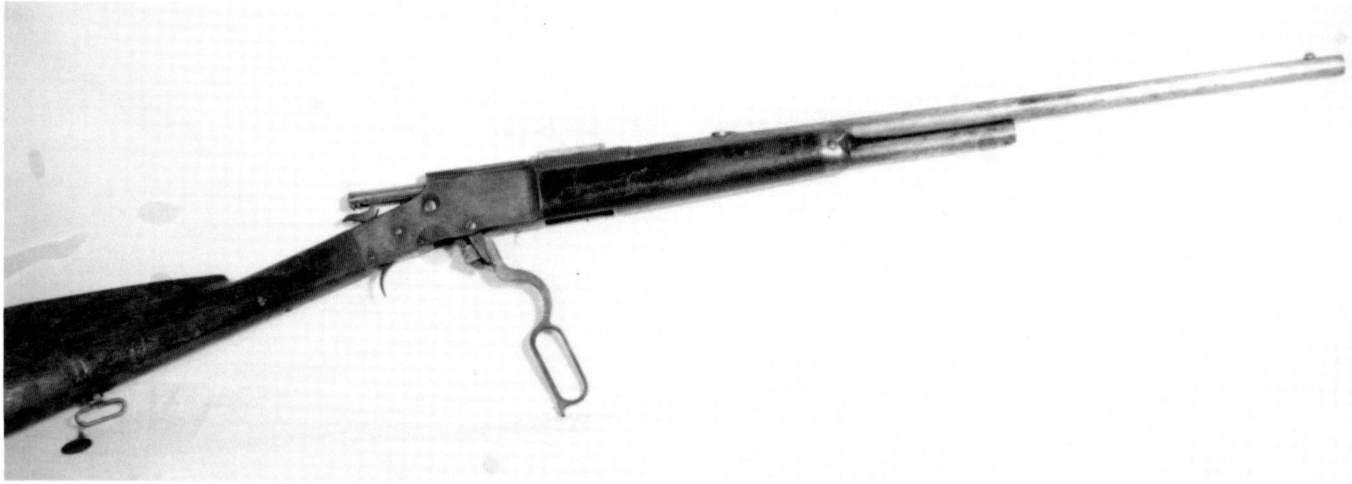

Plate 7-22. (SN 2622 cal .50-95 Winchester) Bullard large-frame. Both forearm and buttstock are beechwood. Has C.I. stamped near caliber markings and VP with two crown stamps also marked on receiver. The buttstock wrist does not have the customary flare-out from the receiver to thicken the diameter of the wrist. The comb height on the buttstock is definitely lower than standard and the wrist longer, but whether this was done in Turkey or at the factory is unknown. From the photographs the wood appears to be matched with both forearm and buttstock showing the same wear and color. Although not shown in the photo, this rifle has a military steel buttplate. The history behind this rifle is unknown. It was definitely not a presentation piece. Was it part of an order for military arms or merely a captured piece from one of Turkey's many military adventures? *Courtesy Askeri Muzesi, Istanbul, Turkey, and Infantry Colonel Muhlis Ozmen, Chief of the Military Museum*

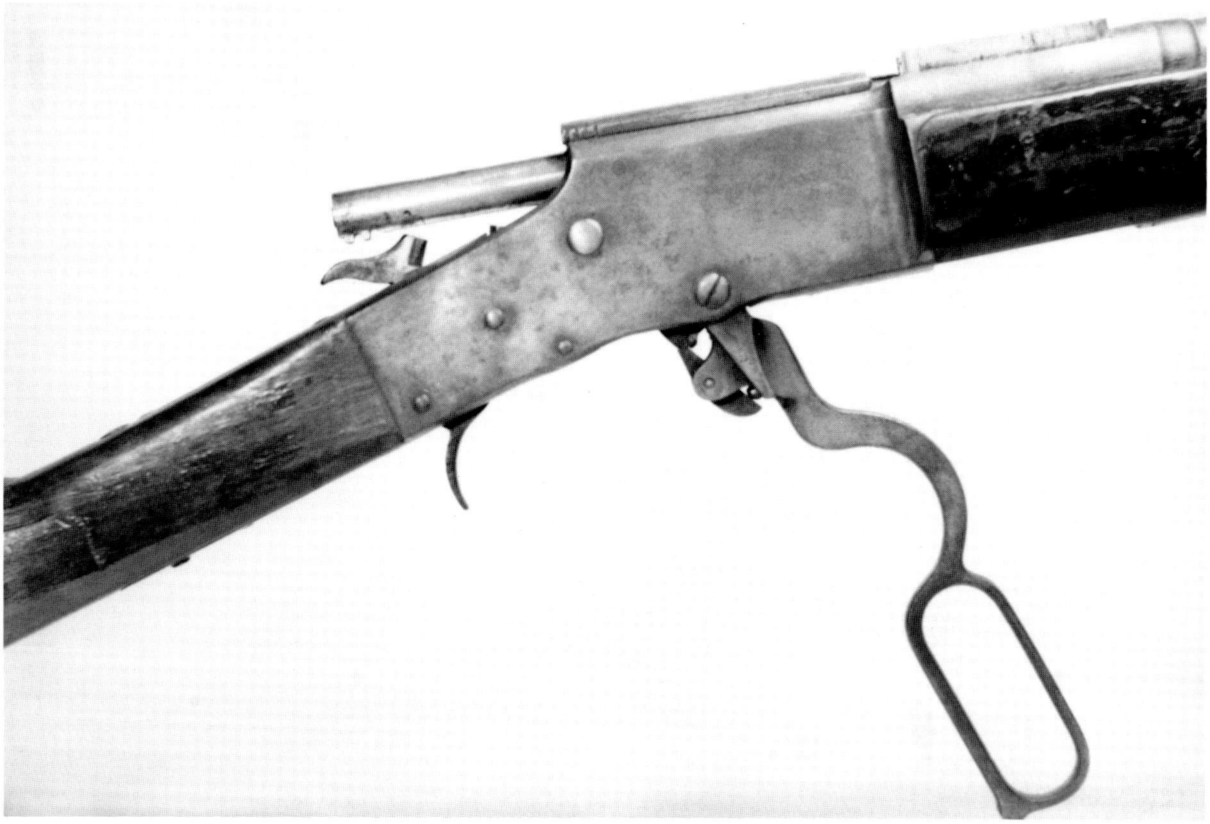

Plate 7-23. (SN 2622) Top: Thin lever pawl design. Right side is marked 'model 1886.' *Courtesy Askeri Muzesi, Istanbul, Turkey*

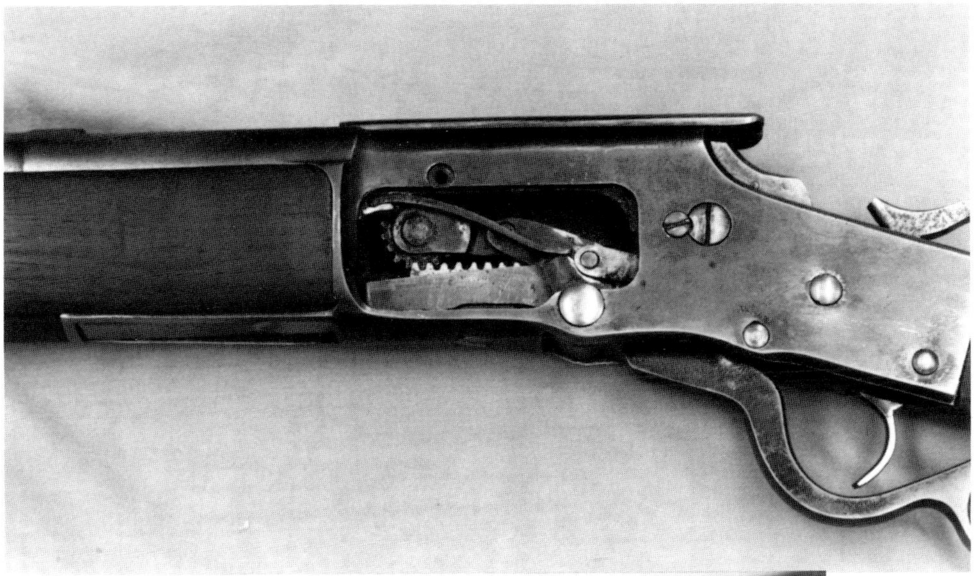

A.

B.

Plate 7-24. (SN 825 cal .45) The author has included this rifle in the experimental chapter for several reasons. Top: note the 7-screw receiver, but only the cover screw and breechblock pin and lock have slot heads. All other pins are just that, pins. The owner says when all pins are in place action is just as tight and crisp as the slotted-pin type. At first glance the overlength dustcover looks like a replacement, but in conjunction with the other differences on this rifle it is believed to be factory original. Bottom: note the larger than standard screw head in bottom of receiver to hold lower rack in place. *Courtesy Kenneth Howell*

Plate 7-25 (SN 825) Right view showing all pin heads and no slots — very unusual yet must have been factory produced because of the close tolerances involved. For what purpose was this rifle made? Another of Mr. Bullard's machine shop exercises? Unfortunately no literature survives to enlighten us. Possibly it was a move to eliminate the process of threading four holes in the receiver. However no other repeaters have been seen with this many unthreaded, unslotted pins. *Courtesy Kenneth Howell*

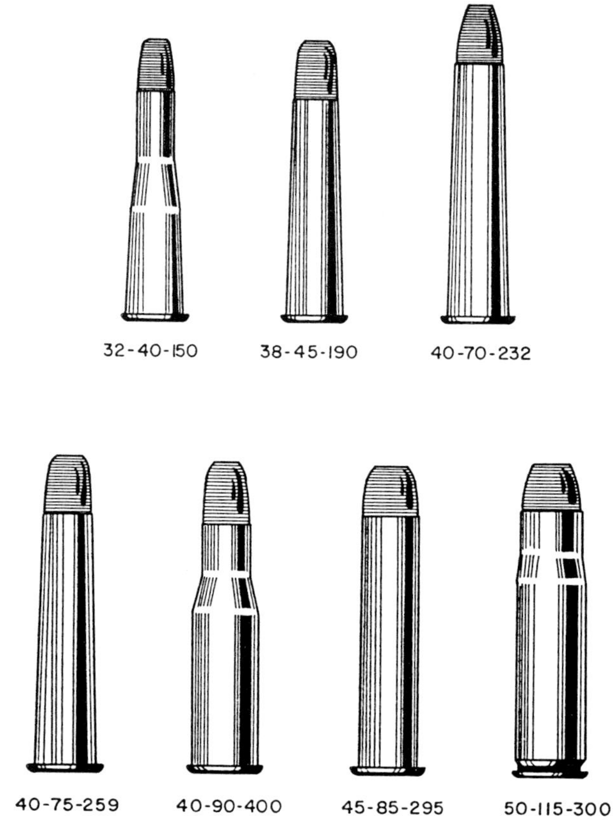

Drawing by Michael E. Whiteley

BULLARD Cartridges

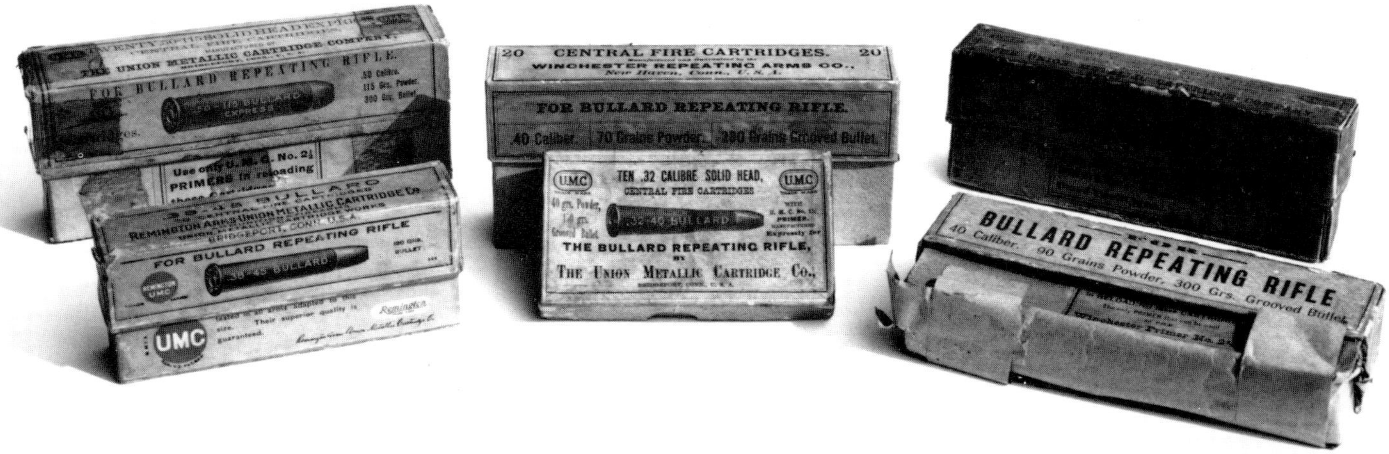

Courtesy Cliff Schisler

8

CARTRIDGES: DEVELOPMENT, PATENTS CALIBERS AND RELOADING TOOLS

James H. Bullard was a proponent of heavy charges and light bullets. To further these ends he developed a complete line of cartridges, seven in all. These were the .32-40-150, .38-45-190, .40-70-232, .40-75-258, .40-90-300, .45-85-290, and the .50-115-300. However Bullard rifles could be had in just about any caliber then available, as the tables and catalogs attest. It should be noted that very few Bullard single-shot rifles were chambered for Bullard loads.

Of these seven loadings, only three differ substantially from other products then on the market. They were the .32-40, a bottleneck cartridge, the .40-90, another bottleneck cartridge, and the .50-115, the most original and important of them all.

James Bullard took out two cartridge patents, both of which involved improvements in the art of cartridge manufacturing: No. 192676 of July 3, 1877, and No. 197823 of December 4, 1877 (Plates 8-1, 8-2). As far as the author has been able to ascertain, no patents were taken out for any specific Bullard cartridge. The .50-115, being of true solid-head design and probably one of the first, if not *the* first, semi-rimless (rimless) cartridge to be mass-produced anywhere in the world, surely deserved a patent on one or both of its features, but this does does not seem to have happened.

Although it has proven impossible for the author to date precisely the introduction of each Bullard cartridge, it can be shown through first-hand accounts and advertisements that as early as September 1884 all the Bullard calibers were available save the .40-90, which is not mentioned in the September 1884 article.

The following article was written by a "visitor" (more likely a publicity man) to the Bullard factory on September 13, 1884. While the following article is full of colorful adjectives extolling the features of the Bullard Company — which amount to blatant advertising — the article is a very good one for insight into introduction dates for the cartridges and reloading tools, as well as some of the background on the .45-85 cartridge. This article was taken from Gerald O. Kelver's book, *100 Years of Shooters and Gunmakers of Single Shot Rifles*:

Being personally unknown to Mr. Bullard my reception was most gratifying in its cordiality, and a request to be allowed to look around the manufactory was met with a prompt compliance, and for some hours I went around looking at the various processes of manufacture of what is generally conceded, even by its competitors, as the finest magazine gun in America or any other country. The machinery and tools for the various processes are all of the latest model, and all first class, and with the fine workmen employed, insure the fine and perfect fitting work for which the Bullard is so justly celebrated.

One of the most essential things in any rifle, and particularly in a repeater, is a strong and simple breech mechanism, as nearly removed from breaking down from weakness or derangement as it is possible for human hands to make it; this the Bullard possesses in an eminent degree; the material used is of the best, being all forged steel, while the design is such as absolutely precludes the idea of a break down through any inherent fault or error in its conception or in its use. Briefly stated, the design may be called a combination of a sliding bolt actuated by a lever, a falling breech block, firmly locked into place when the gun is ready to fire, similar to the breech block so widely known in the earlier Remington guns, and a carrier block to bring up the cartridge from the magazine, all these parts are moved by a simple motion of the lever by positive action, and none depends upon springs for their movement or position.

The bolt is a heavy round piece of steel sliding in a long bearing, and free from any 'side shake'. It is squared on its lower surface, where it is cut with heavy teeth, which teeth rush into a rolling pinion attached to the end of the lever by a strong connection. This bolt is pierced from end to end for the firing-pin, and has the extractor on its upper and forward face. The heavy breech block is so placed as to swing up in the rear of this bolt, and is firmly locked in its position when closed by a heavy strut of steel, which is brought under its rear and lower face by a positive motion of the lever. This breech block closes up in rear of the bolt and is locked as firmly in its position as does the falling breech block in the Remington rifle come up and lock in rear of the cartridge itself, ensuring a perfect certainty against any possibility of a 'blow out', even though the heaviest charge be used. In fact, it is simply impossible for the bolt to blow out; everything locks firmly with its place in the breech frame, the screws being simply pivots upon which the parts rotate when moved by the lever, either in throwing back or closing the bolt.

The action is simple in the movements, not liable to derangement, and is the smoothest and easiest moving mechanism I have ever seen on any gun. It is almost as easily moved with the hammer down as with it cocked, and yet there is a mainspring of the heaviest type to ensure a heavy blow on the primer. This ease of motion is explained by the proper leverage which is brought to bear during all the travel of the lever, and the perfect workmanship of the mechanism, everything being finished and polished to insure smooth working.

The lock mechanism is strong and simple; by simply removing four screws the stock is first detached, and then the lock, all assembled on the lower tang piece, is easily removed. There is no need of jamming of thumb nails or barking of knuckles in dismounting anything; no screws that are removed in taking out the lock release, the mainspring or the hammer, but all remain in place, to be easily dismounted afterwards should occasion require. It is the same with the breech mechanism. All parts fit so nicely and yet so easily that no driving out by 'brute strength and awkwardness" is necessary. The whole breech mechanism, including the lock, can be assembled easily by a non-expert in fifteen minutes, and all the parts be as free from scar or blemish as when first finished, making it possible for the gun owner to take his gun down to clean it and yet get it together again 'looking as good as new', a thing to be prized by all lovers of fine guns.

Perhaps the pleasantest part of my visit was the courtesy and attention with which all remarks or suggestions were received. It is an open secret that Mr. Bullard is desirous of turning out a gun for sportsmen for sporting purposes, and one that will combine any or all the ideas possible to put in a gun, so as to suit the taste and ideas of sportsmen, and any suggestions or ideas receive due consideration and attention, and, after experiment, are adopted if proven beneficial. Certain things neither he, nor any man, can do. For instance, he does not pretend to make a gun that is to serve as a shell reducer or a bullet swage as some would like to have him; but he is willing to do anything in his power, within the range of possibilities of gun construction, to make a perfect weapon for legitimate gun work, so that, if a man behind the gun does his duty by it, the gun will always respond, do its share of the work perfectly and never 'go back on' the sportsman. as an evidence of his ideas and his desire to please and make what has been so long needed, let me invite your attention to the .45-85-285 cartridge now on the market, introduced by Mr. Bullard at his own risk of pecuniary loss, and in the face of the opposition of cartridge makers and gun dealers. He saw conclusive evidence that the .45-70-405 cartridge was not suited for short range work or hunting, but he experimented with and put this new cartridge upon the market, and now it is heralded over the land as the latest and best for heavy work by other gunmakers whose gun; it fits, they having had nothing whatever to do with its conception, making or use until they saw its success and then stole it and advertised it in connection with their guns. Let us give honor where honor is due, and that to the man who stood the risk of pecuniary loss in trying to get what was wanted, and not to those who stood quietly by at no expense or risk of loss, and then jumped at the thing because it was a success.

As another evidence of his advanced ideas let me call attention to the special guns he now makes, which I hope to see soon become the regular standard. Of calibers running through .32, .38, .40, .45, .50, nearly all have a special ammunition of heavy charges of powder and light bullets. Through him and him alone more power has been given our guns than ever before, and the .32 and .38 special, the .40-75, the .45-85, and the .50-115 are the most powerful weapon of their class now to be found, throwing each its proper ball with terrible killing force over a line that is as nearly flat as it is possible to get, and yet keep the necessary accuracy. Taking varying lengths of barrel, of magazine, style of stocks, weight, caliber and finish, he now makes something near sixty or seventy varieties of arms more than any other one gun maker, and certainly enough from which to pick and choose and suit any one. Again it goes without saying that the best gun that can be made will not succeed except with good ammunition suited to its wants and work. When one can afford to buy and use only factory-made cartridges reloading is not needed; but to the many a long felt want has been that of a first-class set of reloading tools. With all due respect to the opinions of others who may differ, I will say that proper reloading tools have never yet been procurable to the general public until those of the Bullard Company were made and put on sale, Many abortions have been made and sold, but never a set that acted as test gauges as well as reloaders while being used, and by which the sportsman as thoroughly inspected his ammunition as do the experts in overlooking ammunition at the factories.

Such a set of reloaders I saw at the Bullard manufactory; they were simple, were easily handled, and reloaded as perfectly as any machine could have

UNITED STATES PATENT OFFICE.

JAMES H. BULLARD, OF SPRINGFIELD, MASSACHUSETTS.

IMPROVEMENT IN METALLIC CARTRIDGES.

Specification forming part of Letters Patent No. **192,676**, dated July 3, 1877; application filed February 24, 1877.

To all whom it may concern:

Be it known that I, JAMES H. BULLARD, of Springfield, in the county of Hampden and State of Massachusetts, have invented a new and valuable Improvement in the Process of Manufacturing Cartridge-Shells; and I do hereby declare that the following is a full, clear, and exact description of the construction and operation of the same, reference being had to the annexed drawings, making a part of this specification, and to the letters and figures of reference marked thereon.

Figure 1 of the drawings is a representation of a plan view of the blank from which the shells are struck, and Fig. 2 is a side view of the same in the second stage. Fig. 3 is a sectional view of my cartridge-shell in the third state of construction; and Fig. 4 is a central sectional view in a further state of construction, showing the headed base. Fig. 5 is a front view of the headed base. Fig. 6 is a sectional view of a modification of my cartridge-shell; and Fig. 7 is a modification of Fig. 1, showing the different layers of metals.

This invention consists in the process of making cartridge-shells from two heterogeneous layers of metal, as will be hereinafter set forth.

My method or process of manufacturing is as follows: I take two loose thin sheets of metal and place one upon the other, without soldering, the lower being preferably brass, such as is ordinarily used for cartridge-shells, and the upper one being copper, or any metal which will not be injured by contact with gunpowder. I then place these sheets under a double-acting press, such as is in common use for making ordinary cartridge-shells, which cuts out the blanks and "cups" them at the same time, giving them the shape shown in Fig. 2. After this first operation the blank is drawn, in the usual manner, into the shape shown in Fig. 3. It is then "headed," so as to form the base, and the cartridge is complete, except such changes as may be necessary to adapt it to the chambers of certain fire-arms.

In the accompanying drawings, A designates the brass outer part of the shell, and B the protecting inner layer or lining, of material which is unaffected by gunpowder. C designates the base of the shell; and in heading or forming said base an annular rim, B′, of the copper or similar lining B is crimped within the annular expansion C′ of said headed base, which prevents it from any possibility of being detached or pulled out. This rim or bead B′, extending inward from the walls of the shell, acts as a re-enforce, which strengthens the base of the shell and prevents the same from being swollen or burst by the action of gases.

No permanent fastening is required to make heterogeneous layers A and B "draw" perfectly even. I prefer, however, to "tin," or coat with tin, either the upper side of the brass plate or sheet, or the lower part of the copper or similar plate or sheet, or the contiguous sides of such sheet of metal.

In the process of annealing the outer brass layer A and inner copper layer B will be soldered together by the action of the heat on said interposed layer of tin. This layer of tin (designated by D) is shown in Figs. 6 and 7.

The advantages of my method of construction above described are as follows: First, I am able to tell exactly how much of each metal is in each shell, and thereby to regulate the manufacture so as to obtain the greatest degree of elasticity, and also the requisite protection from the action of the powder. Secondly, my process does not require any special preparation of the metals, as in the case of soldering or other permanent fastening. Thirdly, I am enabled to separate the scrap without recourse to heat or any other agency. Fourthly, the only expense attending the manufacture, beyond that attending the manufacture of an ordinary brass cartridge-shell, is the cost of the copper or other material of which the lining is made, and that of the interposed tin when the latter is used. Finally, these cartridge-shells can be stored filled with powder for an indefinite length of time without injury.

What I claim as new, and desire to secure by Letters Patent, is—

1. An improvement in the art of manufacturing cartridge-shells, which consists in placing a sheet of copper upon a sheet of brass, cutting a blank from the double sheet, and

drawing and heading the blank, substantially as described, and for the purpose set forth.

2. A cartridge-shell having a brass body, A, and headed base C, and a copper lining, B, provided with a re-enforcing rim or filling, B′, substantially as specified.

In testimony that I claim the above I have hereunto subscribed my name in the presence of three witnesses.

JAMES H. BULLARD.

Witnesses:
W. S. GREENE,
F. E. BECKWITH, Jr.,
M. G. BULLARD.

J. H. BULLARD.

METALLIC CARTRIDGE.

No. 192,676. Patented July 3, 1877.

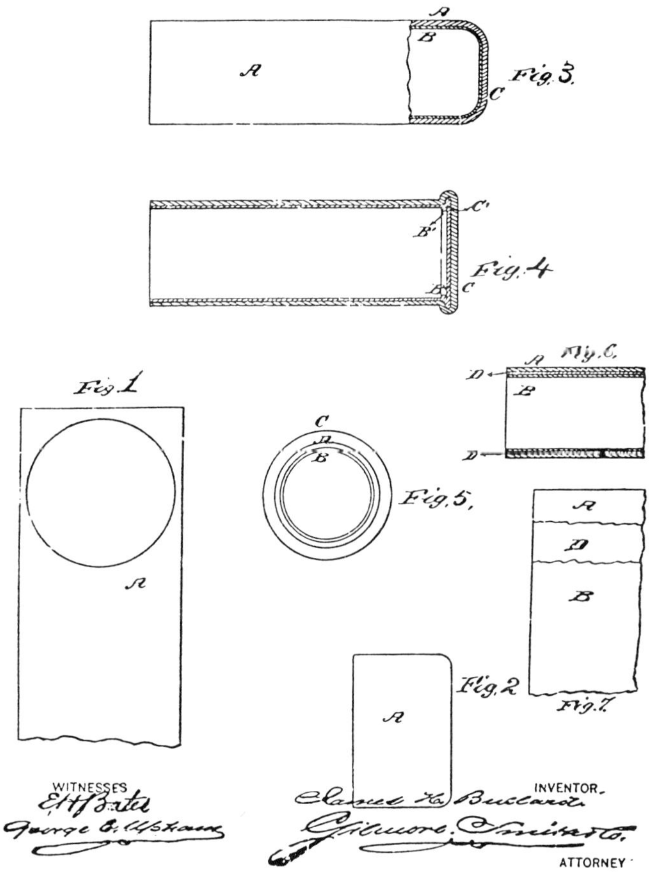

Plate 8-1.

J. H. BULLARD.
Metallic Cartridge-Shell.

No. 197,823. Patented Dec. 4, 1877.

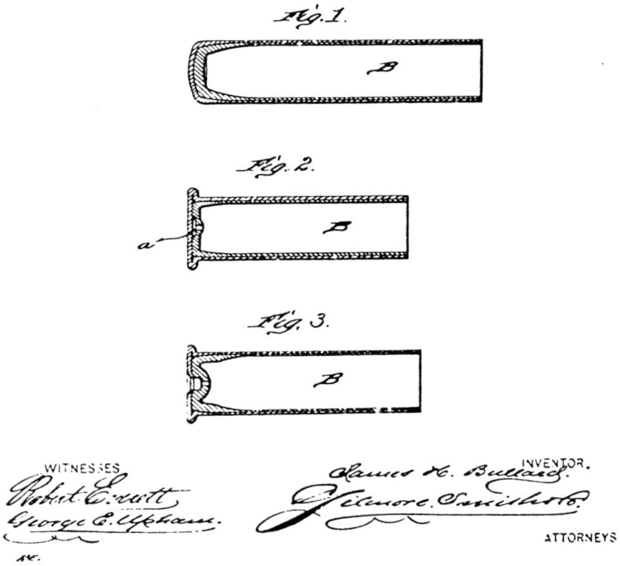

UNITED STATES PATENT OFFICE.

JAMES H. BULLARD, OF SPRINGFIELD, MASSACHUSETTS.

IMPROVEMENT IN METALLIC CARTRIDGE-SHELLS.

Specification forming part of Letters Patent No. **197,823**, dated December 4, 1877; application filed September 29, 1877.

To all whom it may concern:

Be it known that I, JAMES H. BULLARD, of Springfield, in the county of Hampden and State of Massachusetts, have invented a new and valuable Improvement in Cartridge-Shells and Process of Making; and I do hereby declare that the following is a full, clear, and exact description of the construction and operation of the same, reference being had to the annexed drawings, making a part of this specification, and to the letters and figures of reference marked thereon.

Figures 1, 2, and 3 are longitudinal central sectional views of my cartridge-shell in different stages of construction.

My invention relates to the manufacture of cartridge-shells for small-arms; and the nature of my invention consists, first, in the process of making cartridge-shells with a copper or other metallic lining put in at any time after the first drawing, for the purpose of making a re-enforce to the butt of the shell, and extending the whole length of the shell, for the additional purpose of protecting the outer case from the action of the gunpowder; and, second, in a cartridge-shell made with a copper or other metallic lining, the base of which lining shall have a pocket or recess for the fulminate, all as hereinafter more fully set forth.

The annexed drawing, to which reference is made, fully illustrates my invention.

I first make a cup for the future shell, of brass, for the outside of the shell, and also another cup, of copper, nickel, or other metal that will not be acted upon by gunpowder, this latter cup being just enough smaller than the outside or brass cup to fit into the same. After these cups have been made, they are put together—that is, one within the other—and "drawn" in the usual manner, until the shell is completed.

In Fig. 1, the lining is put in after the third drawing, and makes what is called a "solid head."

In Fig. 2, I represent a shell with the recess *a* formed for the fulminate, and which is commonly known as an "inside primer," the lining in this case being put in after the head is formed.

Fig. 3 represents what is called a "folded head" and an "outside primer," the head and pocket for the primer being formed at the same time, after the lining is put in the outside case.

By this method of cupping each metal separately, I am enabled to make a lining of equal thickness, or thicker than the outside of the case, and I can also make the rear part of the shell much stronger than can be made of a single piece, for the reason that beyond a given thickness the metal cannot be drawn successfully, as it will tear apart under the press. It is obvious that two pieces drawn separately, as thick as can be in the first place, will be twice as strong as one single piece of half the thickness. This is of great importance in shells used in machine-guns, like the Gardner, Gatling, and other guns of the same class, as the shell remains only for an instant in the chamber of the gun, and if the charge hangs fire from any cause, the shell will not explode until partly withdrawn from the chamber, and, unless the case or shell is unusually strong, it will burst, and inevitably injure and damage the mechanism of the breech, rendering it unserviceable for the time at least. Such accidents have happened in just this way, and, unless a shell is fired the instant the fulminate is struck, are liable to happen at any time with the shells now in use.

Another advantage in having the rear part of the shell thick is the fact that most guns have a recess or slot at the rear end of the chamber, for the purpose of allowing the extractor to engage with the head of the shell, and there is always more or less of a vacant space in the chamber. When a shell is exploded, the force of the explosion drives the shell (as usually made) into this space, rendering it difficult, and in some instances almost impossible, to extract. Neither will the shell be as liable to swell as shells made in the usual way, and stick from that cause.

Another advantage lies in the fact that, by putting this lining in after the head is formed or folded, it completely shuts out the possibility of any gas getting between the folds of the head and causing it to swell when fired, thus making the most perfect re-enforce known; and when this lining is made of copper or other non-corrosive metal, it will thoroughly protect

the brass outside from the action of the powder, making a most perfect shell in every particular.

By my method of making the shells, what is known as an "inside primer" can be used with great advantage, as the pocket or recess *a* can be made in the base of the lining, and the fulminate placed therein, and then said lining put in the outside case or shell A the last thing before the head is formed, or even after, if thought best; or it can be delayed until the time immediately preceding the reducing of the muzzle end of the shell to fit it to the chambers of different guns.

By placing the fulminate in the lining, as described, the piece called a "cup," usually employed for the twofold purpose of holding the fulminate and acting as a re-enforce, is not used, and there is no necessity of pinching or pressing a groove in the sides of the shell to hold it in, thus leaving the outside of the shell smooth and unbroken, and much easier to withdraw from the chamber of the gun, and is also not as liable to break in two as has been the case with shells made in the usual way. Also, by this construction of a cartridge-shell, the outer brass shell is protected from the action of the powder by the inner non-corrosive lining, the latter extending out to the end of the brass outer shell, whereby all liability of the entrance of the gases generated by the explosion of the powder between the shells is prevented, which is liable to occur when the inner lining extends but a short distance from the base. This extension of the inner lining stiffens and strengthens the cartridge-shell, and if there should be, from any cause, a flaw or imperfection in the outer shell or case, the inner lining will prevent any escape of gas into the mechanism of the gun, thereby doing injury to the fire-arm. If the inner lining should have a flaw or imperfection, the outer case would perform the same duty, thus making a much safer shell than has heretofore been used. The possibility of a flaw in the same place in both cases of metal would be very rare indeed.

I do not broadly claim, in this application, a copper-lined brass cartridge-shell, as that is fully shown and described in my Letters Patent dated July 3, 1877.

What I claim as new, and desire to secure by Letters Patent, is—

1. The improvement in the art of manufacturing cartridge-shells, which consists in first forming two separate cups of different metals, one of which is re-enforced at the head by inserting the latter within the outer cup at any time after the first drawing operation, and subsequently completing the same, forming a cartridge-shell with an interior re-enforced casing of different metal from the outer, substantially as and for the purposes set forth.

2. A cartridge-shell consisting of an outer shell, an inner lining of non-corrosive metal, extending to the outer end of the outer shell, said inner lining being re-enforced at the butt, and provided at its base with a pocket or recess for the fulminate, substantially as and for the purposes set forth.

In testimony that I claim the above I have hereunto subscribed my name in the presence of two witnesses.

JAMES H. BULLARD.

Witnesses:
I. NELSON,
L. PRESTON.

Plate 8-2.

done, though of course not so rapidly. These tools consisted of a resizing die and anvil, this latter being the one anvil for all the operations of reloading. The resized shell was driven out by a punch that extracted the primer at the same time, the shell having been resized before the primer was extracted. The new primer was easily seated by another punch, and then the shell filled with powder and inserted in the reloading die, the ball being seated in its place to exactly the proper depth and then crimped all in one operation, and the perfectly reloaded shell then forced from the die. This latter die serves as the gauge, as a swelled-headed shell is instantly detected, the ball is accurately seated and to the exact and proper depth, so there can be no doubt but that the cartridge will feed up properly, enter the chamber and then extract as it should.

THE .32-40-150

The .32-40 Bullard was a bottleneck cartridge of sub-caliber dimensions, which made it unusable in any other .32-40 rifle chamber of that time.[1] This fact in itself would not necessarily have doomed this loading, but the fact that it was no better ballistically than its competition precluded its commercial success. A.C. Gould had this to say about the .32-40 Bullard: "The .32-40 Bullard cartridge is a good one; but the .32-40 Winchester and Ballard being more used and higher esteemed, the Bullard cartridge is seldom found in gun stores."[2] (Plates 8-3A, 8-4B.)

Almost all the single-shots stamped .32-40 are for Winchester loadings, not Bullard, but as can be seen from the various plates showing caliber markings, this is never made clear on the stampings. The original owner knew the loading, but subsequent owners would only know by experience that a Bullard rifle was chambered for .32-40 Winchester and not .32-40 Bullard.

The following comments are from *The Breech-Loading Single-Shot Match Rifle* by Major Ned Roberts and Kenneth Waters. Roberts and Waters' comments about the .32-40 Bullard once again point to the possibility that Mr. Bullard may have been anticipating smokeless powders when he brought out his bottleneck cartridges in .32-40 and .40-90.

Then, about 1888 or late 1887, they commenced making this rifle with detachable and interchangeable barrels. Also, a Special Bullard Schuetzen rifle was brought out with either solid frame or detachable barrel. This Schuetzen model was a heavier rifle, weighing from 12 to 13 pounds, with half-octagon barrels only. Probably the most important difference between the models was that the Schuetzen was furnished with interchangeable double-set and plain single triggers, which I believe is the only rifle ever furnished with two triggers which could be easily and quickly interchanged by the shooter. The standard Target model was not furnished with double-set triggers.

A whole new series of cartridges was brought out for the Bullard rifles, of solid head reloadable type, all of which were manufactured by the Union Metallic Cartridge Company. These included the 32-40-150 and 40-90-300 Bullard bottlenecked; also the 38-45-190, 40-75-258, 45-85-290 and 50-115-300 straight-taper cartridges.

The 32-40 Bullard bottlenecked cartridge was an especially good one, but as the bullets weighed only 150 grains, the majority of riflemen considered it less accurate than the 32-40 Marlin-Ballard cartridge using 165 and 185 grain bullets. The bore diameter of the 32-40 Bullard barrels was .306" with a groove diameter of .311", as compared with the .319" groove diameter of 32-40 Marlin-Ballard caliber bullets. The Bullard target rifle in 32-40 Bullard caliber that I owned gave very nice accuracy loaded with 165-grain paper-patched bullets at 200 yards rest, as well as being far more accurate than I could hold offhand at that range. I also used both 165- and 153-grain lubricated bullets, breech seated, with loads of either Du Pont No. 1 Rifle Smokeless or Schuetzen Smokeless in order to avoid the need for cleaning the bore between shots and found that I could make 1½-inch groups at 100 yards or 3½-inch groups at 200 yards, shooting from rest.

I think it is unfortunate that the 32-40 Bullard cartridge is no longer on the market, as I believe it would prove highly accurate at 200 yards when loaded with some of our present-day smokeless powders, such as Du Pont I.M.R. No. 4227, which give better ballistics in bottlenecked cases than in straight-taper ones. Also, the standard 32-40 Marlin-Ballard case has too great a powder capacity when used with the modern smokeless powders adapted for use in single-shot target rifles. My rifle in this caliber had a 28-inch half-octagon barrel, double-set triggers, and pistol-grip stock, and of course vernier and wind-gauge target sights. It was a fine rifle and a favorite for some time.

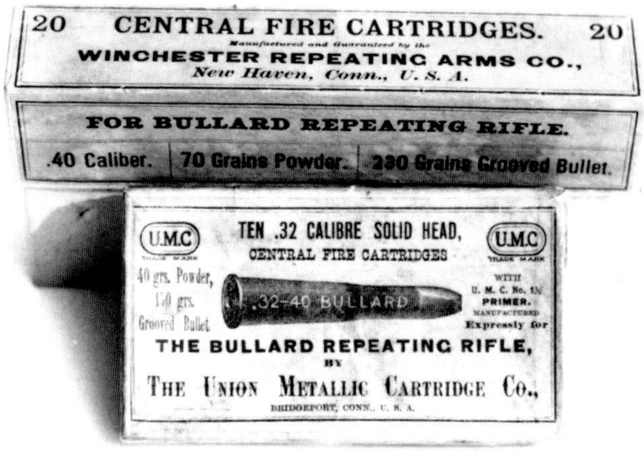

A.

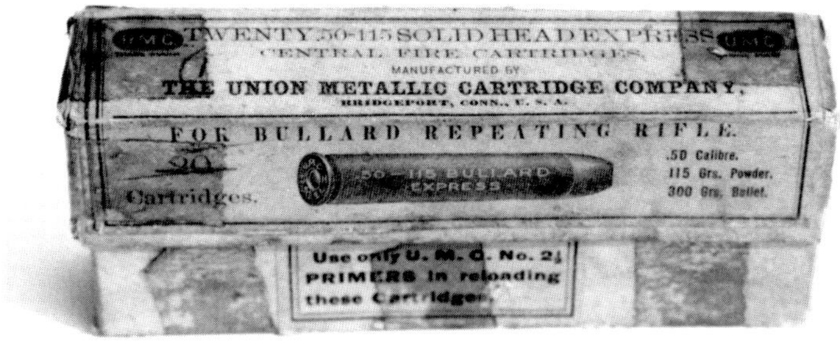

B.

Plate 8-3. Top: Winchester-made .40-70 Bullard loads, UMC .32-40. Bottom: UMC .50-115-300 box of cartridges — see note in chapter section about this headstamp. *Courtesy Cliff Schisler*

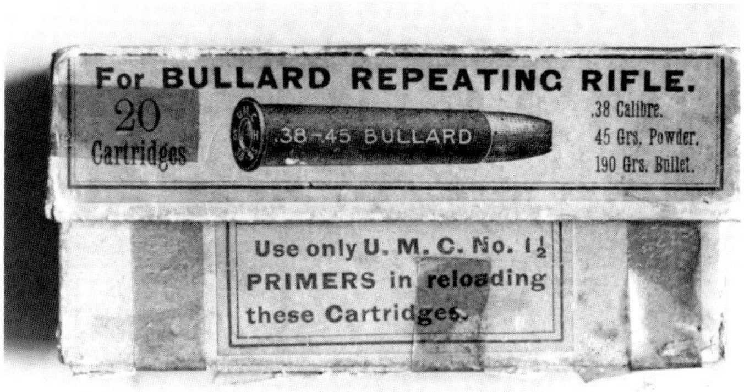

A.

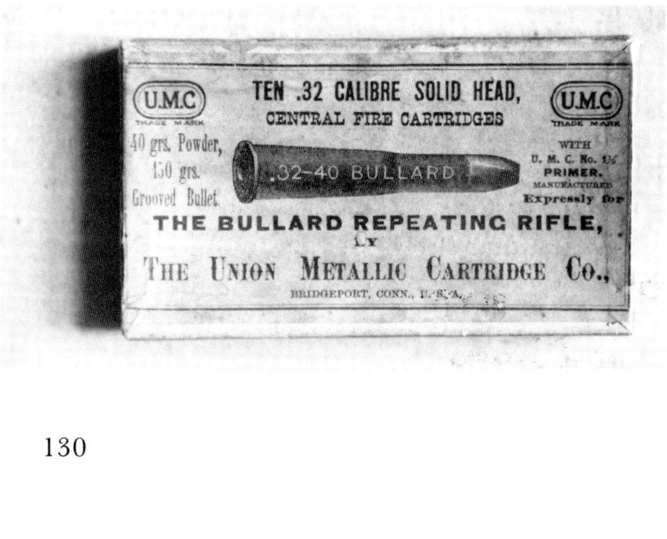

B.

Plate 8-4. Top: Compare this earlier UMC box of .38-45 Bullard cartridges with that in Plate 8-6. Bottom: Box of UMC .32-40 Bullard cartridges. *Courtesy Donald M. Blyth*

Plate 8-5. Box of .38-45 Bullard cartridges. The UMC label is colored red. The guarantee seal on the back is green and stamped with a 235 code number. The front of the box has a pink-colored label. This box could not have been manufactured prior to 1911 as Remington Arms and the Union Metallic Cartridge Company amalgamated in that year to form one company. *Courtesy Cliff Schisler*

Plate 8-6. Reproduced from the May 1980 issue of the *Gun Report*. *Courtesy Charles R. Conn*

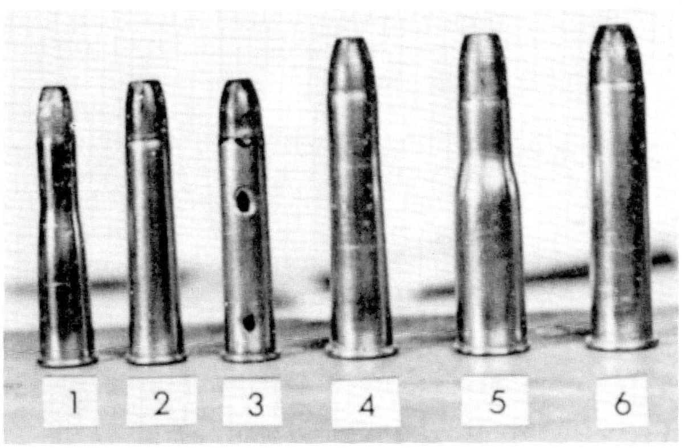

Plate 8-7. (1) WRACO 32-40 BUL; (2) WRACO 38-45 Bul; (3) WRACO 38-55 cut to 38-45 BUL; (4) WRACO 40-70 BUL; (5) WRACO 40-90 BUL; (6) WRACO 45-85 Bul. Reproduced from the July 1974 issue of the *Gun Report*. *Courtesy Charles R. Conn*

Plate 8-8. From the style of the label and the April 1884 Hobbs patent date on it, this UMC box was manufactured sometime after mid-1884. See Plate 8-3 for comparison to this label. *From the October 1969 issue of the Gun Report*

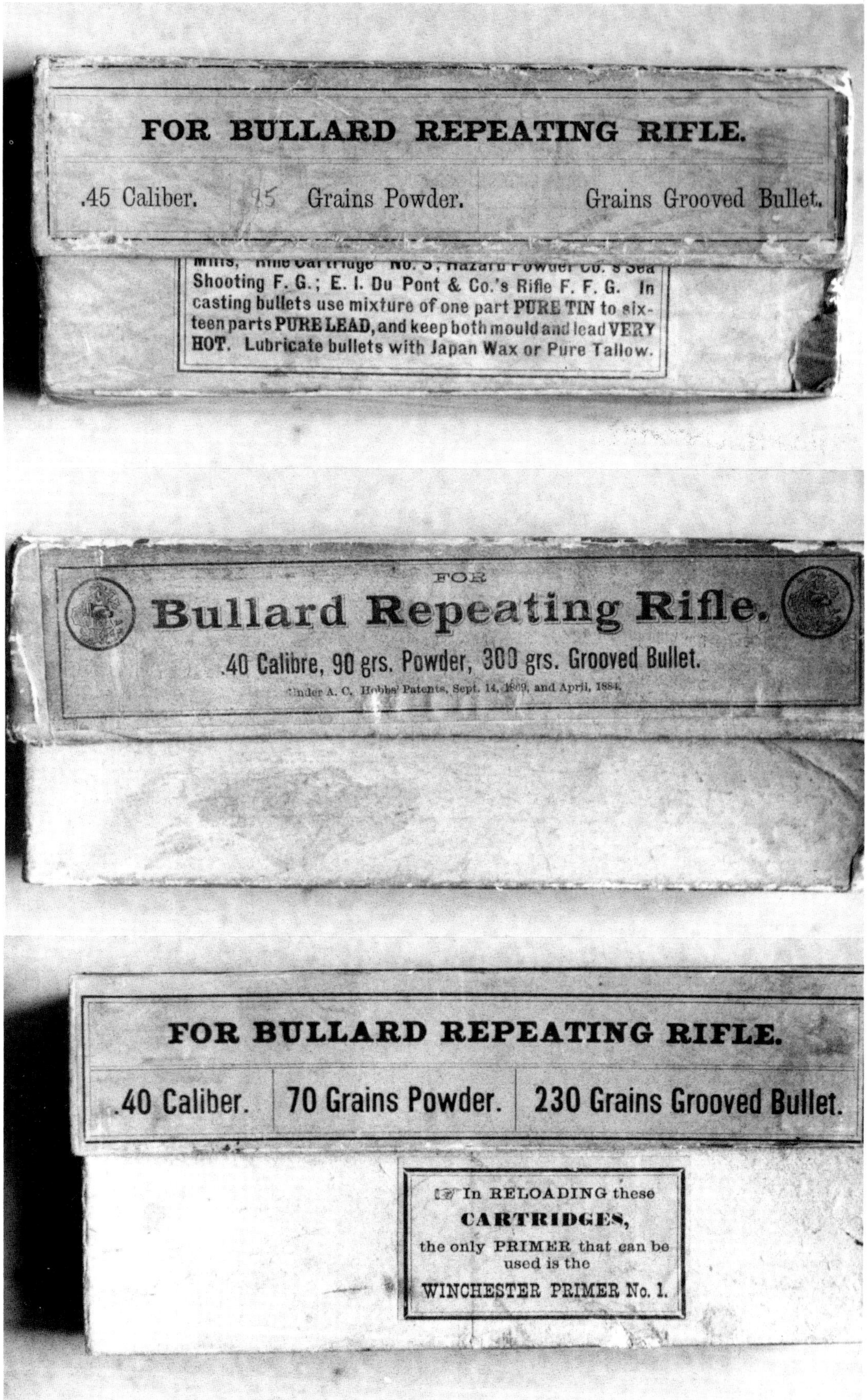

Plate 8-9. Top to bottom: .45-85 Winchester made, .40-90 UMC made, .40-70 Winchester made. *Courtesy Donald M. Blyth*

Major Roberts also wrote about the first rest matches. As always, his comments were interesting and informative:

> The caliber of the rifles used in those first rest matches ranged from 40 to 32, but none of the exclusively rest shooters used a smaller caliber than 35, and 38-50, 38-55, and 40-70 calibers were preferred by many of these experts for several years. However, by about 1895 the Maynard 32-35, 32-40, and 35-50; the Ballard, Bullard, Remington-Hepburn, and Winchester single-shot match rifles in 32-40 and 38-55 calibers; and the custom-made Pope, Schoyen, Zischang, Wurfflein, and Farrow rifles in 32-40, 33-40, and 38-55 calibers had been adopted by practically all of the noted expert offhand and rest shooters for 200-yard work, as a multitude of bench-rest and machine-rest tests had proven conclusively that these calibers were more accurate than the 40 or larger calibers.

The specifications of this round are as follows:

cartridge length:	2.26 inches
case length:	1.85 inches
bullet diameter	.315 inches, 3-groove solid lead
case type:	bottleneck
neck diameter:	.332 inches
base diameter:	.455 inches
shoulder diameter:	.413 inches

Used in single-shot and small-frame repeaters

Many of the headstamps shown were supplied by Howard Hoovestol of the International Cartridge Collectors Association Inc.

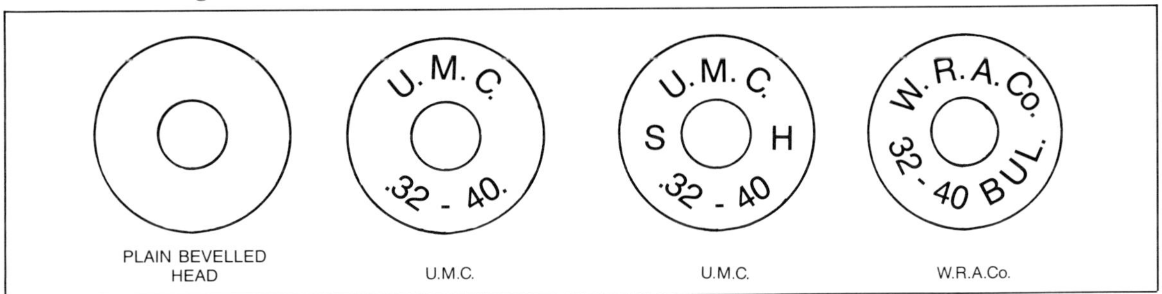

32-40 Bullard Headstamps.

Cal 32	The style in which repeaters of this caliber are marked.	32 40	The style in which single-shot rifles of this caliber are marked.

THE .38-45-190

This round was again a special Bullard round used in the single-shot and small-frame repeaters. According to A.C. Gould, "The .38 Bullard is but little known, and but few rifles are made to take it."[3] This round was loaded by the Union Metallic Cartridge Company (UMC) and Winchester. However, unlike the .32-40 Bullard caliber cartridge, there was little difference between this round and its .38-caliber contemporaries (Plates 8-4A, 8-6, 8-7). The statement made by Mr. Gould that "few rifles are made to take it" is incorrect unless — and remember he was writing this in the early 1890s — other makers shortly after the demise of the Bullard Company began to make a similar load with very similar case dimensions. Plate 8-6 shows the label of a box of UMC .38-55 cartridges made expressly for Ballard, Marlin and Bullard rifles, which can only mean one thing, that the rounds were the same. Was Mr. Gould incorrect in his assessment of the situation or did Ballard and Marlin copy a good idea from J.H. Bullard?

The approximate specifications (approximate because everyone has a different feel for a micrometer, and due to the age of the cartridges they may have been reloaded and resized many times, making exact measurements almost impossible or irrelevant) of this round are as follows:

cartridge length:	2.26 inches
case length:	1.80 inches
bullet diameter	.373 inches, 2-groove solid bullet
case type:	straight taper case
neck diameter:	.397 inches
base diameter:	.454 inches; both UMC and WRA Co. measured out at this figure on the micrometer

Used in single-shot and small-frame repeaters

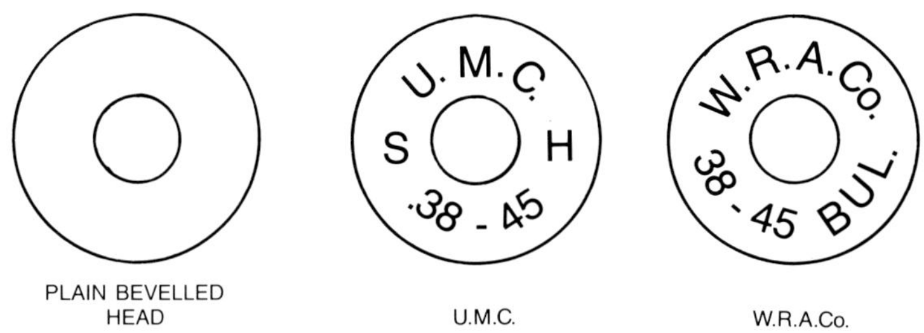

Three known headstamps for the .38-45 Bullard loads.

Cal 38	See Plate 4-15C The style in which repeaters of this caliber are marked.	38 45	The style in which single-shot rifles of this caliber are marked.

THE .40-70-232

This Bullard loading was to find its way into the single-shot rifle range and the large-frame repeaters.

Some of the first 100 or so large-frame repeaters made with the Bullard Repeating Arms Association address handled either the .45 or .40 caliber cartridges then available, and as the special loadings that Bullard made also fit these rifles, its quite conceivable that he fully intended to build this duality into his rifles right from the beginning. He was the guiding force and designer behind these cartridges. His cartridge work would have been completed when he left the Bullard Company after the summer of 1885. As can be seen by the plates showing caliber markings, no Bullard rifles seem to have been marked precisely to indicate whether the rifle handled a Bullard cartridge or a Winchester loading, etc. With many loadings this was not a problem, but with some calibers, such as the .32-40 Bullard, which did not interchange with any other .32-40 cartridge, it could pose a problem.

TRAJECTORY OF BULLETS.

During the past year a series of experiments were carried on at the range of the Massachusetts Rifle Association, at Walnut Hill, Mass., under the direction of Messrs. C. W. Hinman and F. J. Rabbeth, with a view of learning the trajectory of bullets fired from various rifles used both on the range and for hunting. For this purpose, tissue-paper screens were erected every 25 yards up to 200 yards. Bullets were fired through these, and the height of the bullet-holes above a fixed plane was carefully measured. A simple calculation gives the height of the bullet at each distance above a line drawn from the muzzle of the rifle to where the bullet pierces the 200 yards' screen. The plane was set out with an engineer's instrument, and the results given are believed, in most cases, to be correct to a tenth of an inch.

Where no particular powder is specified, Hazard's Sea Shooting F. G. was used.

Rifle.	Calibre.	Length of Barrel.	Powder.	Bullet.	Height of Trajectory in inches at								
					0 yds.	25 yards.	50 yards.	75 yards.	100 yards.	125 yards.	150 yards.	175 yards.	200 yds.
Remington. — Military	.40	34	*95	405P	0	2.9	5.4	7.2-	7.6	7.4-	6.0+	3.7-	0
" "	.40	34	*95	370P	0	3.5	6.3-	8.0+	8.7+	8.3†	6.6	4.1-	0
" Sporting	.25	28	*20†	67N	0	3.6-	6.4	8.3	9.4	9.2+	7.6+	4.7+	0
" "	.32	28	*35†	147N	0	4.1	7.3-	9.3+	10.2	9.8†	8.0	4.7-	0
" Military	.44	34	90†	520P	0	4.0	7.2	9.3	10.4	9.7	7.9	4.7	0
Bullard. — Sporting	.40	28	75‡	230N	0	3.7	6.8	9.1	10.2	10.0	8.3	4.9	0
" "	.50	26	115‡	300N	0	3.9+	7.1+	9.4-	10.6-	10.4-	8.6†	5.3	0
Winchester	.40	28	62‡	210N	0	4.1	7.3	9.4+	10.5	10.3	8.6	5.2	0
"	.32	24	20‡	115N	0	5.8	10.3-	13.3	14.5	13.9+	11.3	6.6+	0
"	.44	24	40‡	200N	0	5.9	10.8	14.2	15.5+	14.9	12.1	7.1+	0
Maynard	.32	28	35	165P	0	4.5	8.0	10.2	11.2	10.9	8.8	5.2	0
"	.35	28	60	245P	0	4.7	8.1+	10.4+	11.3	10.9-	8.8	5.2-	0
"	.40	28	60	265P	0	4.5-	7.9	10.2	11.2-	10.8-	8.8	5.2+	0
"	.40	28	60	330P	0	5.2	9.3	11.8-	12.7+	12.2-	9.8+	5.9	0
Springfield. — Military	.45	32.6	70‡	405N	0	4.8-	8.3+	10.7+	11.7	11.3	9.2-	5.6	0
Remington. — Sporting	.38	28	50	253P	0	6.1	10.5	13.4	14.5	13.6	10.9	6.4	0
" "	.38	28	50	330P	0	6.6	11.4	14.5	15.5	14.6	11.7	6.6	0
Wesson	.32	24	13‡	90N	0	7.5	13.5	17.5	19.2	18.4	15.0	8.9	0
Ballard	.22	28	5‡	30N	0	8.3	14.5	18.7	20.4	19.8	16.2	9.4	0

* About one sixth the whole volume was "Brackett" powder.
† Hazard "Ducking" powder.
‡ Factory cartridges (P—patched, N—naked.)

Plate 8-10. Article showing Bullard ballistics for the .40-75-230 and the .50-115-300, further proof that the .50-115 was well established long before 1886. Both loads were listed as factory cartridges. *The Rifle, Vol. 1, No. 1, May 1885*

This round was interchangeable with the Marlin .40-60-260, as according to several Bullard catalogs the cartridges were of the same length.

The cartridge specifications are as follows:

cartridge length:	2.54 inches
case length:	2.09 inches
bullet diameter	.413 inches, single wide-groove solid lead
case type:	straight with full-length taper
neck diameter:	.432 inches
base diameter:	.505 inches

Used in single-shot and small-frame repeaters

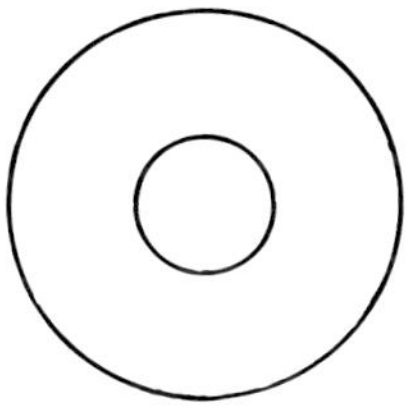

PLAIN BEVELLED HEAD

.40-70 Bullard.

W.R.A.Co.

Cal
40

The style in which repeaters of this caliber are marked.

THE .40-75-258

This round used the same brass as the .40-70 Bullard and, except for a difference in ballistics caused by the extra five grains of powder and heavier bullet, it was a carbon copy of its less powerful brother. Cartridge collectors may wish to lump this and the aforementioned cartridge under one heading, as they are so similar. The author has chosen to separate them into two classifications simply to aid their study.

The cartridge specifications are as follows:

cartridge length:	2.54 inches
case length:	2.09 inches
bullet diameter	.413 inches, single wide-groove solid lead
case type:	straight with full-length taper
neck diameter:	.432 inches
base diameter:	.505 inches

Used in single-shot and large-frame repeaters

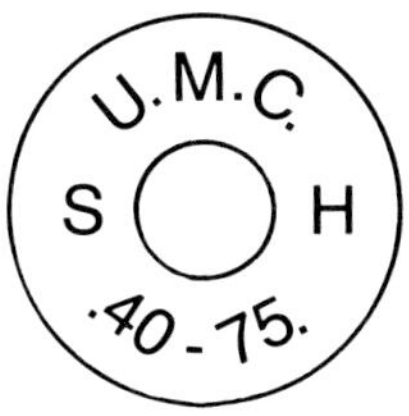

U.M.C.

One known headstamp for this round.

Being the same proportions as the .40-60-260 Marlin, this round is interchangeable with it, and according to Frank C. Barnes[4]: "The old Bullard catalog states that the .40-60 Marlin can be fired in guns of the above caliber. If so, then the .40-66 WCF could also be substituted since it is the same case as the Marlin round."

40
75

The style in which single-shots of this caliber are marked.

THE .40-90-300

This cartridge is the second Bullard cartridge of the previously mentioned three that differs substantially from its contemporaries. It is of a bottleneck design and was loaded by UMC and Winchester for Bullard. As stated in one catalog, "The .40-90 Bullard cartridge is loaded with great care. A highly compressed, grooved and lubricated bullet is used, weighing 300 grains. It has a flatter trajectory than any other 40 cal cartridge in the market." Frank C. Barnes adds, "The .40-90 Bullard was undoubtedly an effective big game caliber, particularly if the now rare 400-gr. loading was used."[5]

Reference Page.	Description.	Cartridges, per 1,000.			Shells, per 1,000.			Bullets, per 1,000.		
		£	s.	d.	£	s.	d.	£	s.	d
23	·22—13—45	2	12	6	1	9	2	–	13	4
24	·25—21—86	4	15	–	3	13	4	–	18	4
24	·25—20—85	2	12	6	1	9	2	–	13	4
24	·25—36—85	4	17	6	2	15	–	–	13	4
24	·25—36—117	6	5	–	3	2	6		—	
28	·32—20—115	2	12	6	1	9	2	–	13	4
28	·32—40—165	4	17	6	2	15	–	1	1	–
30	·38—40—160	3	2	6	2	5	–	1	1	–
31	·38—55—255	5	5	–	2	15	–	1	6	8
31	·38—56—255	5	15	–	2	17	6	1	6	8
31	·40—60—210	5	15	–	2	17	6	1	1	–
31	·40—60—260	6	2	6	2	17	6	1	6	8
31	·40—65—260	6	2	6	2	17	6	1	6	8
31	·40—82—260	6	15	–	3	10	–	1	6	8
33	·425—30—230	3	15	–	2	5	–	1	1	–
33	·430—20—200	3	8	4	1	17	6	–	16	–
34	·44—40—200	3	2	6	2	4	2	1	1	–
35	·442—20—220	3	8	4	1	17	6	–	16	–
35	·450—18—225	3	8	4	1	17	6	–	16	–
35	·450—28—270 (450 No. 1) ..	4	15	–	2	15	–	1	11	8
35	·45—60—300	5	15	–	2	18	4	1	11	8
36	·45—70—350	6	8	4	3	–	–	1	11	8
36	·45—70—405 (U.S.)	6	8	4	3	–	–	2	2	6
36	·45—70—500	6	15	–	3	–	–	2	8	4
36	·45—70—405 (Marlin)	6	8	4	3	–	–	2	2	6
36	·45—75—350	6	12	6	3	8	4	1	11	8
36	·45—85—300 B.	6	8	4	3	–	–	1	11	8
37	·45—90—300	6	14	2	3	8	4	1	11	8
37	·45—85—350	6	14	2	3	8	4	1	11	8
37	·45—82—405	6	14	2	3	8	4	2	2	6
37	·45—85—300 (Express)	6	14	2	3	8	4	1	11	8
41	·50—95—300	7	2	6	3	10	–	1	16	8
41	·50-110—300	7	12	6	3	15	–	1	16	8

CENTRAL FIRE CARTRIDGES

FOR

COLT'S, MARLIN, WINCHESTER, STEVENS', WEBLEY, BALLARD, BULLARD and KANGAROO RIFLES;

also for

COLT'S FRONTIER REVOLVERS.

Plate 8-11. Note entry under page 36: .45-85 with 300B-grain bullet. *Courtesy Eley Ltd.; IMI PLC*

The author would be most interested in obtaining a photograph of a cartridge box with this loading, as all that have been seen are 300-grain loadings (Plate 8-9).

The cartridge specifications are as follows:

cartridge length:	2.55 inches
case length:	2.04 inches
bullet diameter	.413 inches, 3-groove solid lead
case type:	bottleneck
neck diameter:	.430 inches
shoulder diameter:	.551 inches
base diameter:	.569 inches

Used in single-shot and large-frame repeaters

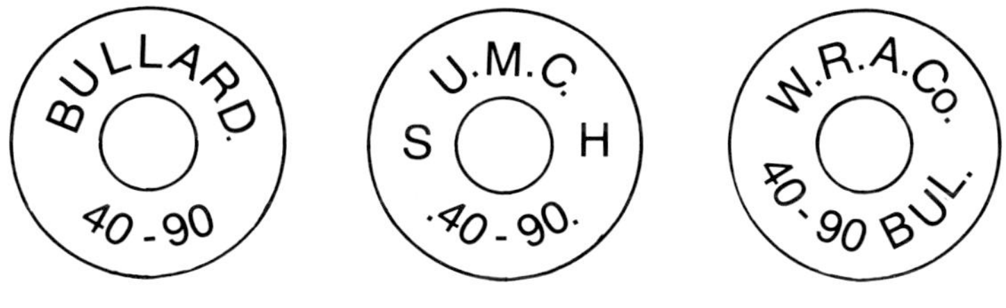

U.M.C. U.M.C. W.R.A.Co.

UMC WINCHESTER
Three known headstamps for this round

Cal
40
90

The style in which repeaters of this caliber are marked.

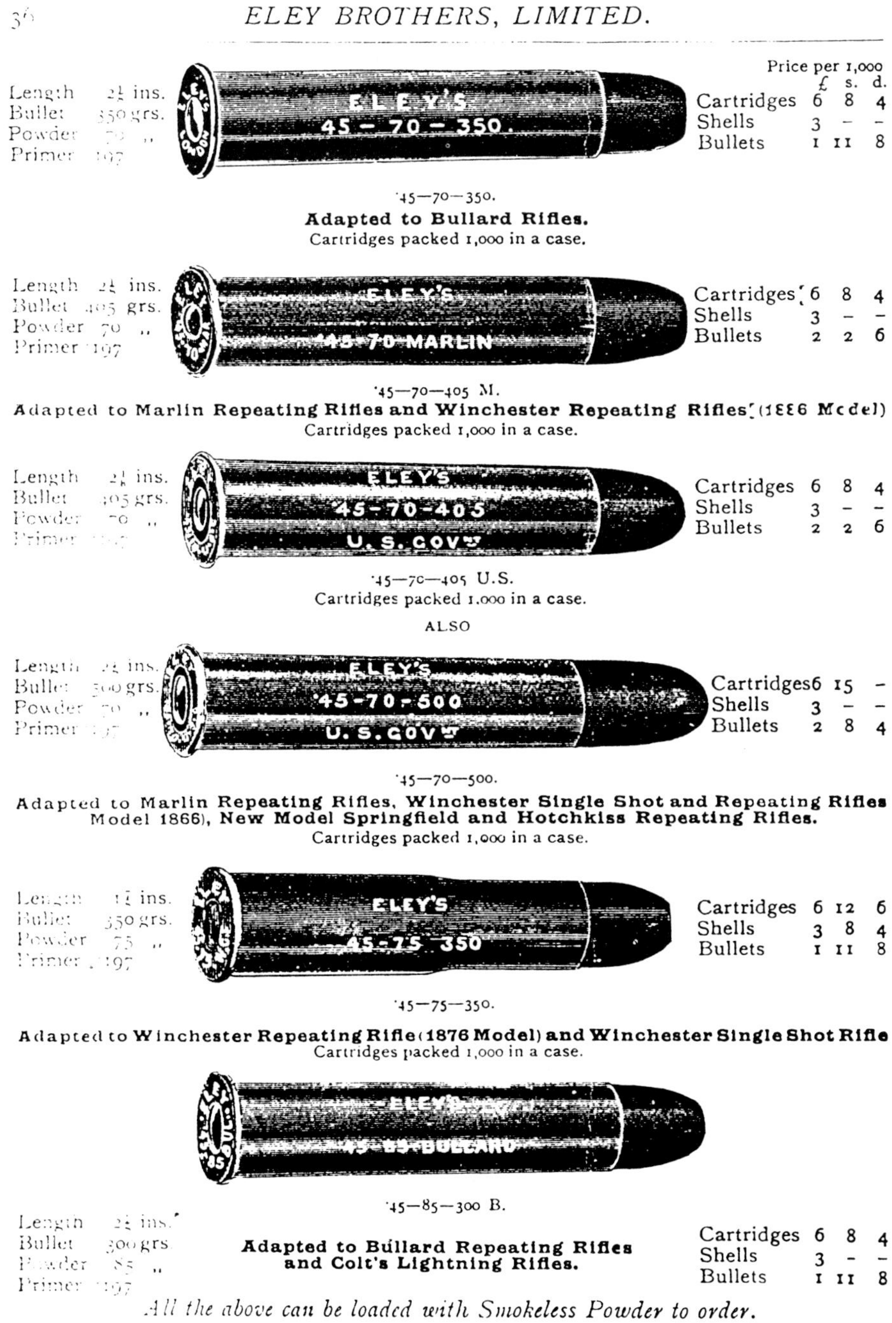
36 ELEY BROTHERS, LIMITED.

Length 2¼ ins. Bullet 350 grs. Powder 70 ,, Primer ·197

	£	s.	d.
Price per 1,000			
Cartridges	6	8	4
Shells	3	–	–
Bullets	1	11	8

·45—70—350.
Adapted to Bullard Rifles.
Cartridges packed 1,000 in a case.

Length 2¼ ins. Bullet 405 grs. Powder 70 ,, Primer ·197

Cartridges	6	8	4
Shells	3	–	–
Bullets	2	2	6

·45—70—405 M.
Adapted to **Marlin Repeating Rifles and Winchester Repeating Rifles (1886 Model)**
Cartridges packed 1,000 in a case.

Length 2¼ ins. Bullet 405 grs. Powder 70 ,, Primer ·197

Cartridges	6	8	4
Shells	3	–	–
Bullets	2	2	6

·45—70—405 U.S.
Cartridges packed 1,000 in a case.
ALSO

Length 2¼ ins. Bullet 500 grs. Powder 70 ,, Primer ·197

Cartridges	6	15	–
Shells	3	–	–
Bullets	2	8	4

·45—70—500.
Adapted to **Marlin Repeating Rifles, Winchester Single Shot and Repeating Rifles** Model 1866), **New Model Springfield and Hotchkiss Repeating Rifles.**
Cartridges packed 1,000 in a case.

Length 1⅞ ins. Bullet 350 grs. Powder 75 ,, Primer ·197

Cartridges	6	12	6
Shells	3	8	4
Bullets	1	11	8

·45—75—350.
Adapted to **Winchester Repeating Rifle (1876 Model) and Winchester Single Shot Rifle**
Cartridges packed 1,000 in a case.

·45—85—300 B.

Length 2¼ ins. Bullet 300 grs. Powder 85 ,, Primer ·197

Adapted to Bullard Repeating Rifles and Colt's Lightning Rifles.

Cartridges	6	8	4
Shells	3	–	–
Bullets	1	11	8

All the above can be loaded with Smokeless Powder to order.

Plate 8-12. Note .45-70-350 loading at top (adapted to Bullard rifles) and headstamp detail on .45-85-300B at bottom. *Courtesy Eley Ltd.; IMI PLC*

THE .45-85-290

This round is interchangeable with the U.S. government .45-70 round, both of which use a case 2.1 inches long.[6] The Marlin .45-85-285 is very similar to the Bullard round and is also interchangeable. This cartridge was loaded by UMC, Winchester and Eley Brothers. Eley Brothers loaded their Bullard cartridge with a 300-grain bullet (see Plates 8-12, 8-13).

As far as can be ascertained, this load came with the option of a 285, 290, 295, and 300 grain bullet. The 285 loading is mentioned in the article about the "visitor" to the Bullard plant on September 13, 1884; the 290-grain bullet weight is mentioned in the catalogs; and the 295 load is shown by Michael E. Whiteley in his artwork for this chapter's frontispiece. The 300-grain load is shown in the Eley catalog.

There does seem to be some controversy about who developed the .45-85-285 loading first, Marlin or Bullard. Graham Burnside, in his article "Marlin's .444 Starts Fire Under Cartridge Fans," says:[7]

> The .45-85 Marlin Cartridge is something of an enigma. This round employs the same case as the .45-70. It is a .45-70 with more powder and a lighter bullet. Normal .45-70 bullets ranged from 405 grains (lead) on up to the 500-grain ball, but this .45-85 Marlin has a bullet of 285-grains — again, such a combination makes good sense in a case of only 2.1 inches in length.
>
> There is no problem of identification here. The cartridge is plainly marked W.R.A. Co. .45-85 MAR. But oddly enough, if the round were not head-stamped it would be identified as a .45-85 Bullard cartridge!
>
> In checking the loading specifications, it is discovered that the .45-85 Marlin and the .45-85 Bullard are the very same thing! They both employ the .45-70 case of 2.1 inches, they both have 85 grains of powder, and they both utilize a lead bullet of 285 grains. The exactness continues when it is noted that the style of the two lead bullets is identical. This last is strange in that Marlin usually used a bullet style which was characteristically of their design.

In his book on Marlins, Bill West states, "In 1884 the .45-85-285 cartridge became available. This had the same case as the .45-70."[8]

We may never know who had this round out first, but we do know this round was available from Bullard in September of 1884. Was the sentence in which the visitor to the Bullard factory mentions the Bullard .45-85-285 alluding to the outright copying by other companies of the cartridge Mr. Bullard designed?

Bullard's record of loading cartridges of his own design with a heavy powder charge and light bullet was well established and his .45-85-285 was in tune with that philosophy. While it is possible that both Marlin and Bullard arrived at the same conclusions at the same time, it would seem that Mr. Bullard, by virtue of his previous cartridge work, should get the credit for the design of this round.

Cal 45	Used in single-shot and large-frame repeaters. The style in which repeaters of this caliber are marked.

The bullet itself is a single, wide-groove, solid lead type. Unfortunately the author does not have an example of this cartridge and therefore cannot give its dimensions.

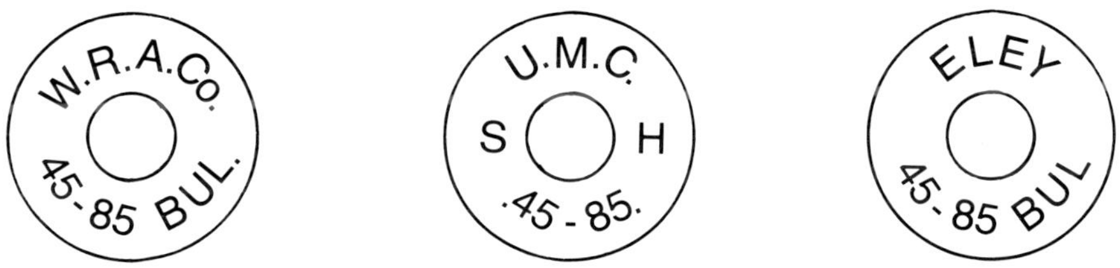

W.R.A.Co. U.M.C. ELEY TAKEN FROM ELEY CATALOG ARTWORK

Detail of the headstamp shown in the Eley catalog.
The author has not personally seen this headstamp.

ELEY BROTHERS, LIMITED.

ELEY CENTRAL-FIRE CARTRIDGES for Winchester, Marlin & similar rifles. Continued.

Description.	Cap	Powder	Bullet	Price per 1,000		
				Cartridges	Shells	Bullets
			Grains	£ s. d.	£ s. d.	£ s. d.
ELEY ·40-82-260 Winchester 	33	Black ...	260	**7 10 –**	3 16 8	– 17 6
ELEY ·405 Winchester 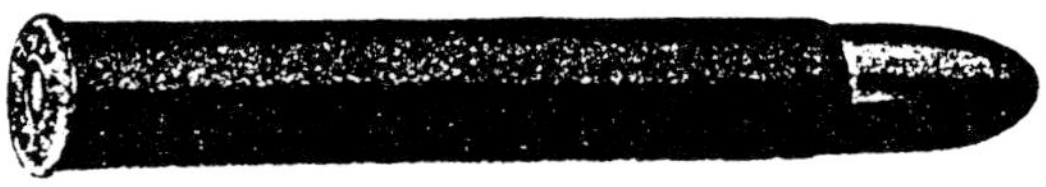	33	Smokeless	300	**9 10 –**	4 10 –	1 17 6
ELEY ·44-40 Winchester  (In boxes of 50)	19	Black ...	200	**3 2 –**	2 7 6	– 15 –
ELEY ·44 Winchester Shot  (In boxes of 50)	19	...	...	**3 5 3**	...	...
ELEY ·45-60 Winchester 	33	Black ...	300	**6 6 8**	3 4 2	1 2 6
ELEY ·45-75-350 Winchester 	33	Black ...	350	**6 10 –**	4 2 6	1 5 –
ELEY ·45-70-350 Winchester & Bullard 	33	Black ...	350	**7 1 8**	3 6 8	1 5 –
ELEY ·45-85 Bullard 	33	Black ...	300	**7 1 8**	3 6 8	1 5 –

38

Plate 8-13. *Courtesy Eley Ltd.; IMI PLC*

This information was contained in an article entitled "Ammunition for Indoor Shooting" from the June 1885 edition of *The Rifle* and involved .45 caliber cartridges: The Bullard Repeating Arms Company of Springfield, Massachusetts have a novel way of adding a grooved circle to a rifle shell which is shown by the accompanying illustration enabling one to use a light charge of powder for shooting. This groove prevents the bullet from being seated beyond a desired point of the shell which is particularly desirable in shooting this ammunition in a magazine rifle. The shell creased in this manner may be loaded with a small quantity of black or wood powder. In the shell illustrated which is the government 45 caliber for 70 grains of powder there are placed 7 grains of wood powder. At the factory of this company a representative of the rifle shot a quantity of this ammunition at a target about 50 yards away. There was a very light report of about the same as a 22 caliber cartridge." (This information provided by Joseph B. Roberts, Jr. of the National Rifle Association.)

Unfortunately this article had to be put on tape because of the fragile nature of the magazine, so the accompanying illustrations mentioned in the article cannot be provided here. An example of one of these cartridges would be most rare.

THE .50-115-300

The most unique Bullard cartridge of all was also the largest of the seven. It was the first semi-rimless (rimless) cartridge produced in the United States and it may have been first in the world. Most references list this cartridge as appearing around 1886 or 1887, but in actual fact its introduction was much earlier. This earlier than generally accepted date of introduction would definitely make the Bullard round a front-runner for the honor of being the first *commercial* version of a semi-rimless (rimless) cartridge produced anywhere. Is it possible that Bullard envisioned box and clip magazines and began preparing for this eventuality? Unfortunately, to the author's knowledge, no records exist that would indicate Bullard's reason for developing a semi-rimless cartridge; his repeaters were of tubular magazine style and Bullard himself had nothing to do with the later Hindley magazine experiments.

The second innovation on this cartridge was the use of a true solid-head construction much the same as that used today. "The primer pocket did not protrude into the case interior as did all other contemporary 'Solid Head' cases," states Capt. George C. Nonte, Jr. in his magazine article entitled "Big Fifty Cartridges" (publication date and publisher unknown).

Another magazine article also elaborates upon the uniqueness of the round. Bob Hagel, in an article entitled "Head Designs for Centerfire Cases" in the March-April 1979 issue of the *Handloader*, states, "It may come as a surprise to many gun buffs, but the first American-made cartridge of what might be termed a rimless design was actually a semi-rimmed number known as the .50-115 Bullard. To my knowledge, this was also the first American cartridge of solid-head design. It may also be the first semi-rimmed, solid-head case made anywhere in the world. It was apparently designed for the Bullard repeating rifle, but I am not familiar with that rifle and do not know whether the action required a semi-rimmed case to function properly. This cartridge appeared in the mid-1880s."

The Bullard action did not require the use of a semi-rimmed cartridge, therefore Mr. Bullard's thoughts on ballistics may have been at the forefront of contemporary technology. One can only guess at the reasons for making the other six Bullard cartridges conventional rather than completely new in concept like the .50-115, as they were all available at about the same time.

A quick look at pages in the following chapter containing sectioned views of the

.50-115 cartridge will indeed confirm the fact that it is a true solid-head cartridge.

This cartridge was never intended for anything but hunting purposes, and both the large-frame repeaters and single-shot rifles were chambered for it.

The .50-115 came with two bullet weights. These were a .50-115-300 express with a hollow bullet (copper tube inserted in bullet) and a .50-115-346 solid ball. A slight variation on this loading shows up on page 11 of the second catalog in the next chapter. It shows a .50-105 version, but the author has never seen a box label to confirm if this loading was ever sold in any quantity. It does not seem to appear in any other catalog or advertising. It too is of solid-head construction but does not appear to have a slight shoulder or the rimless case of the .50-115. Possibly the drawing is inaccurate. The .50-105 Winchester round is almost a quarter inch longer and was not introduced until about 1899.

This cartridge was loaded by UMC and possibly by others, but to the author's knowledge no headstamp has ever been seen on a .50-115 cartridge.

Further proof that this caliber was on the market by mid to late 1884 is contained in Theodore Roosevelt's statement, "When I first came to the plains I had a heavy Sharps rifle, 45-120, shooting an ounce and a quarter of lead, and a 50-caliber, double-barrelled English express. Both of these, especially the latter, had a vicious recoil; the former was very clumsy; and above all they were neither of them repeaters; for a repeater or magazine gun is as much superior to a single- or double-barrelled breech-loader as the latter is to a muzzle-loader. I threw them both aside; and have instead a 40-90 Sharps for very long range work; a 50-115 6-shot Bullard express, which has the velocity, shock, and low trajectory of the English gun; and better than either, a 45-75 half-magazine Winchester."[9]

As a further point or two concerning this round, Capt. George C. Nonte, Jr., in his article "Big Fifty Cartridges," states, "The cartridge was available loaded with both 312- and 300-grain bullets ahead of 115 grains of black powder." The author has seen no information to confirm what may be a third bullet weight for this round but would like very much to hear from someone who can provide photographic proof of its existence. This cartridge was probably the most expensive commercially made cartridge of its time. Check prices in Chapter 9.

The cartridge specifications were as follows:[10]

cartridge length:	2.56 inches
case length:	2.19 inches
bullet diameter	.512 inches, 3-groove 300 grain, hollowpoint
case type:	semi-rimless with a very slight neck
neck diameter:	.547 inches
base diameter:	.585 inches
shoulder diameter:	.577 inches
Used in single-shot and large-frame repeaters	

.50-115-300 THIS REPRESENTATION TAKEN FROM CARTRIDGE BOX ARTWORK

Detail of headstamp shown in box top art of Plate 8-3B, but no one the author has spoken to or corresponded with has seen a .50-115 with a headstamp. Even the box shown in Plate 8-3B had a lone cartridge in it but no headstamp.

Possibly the box art is just that, box art! For what it's worth, here are the headstamp designs. The second box-top artwork headstamp (right) seen on a UMC box (after 1911). Like the one on the left it has not been seen on an actual cartridge.

REMINGTON U.M.C. .50-115-346 THIS REPRESENTATION FROM TAKEN CARTRIDGE BOX ARTWORK

Two headstamps from .50-115 boxtop art.

Express 50 115	See Plate 4-15C The style in which repeaters of this caliber are marked.

For all those people who would like to fire their Bullard rifles again without taking out a second mortgage, the following information is made available through the kindness of Petersen Publishing Company. WARNING: if in doubt about the safety or functioning of your rifle, have it checked out by a competent gunsmith, otherwise you could be taking a grave risk!

Here then is the conversion data needed to make new brass for old charcoal burners.[11] The .32-40 Bullard is the only cartridge left out of the lineup, either because it cannot be made out of anything else or the demand for its use is too low.

CONVERSION DATA

CALIBER	BULLET	GROUP	CASE
.38-45	190 gr #375164	B	RS, T, .30-40
.40-70	260 gr #412174	B	RS, .45-70
.40-75	260 gr #412174	B	RS, .45-70
.40-90	300 gr #415175	B	NE, T, .348 Win
.45-85	300 gr #457191	B	.45-70
.50-115	300 gr #515139	B	NE, .348 Win

CODES: RS = RESIZE; NE = EXPAND NECK; T = TRIM TO LENGTH

GROUP B indicates that the Bullard action is *moderately* strong.

The section of this chapter devoted to reloading tools was written for this book by James Zupan, a noted authority on reloading tools and the author of many articles and books on this subject. His latest book on the Frankford Arsenal is entitled *Tools, Targets, and Troopers*. The balance of this chapter, then, are his words; with photographs supplied by Mr. Zupan, except where noted. His footnote numbers are also retained.

D. BROWN, Jr.

Tool for Capping, Uncapping, and Loading Cartridge Shells.

No. 232,173. Patented Sept. 14, 1880.

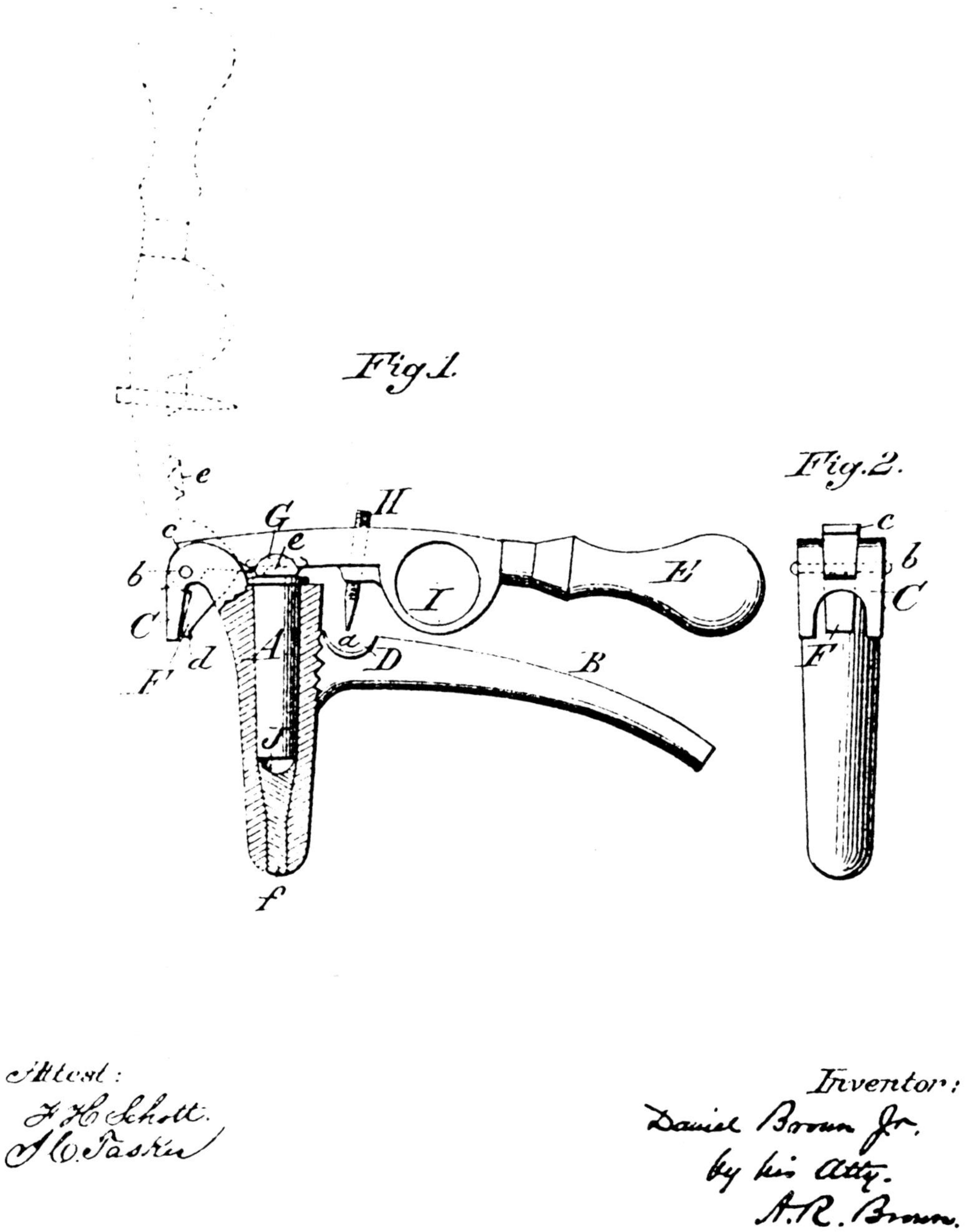

Plate 8-14. D. Brown's first model loading tool.

BULLARD RELOADING TOOLS

by James Zupan

Bullard Repeating Arms Company, in keeping with the customs of the last quarter of the 19th century, offered reloading tools with their fine selection of rifles. Ballard, Maynard, Remington, Sharps, and Winchester, to name the industry leaders, offered complete lines of reloading tools and bullet molds. Some of these companies offered only one type of tool and others several different styles. Some manufactured their own and others purchased them from firms that specialized in implements, like the Bridgeport Gun Implement Company.

Bullard initially sold the Brown Variform Reloader, then began manufacturing their own unique set of hand tools. Both types represented the two distinct designs that were used by the other arms companies.

As the 1880s began, two major types of hand tools were available from the various makers. Each type had their own proponents and a firearms maker rarely offered both. The oldest type consisted of individual dies designed to perform one of the tasks necessary in reloading centerfire ammunition, i.e. decap the spent primer, set a new primer, and seat the bullet. Some sets added resizing dies, powder chargers, wad cutters, etc. Remington, Sharps, the early Ballard, and Maynard tools represent this form.

Winchester, a later Ballard, and Ideal produced a tool that combined the basic operation into one hand tool. These usually looked like and were called "nutcracker" tools.

The first reloading tool offered by Bullard was the Brown Variform Combination Reloader. What the company was offering was the final product that originally began as the invention of Daniel Brown, Jr. of Johnson, Rhode Island. Brown's first patent was number 232173, filed on January 7, 1880. It described a "tool for capping, uncapping and loading cartridge shells." Brown claimed:

> My invention relates to tools for loading and reloading central-fire cartridge-shells for rifles; and it consists of a combination of suitable parts in one implement, as hereinafter described, to form a device for seating the bullet in the shell, a device for setting the primer, and a device for extracting the exploded cap.[1]

The tool functioned as simply as it looked. The patent included this description:

> To seat the cap or primer I inset the cap in its place within the primer-pocket of the shell, and place the shell between the prongs of the seat C, the flange resting against the inner sides thereof. The handle E must be slightly raised for this purpose. I then press down the handle E, which brings the cap-setter F forcibly against the head of the cap and crowds it home upon the anvil of the shell.
>
> To seat the bullet in the shell I place the bullet into the mouth of the shell and insert both, the bullet downward, into the chamber A. As I bring down the handle E the presser G crowds the shell down into the

chamber and drives the bullet into the shell to the desired distance.

If it is desired to seat a round ball instead of a conical one, I use a plug, J, which is movable and fits within the chamber A. This plug fills the bottom of the chamber to such an extent as to furnish a seat for the round ball. This plug has a stem, F, which extends downward through the bottom of the chamber, and is dislodged by the pressure of the finger outside, when it readily falls out through the mouth of the barrel A.

To lift the cartridge out of the chamber I raise the handle E into the position indicated by dotted lines, which brings the lip 'd' up under the flange of the shell and raises it out of the chamber.

To decap the shell I lay it in a slanting direction in the groove D, the ridge 'a' serving as a rest for the flange of the shell. I bring down the handle E, and cause the spur H to enter the indentation in the primer made by the hammer in exploding it, and bearing down I cause the spur to cut through the cap, and then, by a twist of the shell, I draw off the cap from the anvil.[2]

One of the eyewitnesses of this patent was William B.W. Hallet, a Captain in the Rhode Island Infantry. His recommendation appears on the advertising flyer pictured. This model was manufactured by George W. Stafford and Company. Brown's office was two doors down on Clifford Street.

Brown filed another patent, number 239084, on February 1, 1881. This time he called it a "Cartridge-Loading Implement." He claimed:

a novel combination of the devices hereinafter described, adapted to press the primer into position upon the anvil, to seat the ball in the shell, and to remove from the shell the primer after it is exploded, all these operations being performed by a single implement, instead of the several tools heretofore commonly used for such purposes.[3]

The patent diagram shows the basic nutcracker design, but it is noticeably heavier and different in appearance. The patent contained this description of its functioning:

To set the primer, I place the primer loosely in the pocket of the shell, and, opening the handles A and B, I place the cartridge-shell within the bifurcated seat K, as shown in Fig. 4. By closing the handles again the cap-seater 'g' presses the primer into the pocket and upon the anvil of the shell.

To seat the ball in the loaded shell, I open the handles A and B and insert the ball loosely in the shell and place them in the bore of the chamber C, the ball entering foremost. By bringing the handles together the presser F forces the shell downward within the bore of the chamber and upon the ball to the required extent.

To withdraw the loaded shell from the chamber, I open the handles A and B, thus bringing the finger L of the lever A into engagement with the lip 'f' of the cylindrical block G. This imparts a vertical motion to said block in a direct line, and the retractor H, which extends under the flange 'b' of the shell D, lifts the shell out of the chamber.

To remove the exploded primer preparatory to reloading the shell, I open the handles A and B and slide the empty shell D upon the cylindrical block G, as shown in Fig. 5. By closing the handles again I drive

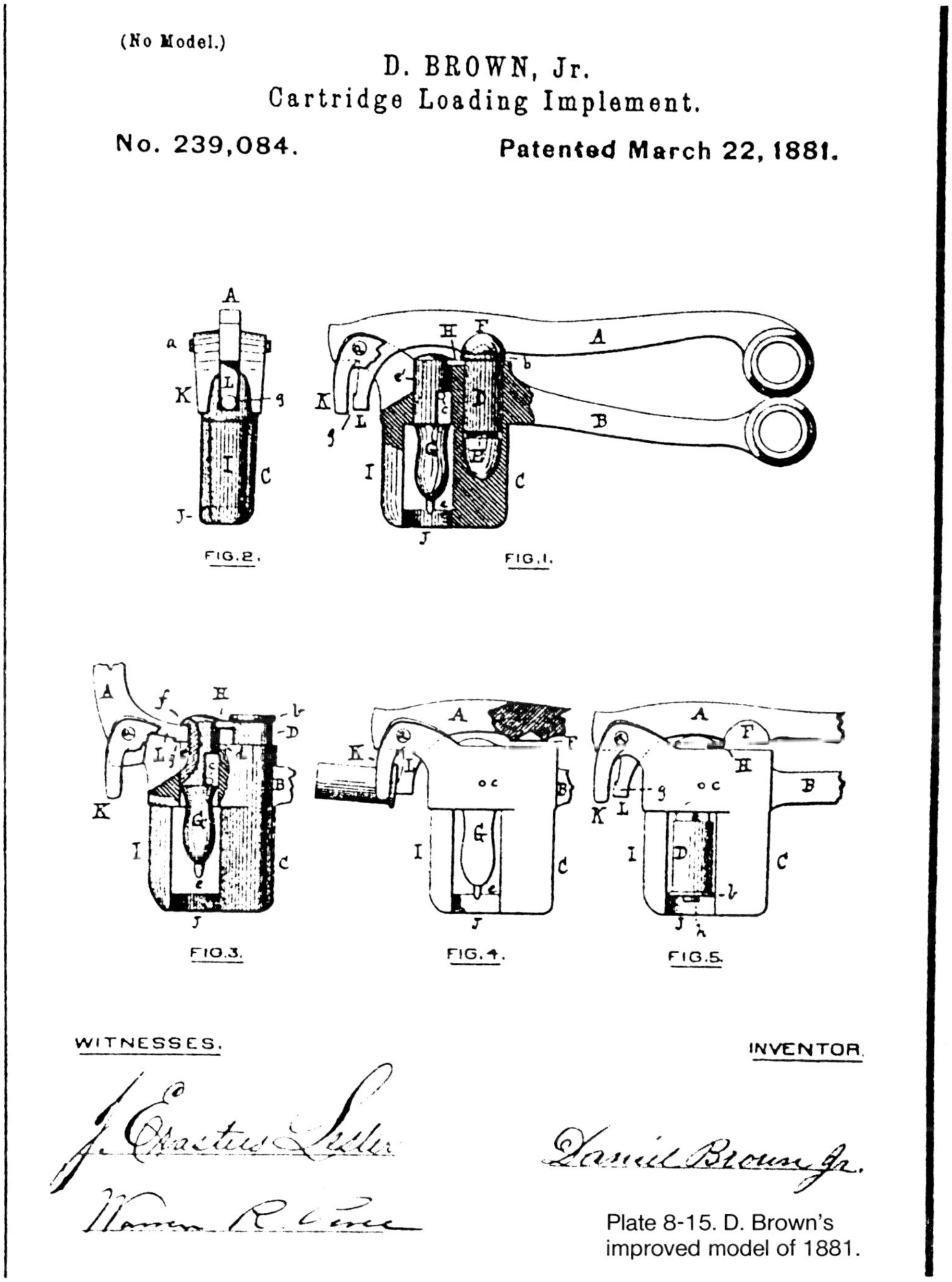

Plate 8-15. D. Brown's improved model of 1881.

> the block G into the shell, and the expelling pin 'e', being exactly guided as the block descends into the shell, enters in and through the central-fire vent of the anvil, and forces the primer off the anvil and out of the primer-pocket, as shown at 'h' in Fig. 5. I bevel and enlarge the seat J, and also reduce the diameter of the central portions of the block G, as described, to enable me the more readily to slide the empty shell upon the block and to withdraw it therefrom; otherwise these movements would require to be made in a direct line and with more force.[4]

Brown claimed the uniqueness of his design lay in the combining of the decapping with a shell-retractor.

These Reloaders as now constructed will be found superior to any upon the market. The leverages have been greatly **increased and all parts made as light and strong as** possible. Extra chambers can be supplied for any size cartridge. All parts are made upon the interchangable plan and can be exchanged in case of breakage. We are now prepared to fill orders for any size, for rifle or revolvers, at reasonable prices. Address all orders to

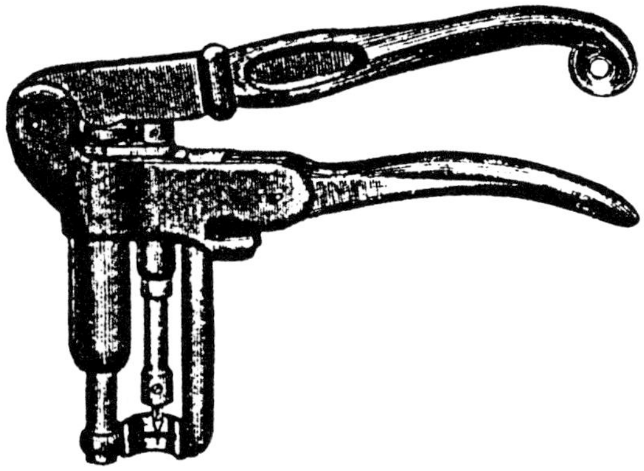

GEO. W. STAFFORD & CO.,

OR BULLARD REPEATING ARMS CO. PROVIDENCE, R. I., U. S. A.

BROWN'S "VACIFORM" COMBINATION RELOADER

Is indorsed by the United States Government as being very ingenious, and well adapted to its purposes. Combines in the most convenient form all the requirements for doing the work of reloading quickly and wel . The accuracy of all rifles is greatly enhanced by having suitable means for reloading the shells by which, about two-thirds the cost of ammunition is saved, and that too when the bullets are purchased ; and as they are readily obtained at the gun stores, the labor of reloading is very trifling.

AMONG THE MANY INDORSEMENTS LATELY RECEIVED, IS THE FOLLOWING :

D. BROWN, 67 CLIFFORD STREET, PROVIDENCE, R. I. NEWPORT, R. I., July 23d, 1883.

DEAR SIR :—I have practically tested and used the VACIFORM RELOADER, 45 Cal., and pronounce it the best tool of the **kind I have ever seen. It is just perfect** in its action, seating both bullet and primer to its proper depth with uniformity.

(Signed) W. MILTON FARROW, Champion Military Marksman, U. S. A.

In his "How I BECAME A CRACK SHOT," he says :—" I would hardly trust anyone to load cartridges for me.

In reloading discard all cartridges that will not enter the chamber of rifle with perfect freedom.

Plate 8-16. Advertisement in early Bullard Arms Company catalog. It probably dates to 1883-84.

Variform Reloader

FOR RIFLE CARTRIDGES.

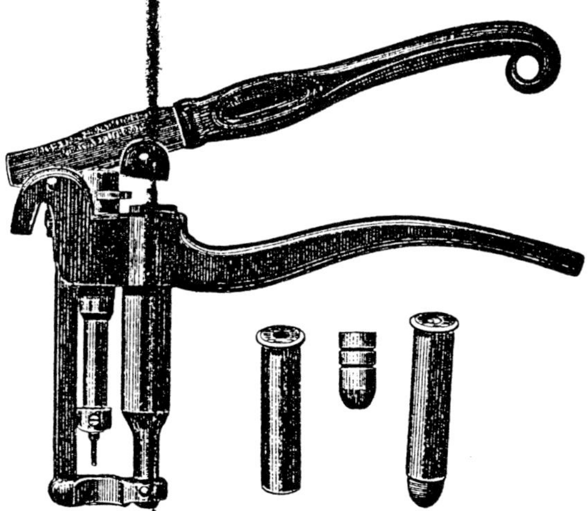

Combines in the most convenient form all the requirements for doing the work of Reloading quickly and well.

The value of the breech-loading system in rifles is greatly enhanced by having suitable means for reloading the shells, by which, about two-thirds the cost of ammunition is *saved*, and that too when the bullets are purchased; and as they are readily obtained at the gun stores, the labor of reloading is very trifling.

Plate 8-17. Cover of Brown's catalog.

Some of Brown's correspondence with the United States Army helps to date the introduction of these tools and their relation to the Bullard firm. He first wrote the Chief of Ordnance on July 6, 1883, requesting a trial of his first tool. He claimed to have had many proposals for the manufacturing of the tool by "some of the leading firms."[5] The Ordnance Department had tested the Winchester tong tool in 1878 and was utilizing some of the inventions developed by Army Officers in their current reloading tools being issued to the troops. They were preparing a new model (1883) that could reload all the service cartridges and another one to reload the 20-gauge "Forager" shot shell. Production of these sets was under the control of the Chief of Ordnance at Frankford Arsenal in Philadelphia.[6] Brown chose an opportune time to contact them.

Brown wrote to F.E. Hobbs on July 12. He stated:

> . . . have sent a sample of my Reloader by mail which although inferior in workmanship may serve to show the practical utility of the invention. A second lot is in process of manufacture and if you desire, a better sample will be sent in a few weeks.
>
> The article has lately been shown to the leading manufacturers, several of whom desire to handle the thing and am now in correspondence with two such parties; one desiring to buy the patents and the other desiring to manufacture on royalty. It is acknowledged by everybody to be the *proper* thing.[7]

He again wrote the Chief of Ordnance on July 22 and indicated that it would be desirable to have the model a *trifle* larger, thereby increasing the leverage."[8] The Ordnance Department noted that it would try the tool sent and correspond with Brown if interested.[9] Nothing in the Chief of Ordnance files indicates any trial or results. They did not contact Brown and any hopes of a military contract faded.

Besides the military's lack of interest, the correspondence dates the transitional point for the two tools. It also indicates the options Brown had. The *American Field* of November 10, 1883, announced that George W. Stafford had purchased Brown's patent rights to the Variform Reloader. The disinterest by the military no doubt played a role in Brown's decision to sell out.

By the spring of 1885 Bullard was ready to cease offering the Brown tool — which he had mistakenly called the Vaciform — and only sell his own set.

Neither style of the Brown Variform Reloader is common, but the number found in non-Bullard calibers indicates that production must have exceeded several hundred, if not one thousand. Whether they were made after Bullard dropped them from his catalog could not be determined. Thus passed another short-lived but unique example of Yankee inventiveness.

It would be reasonable to assume that the Bullard Company had always intended to offer a proprietary set of hand tools as they did cartridges. Financing and the tooling necessary to produce them may have postponed their inception. A scarcity of factory correspondence prevents an actual dating of this. The earliest reference to them predates their introduction in a catalog. Part of a report written on September 13, 1884, by an unknown individual after a visit to the Bullard factory states:

> Again it goes without saying that the best gun that can be made will not succeed except with good ammunition suited to its wants and work. When one can afford to buy and use only factory-made cartridges reloading is not needed; but to the many a long felt want has been that

of a first-class set of reloading tools. With all due respect to the opinions of others who may differ, I will say that proper reloading tools have never yet been procurable to the general public until those of the Bullard Company were made and put up for sale. Many abortions have been made and sold, but never a set that acted as test gauges as well as reloaders while being used, and by which the sportsman as thoroughly inspected his ammunition as do the experts in overlooking ammunition at the factory.

Such a set of reloaders I saw at the Bullard manufactory; they were simple, were easily handled, and reloaded as perfectly as any machine could have done, though of course not so rapidly. These tools consisted of a resizing die and anvil, this latter being one anvil for all the operations of reloading. The resized shell was driven out by a punch that extracted the primer at the same time, the shell having been resized before the primer was extracted. The new primer was easily seated by another punch, and then the shell filled with powder and inserted in the reloading die, the ball being seated in its place to exactly the proper depth and then crimped all in one operation, and the perfectly reloaded shell then forced from the die. This latter die serves as the gauge, as a swell-headed shell is instantly detected, the ball is accurately seated and to the exact and proper depth, so there can be do doubt but that the cartridge will feed up properly, enter the chamber and then extract as it should.[10]

The description leaves little doubt that the tools referred to are the same ones listed in the catalogs.

In April 1885, 1,000 copies of a single sheet, "Directions for Reloading Cartridges with the Bullard Reloader," were printed for inclusion in their current catalog. Page 13 of this catalog showed the Brown Reloader. In catalogs issued later, the Brown tool was omitted. This chronology comes as close to dating the two sets of tools as is possible without any corresponding factory records.

Unfortunately no sets or individual tools could be located nor had anyone seen any. The fact that the set was offered in more than one catalog indicates that they were probably produced in limited quantities. It's reasonable to assume this because the rifles were never that common either.

Several points are worth noting about the design. It consists of a round die bored out to act as a ball seater, part of primer extracting, and resizing die. This die rests on a base similar to those used by other reloading tool makers. The set lacks a powder measure, which is unusual considering it was provided with the Brown tool. No patents could be found for this tool. The lack of anything unique in its design probably accounts for this.

Certain aspects indicate that it may have been influenced by the reloading set being issued to the Army at that time. The anvil is similar in design and usage to the Model 1880 45-70 and later 50-70 anvil. No other commercial firm referred to that piece as anything other than a base. The primer seater is also similar. Only Frankford Arsenal and Bullard Arms Company attempted to seat primers by hitting them! They were also the only ones to make sizing dies that could be used as resizing gauges. Frankford Arsenal began producing their tools in 1878, only a few years prior to Bullard. If there was a connection Daniel Brown may have been it. Most arms companies sought the lucrative contracts that the Ordnance Department issued as supplements to their own arms and ammunition production.

DIRECTIONS

FOR RELOADING CARTRIDGES

WITH THE

BULLARD RELOADER

To prepare shell. Place the fired shell in sizing die. Place sizing die and shell in anvil with the shallow cup upward. Strike the sizing die until the shell is forced completely in. Turn over anvil, put plunger into the shell, which will extract shell from the sizing die and expel the exploded primer

Now, put this reduced shell into the ball seater with the head end up. Place over it the anvil with the long cup upward, and seat the primer by driving the primer seater through the small hole in the anvil as far as it will go.

Now the shell is in its original shape and size, and the primer properly seated. Fill the shell full of powder well shaken down, preferably loaded through a tube so that the powder shall be even full in the shell; if no tube is used, shake the powder well down and fill again even full.

Place the ball evenly on the cartridge with the shell in the anvil, shallow cup upward. Place over the ball seater, and drive down to the shoulder, preferably with a wooden mallet. Use occasionally a drop of oil in the reloader, which will prevent the cartridge and ball sticking.

Great care should be used in preparing cartridges of an even length, and in every case the plunger should be driven down to the shoulder. Shells should be washed in soda water and dried before using. To make lubrication for bullets, use equal parts sperm oil and beeswax.

We will guarantee our Rifles in every case when the cartridges are re-loaded with the above tools, according to these directions.

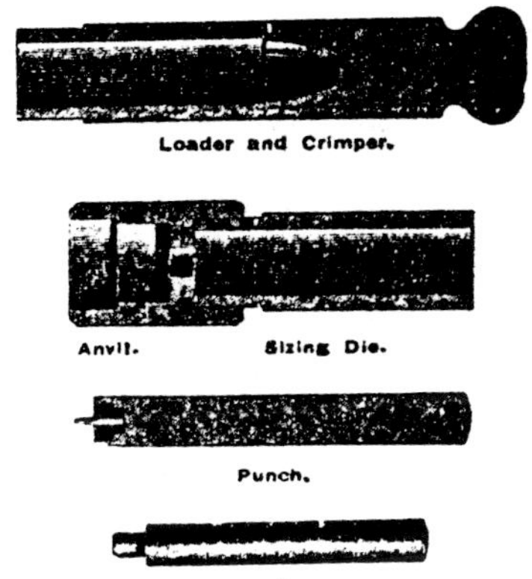

TO DISMOUNT THE GUN.

1. Take out the tang screw.
2. Half cock hammer, and take out lock-frame and hammer screws.
3. Pull out lock frame, down and backwards.
4. Disconnect links from brace.
5. Take out side plate screws, and remove side plate.
6. Remove carrier lever spring.
7. Take out breech-bolt.
8. Take out breech-block.
9. Take out extractor pin and remove extractor; push back pin before removing bolt.
10. Take out guard lever bolt.
11. Draw out guard lever and its connections.
12. Draw out carrier lever.
13. Take out carrier.

Reverse operations in Assembling.

Plate 8-18. This flyer was printed in April 1885 for insertion into the then current Bullard catalog.

28 BULLARD REPEATING ARMS COMPANY.

Price List of Cartridges for Hunting Rifles.

Cartridges for all our arms are manufactured by the U. M. C. Co., and are guaranteed to be of first quality. Solid head and re-loadable shells.

32 Cal.	40	grains Powder,	150	Grains lead,	$31.50	per M.	Shells,	$20	Bullets,	$7.25	per M.
38 Cal.	45	"	190	"	31.50	"	"	20	"	8.00	"
40 Cal.	75	"	258	"	33.00	"	"	18	"	9.00	"
40 Cal.	90	"	300	"	40.00	"	"	22	"	10.50	"
45 Cal.	60	"	300	"	33.00	"	"	20	"	9.00	"
45 Cal.	75	"	350	"	35.00	"	"	20	"	9.50	"
45 Cal.	70	"	405	"	35.00	"	"	22	"	12.00	"
45 Cal.	85	"	290	"	37.00	"	"	22	"	9.50	"
50 Cal.	95	"	300	"	40.00	"	"	23	"	11.00	"
50 Cal.	115	"	300	" Exp.	90.00	"	"	50	"	11.00	"
50 Cal.	115	"	346	"	90.00	"	"	50	"	9.00	"

Primers $2 per M.

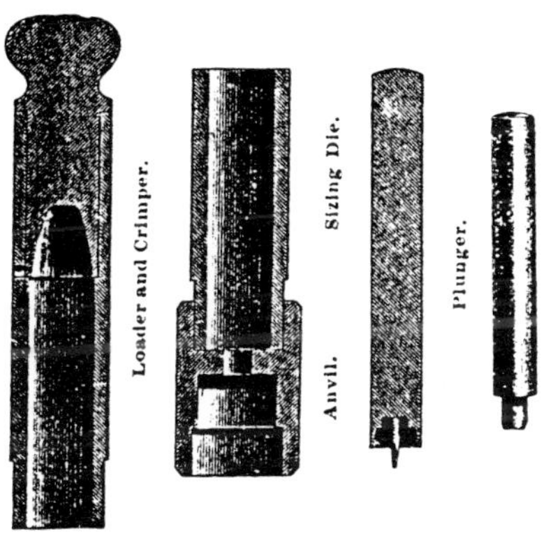

Directions for Re-loading Cartridges with the Bullard Re-loader.

To prepare the shell. Wash the fired shells in strong soap suds or soda water. and dry thoroughly. Place the fired shell in sizing die. Place sizing die and shell in anvil with the shallow cup upward. Strike the sizing die until the shell is forced completely in. Turn over anvil, put plunger into the shell, which will extract shell from the sizing die and expel the exploded primer.

Now put this reduced shell into the loader with the head end up. Put on the primers and seat with the primer seater. Remove the shell, fill with powder, well shaken down, preferably loaded through a tube, so that the powder shall be even full in the shell. Place the ball evenly on the shell, with the shell in the anvil, shallow cup upward. Place over the crimper and drive down to the shoulder, preferably with a wooden mallet.

Great care should be used in preparing cartridges of an even length and in every case the crimper should be driven down to the shoulder.

We will guarantee our rifles in every case when the cartridges are re-loaded with the above tools, according to these directions.

SPRINGFIELD, MASS., U. S. A.

Plate 8-19. Later example of advertisement of Bullard hand tools.

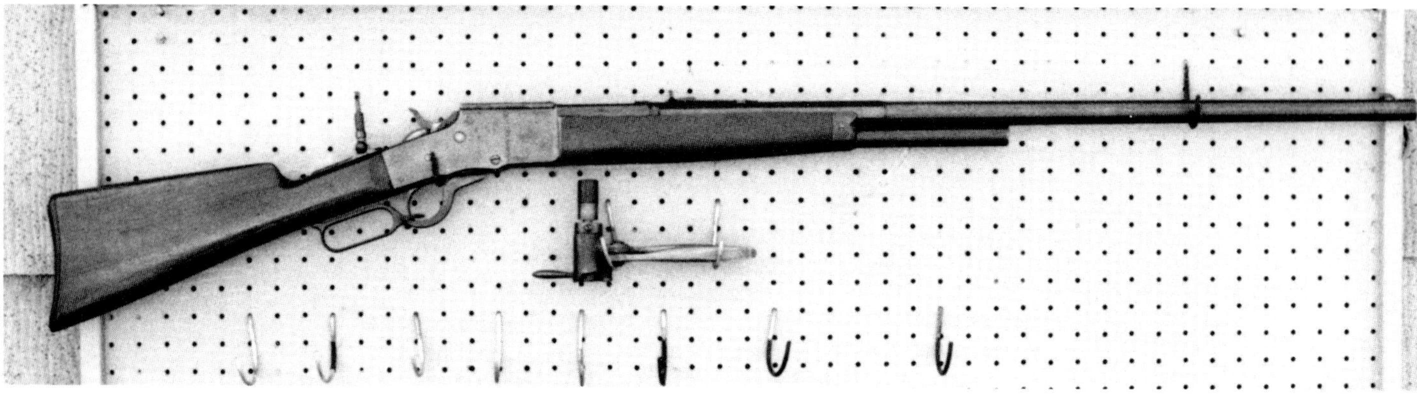

Plate 8-20. Top: (SN 1072 cal .50-115) Breech marked 'model 1886.' Note extractor pin removal hole is drilled right through both sides of breech. Bottom: .50 cal Bullard reloading tool probably manufactured by Ideal. *Courtesy H.A. McCallum*

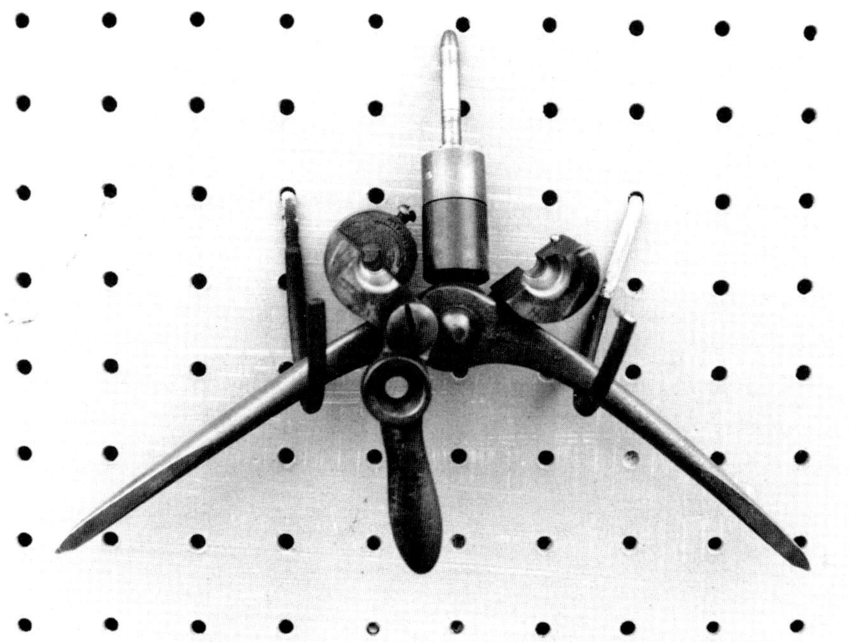

Plate 8-21. Note body of tool is marked 50 Bullard. *Courtesy H.A. McCallum*

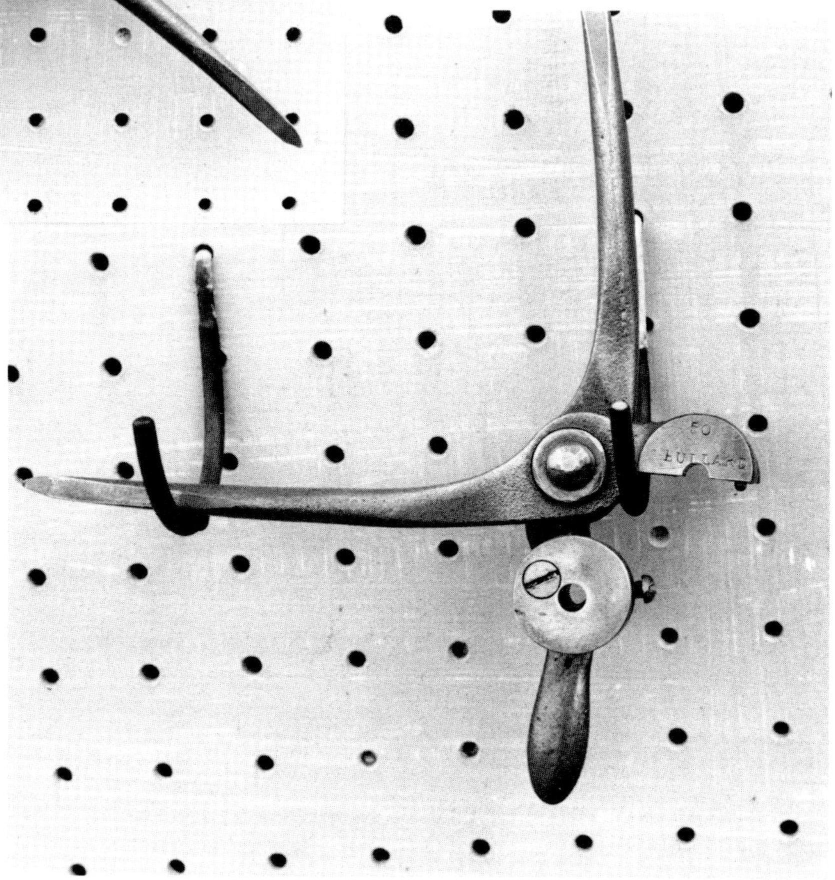

Plate 8-22. Detail of same reloading tool. *Courtesy H.A. McCallum*

Bullet molds are listed in the Bullard catalogs but no information could be found on them. Even the catalog only just listed them. Several Ideal tools have been seen in the Bullard calibers and this may have been another way the Bullard Arms Company sought to supply reloading tools with their single-shot and repeating rifles.

FOOTNOTES FOR BULLARD RELOADING TOOLS BY JAMES ZUPAN

1 Patent Papers, Government Printing Office, 1880.
2 Ibid.
3 Patent Papers, Government Printing Office, 1881.
4 Ibid.
5 National Archives of the United States, Record Group 156: Army Ordnance Department Records, 1883, Entry 3390.
6 Ibid.
7 Ibid.
8 Ibid.
9 Ibid.
10 Gerald O. Kelver, *100 Years of Shooters and Gunmakers of Single Shot Rifles*, p. 100.

FOOTNOTES (INITIAL SECTION OF CHAPTER 8)

1 C.W. Sawyer, *Our Rifles* (Boston: The Cornhill Company, 1920), pp. 114-115. The following quote from this book would seem to indicate that no distinction was made by Bullard in stamping the .32-40 cal on his rifles as to whether it was a Bullard or a Winchester load: "The barrel is stamped .32-40, but it will not take an ordinary .32-40 shell. The accepted standard at that time of barrel boring for .32-40 ammunition was .319, but the Bullard people bored their barrels .311 to force the use of special cartridges of their make." Research has shown that most .32-40 Bullards are in the Winchester loadings, although Major Ned Roberts did own a Bullard in .32-40 Bullard caliber.
2 A.C. Gould, *Modern American Rifles* (Plantersville, South Carolina: Thomas G. Samworth, 1892, republished in 1946), p. 106.
3 A.C. Gould, *North American Rifles*, p. 106.
4 Frank C. Barnes, *Cartridges of the World*, 3rd edition (Northfield, Illinois: Digest Books, 1972), p. 97.
5 Frank C. Barnes. *Cartridges of the World*, p. 98.
6 Jack Behn, *".45-70" Rifles* (Highland Park, New Jersey: The Gun Room Press, second printing, 1972), pp. 135-136.
7 Graham Burnside, "Marlin's .444 Starts Fire Under Cartridge Fans," *Shooting Times*, April 1965, p. 39.
8 William West, *Marlin and Ballard — Firearms and History* (Whittier, California: Stockton Trade Press, 1968), pp. 5-7.
9 Theodore Roosevelt, *Hunting Trips of a Ranchman* (New York: G.P. Putnam's Sons, 1885), p. 188. Since this book was published in 1885, material for the book had to have been gathered prior to or during early 1885.
10 Capt. George C. Nonte, Jr., gives his figures as bullet diameter .508 inches, case length 2.25 inches, cartridge length 2.75 inches. The author found that his own figures matched Frank C. Barnes figures fairly closely, but again a cautionary note must be struck when measuring old cartridges which may or may not have been reloaded many times.
11 George Nonte, "New Brass for Old Charcoal Burners," *Guns and Ammo Annual* (Los Angeles: Petersen Publishing Company, 1973), pp. 268-273.

Courtesy Verne W.L. MacRoberts/photo by Don Reid

9

BULLARD ADVERTISING & CATALOGS

Of the following four catalogs, only one bears an actual date. The rest have been placed in chronological order based on the author's research, which will be discussed with each catalog in progression.

Following the catalog section, the author has put together as extensive a listing as possible at this time, giving publications, dates and subject material of the many ads and articles that were written about the Bullard Company and its products over its short lifespan. These articles are in chronoligical order as to date published.

Following that will be a list of some of the many famous shooters who owned Bullard rifles and who with their skill at arms did as much as any advertising ever did to make Bullard arms more known to the public at large and the shooter in particular.

The catalog chosen to be first was placed there for two reasons. First, even though the cover design was to remain the same for all the other catalogs published (those discovered so far), except for possible differences in coloring on the covers, there is one very significant variation on this cover and this is the use of the word Association in place of Company. Since the Bullard Association was reorganized under Maine laws on October 11, 1883, to form an incorporated company, it is unlikely that they would have retained the word Association much after this time. After October 1883, Bullard Repeating Arms Company would have been the legal name. Second, the line on page 7 of the catalog states, "We shall have ready for the fall trade our . . . to 500 yards." The statistics show that approximately the first 100 large-frame repeaters bore the legend Association, but that after that Company was used in the address. The only calibers listed in this catalog are ones then readily available, not special Bullard numbers. Again that is borne out by the tables at the back. Soon after the first batch of 100 a .50-115 appears. It has already been established that all the Bullard calibers were known by September 1884 and since this catalog makes no mention of these calibers it would be safe to conclude that this catalog should be dated prior to October 1883. This theory also lends credence to the two-factory theory, as land for the new factory was not purchased until July 1883 and those 100 arms may have been produced in a small building elsewhere. As an aside, page 7 also mentions a .38-caliber rifle, which means that there is a good chance that a very few of these small-frame repeaters were marked Association as well on the address. However, this is only theory, as none have ever been seen by the author.

Here then is the first catalog, 16 pages in length, measuring approximately 8½x7 inches, and cover color unknown.

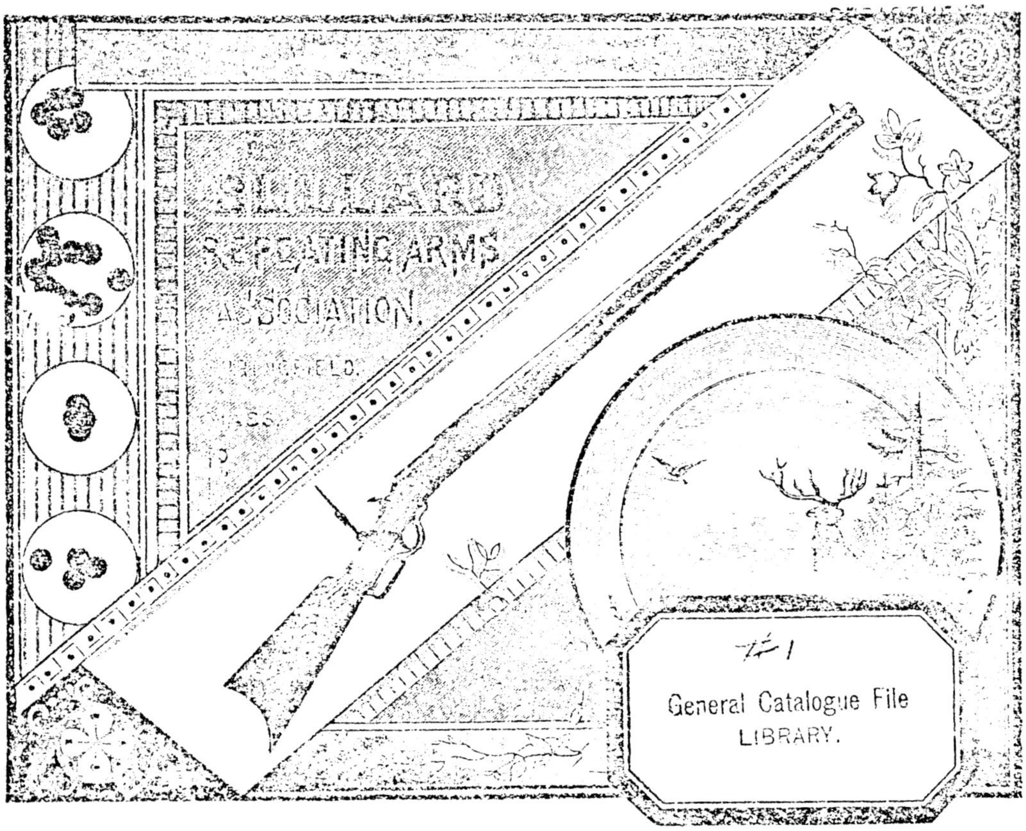

BULLARD
REPEATING FIRE ARMS.

RIFLED MUSKETS,

Carbines, Hunting and Target Rifles

FOR MILITARY AND SPORTING USE.

UNEQUALED

FOR SAFETY, EXCELLENCE OF MATERIAL, WORKMANSHIP, FORCE,
ACCURACY OF FIRING, SIMPLICITY OF CONSTRUCTION,
AND CONVENIENCE OF LOADING.

MANUFACTURED BY THE

BULLARD REPEATING ARMS ASSOCIATION,

SPRINGFIELD, MASS., U. S. A.

Plate 9-1. Note the use of the word "Association." *Courtesy Winchester Gun Museum/T.E. Hall, curator*

THE statement of Napoleon that "Inventions that are before their time remain useless until the stock of general knowledge comes up to their level," is proved by the fact that although breech-loading fire-arms were invented during the reign of Henry II. of France, previous to 1550, they were never practically used on the battle field, until the battle of King's Mountain, North Carolina, in 1780, when a breech-loading gun, the invention of Major Patrick Ferguson, of the 71st Highlanders, was used with good effect.

But not until the invention of Sharps' Rifle in 1857 were the principles of breech-loading of any practical use; and although a repeating breech-loader was invented by S. North, of Middletown, Connecticut, in the year 1825, the principle was not applied in practice until the improvements made by Henry, which resulted in the arm known as the Henry Rifle. The object sought in these and many later inventions was convenience and rapidity in loading, and safety to the user in firing the arm, and these conditions are more fully met in the Bullard repeater than in any other arm in the market, for the following reasons :

1st. Safety.

The action is positive, not dependent upon springs. It is self-locking, with a solid breech-block behind the bolt, which must be in place and securely locked before it is possible for the hammer to reach the firing-pin. Before placing this gun on the market it was subjected to the most thorough tests as to its strength and endurance, and has stood the heaviest charges without the least sign of weakness.

2d. Rapidity.

It is possible to fire the Bullard Rifle with greater rapidity than any other repeating rifle, from the fact that it works easier and smoother by reason of its more direct leverage on the work to be done, the heaviest work being done with the best leverage, as in extracting the cartridge, which is started when the lever is in position to exert the greatest strain. Cocking the hammer is also done by direct leverage inside the receiver or frame instead of a sliding motion of bolt or firing-pin on and over the top of hammer, which is very often liable to grind and always makes the arm work hard and unpleasantly. (See sectional cut, Fig. 2.) The Bullard has been fired twelve shots in five seconds, the best record of any other gun being eleven shots in seven seconds using U. S. cartridges.

Bullard Repeating Arms, Springfield, Mass., U. S. A.

3d. Loading.

The Magazine is charged from the under side, and it can be done with equal facility by a right or left-handed person. And as there are no holes or spring covers on the side, it is not possible to have it clogged by passing through brush or laying it on the ground or in trenches, etc. It is also much easier to load on horseback than any other gun, as there is more choice of position than when the opening is on the side. It can be loaded as a single loader either top or bottom, leaving the magazine full at all times for an emergency.

4th.

The Bullard is the only Lever Repeater that will successfully use the regular United States copper cartridge (45 cal. 70 gr. powder), owing to its absolute certainty to extract the shell, which being made of copper and with a folded head, does not contract after firing (as is the case of brass shells) but often sticks in the chamber, a difficulty more frequently mentioned in reports of trials of breech-loading fire-arms by Government experts, than any other.

The Bullard never fails.

Another advantage in using the U. S. cartridge is, that it is the standard ammunition and can be obtained anywhere in the country. This point will be appreciated by frontiersmen.

—5—

Bullard Repeating Arms, Springfield, Mass., U. S. A.

5th. Workmanship.

The quality of work on the Bullard is pronounced by the best judges to be superior to any other repeating rifle in the world, and the fact that Mr. Bullard, the inventor and manager at our manufactory, was for five years master mechanic at the well known Smith & Wesson works, is a sufficient guarantee that our standard is high and will be kept so. The material used is of the best throughout, no iron or malleable iron being used. All the parts are of forged steel except the carrier, which is composition or gun metal.

6th. Range.

A bullet from the Bullard Rifle using the regular U. S. cartridge (45 cal. 70 gr. powder) has, at a distance of **3,000** *yards* from muzzle of gun, penetrated a *three-inch spruce plank and eight inches of sand.*

7th. Guarantee.

Our confidence in the points enumerated above is such that, in case of failure of any part by reason of defective material or workmanship, a duplicate will be forwarded free of charge upon receipt of the broken part. All parts are sure to fit.

—6—

Bullard Repeating Arms, Springfield, Mass., U. S. A.

8th.

To any responsible party we will send one of our guns with the privilege of a three days' trial and examination on receipt of five dollars as guarantee for express charges.

9th.

We challenge the world to produce a repeater that can compare with ours in ease of operation, safety, quality of workmanship, style, balance, finish, and rapidity of firing.

We shall have ready for the fall trade our (45—70) (45—60,) (50—95 express) (40—60) and a 38 special of 40 gr. powder and a 250 gr. bullet, which is a fine target gun up to 500 yards.

On the pages following will be found cuts and prices which we trust will be satisfactory, and that you will place your orders with us at your earliest convenience.

All orders will be filled in the order received.

Respectfully yours,

BULLARD REPEATING ARMS ASSOCIATION,

SPRINGFIELD, MASS., U. S. A.

Bullard Repeating Arms, Springfield, Mass., U. S. A.

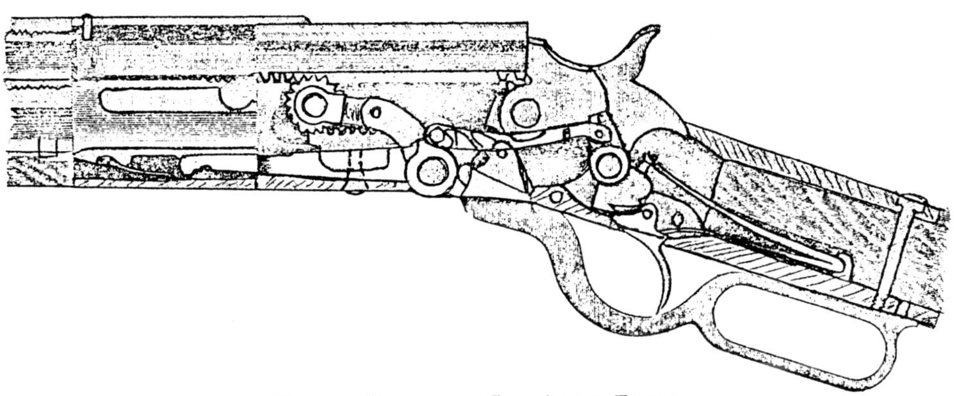

FIG. I. SECTION OF GUN AFTER FIRING.

A, Breech block held firmly in place by brace *C*, which is pivoted and cradled on hammer pin and lock frame. *C* is brought into place and firmly held by links *B* (only one shown), which are connected by strong pins to guard lever *D*. It will be seen by the above cut that it is absolutely impossible to get the hammer to strike the firing pin until the brace *C* is in its proper place. It will also be observed that it is impossible to disarrange the parts as all are pinned and hinged together, and that there are no sliding surfaces.

Bullard Repeating Arms, Springfield, Mass., U. S. A.

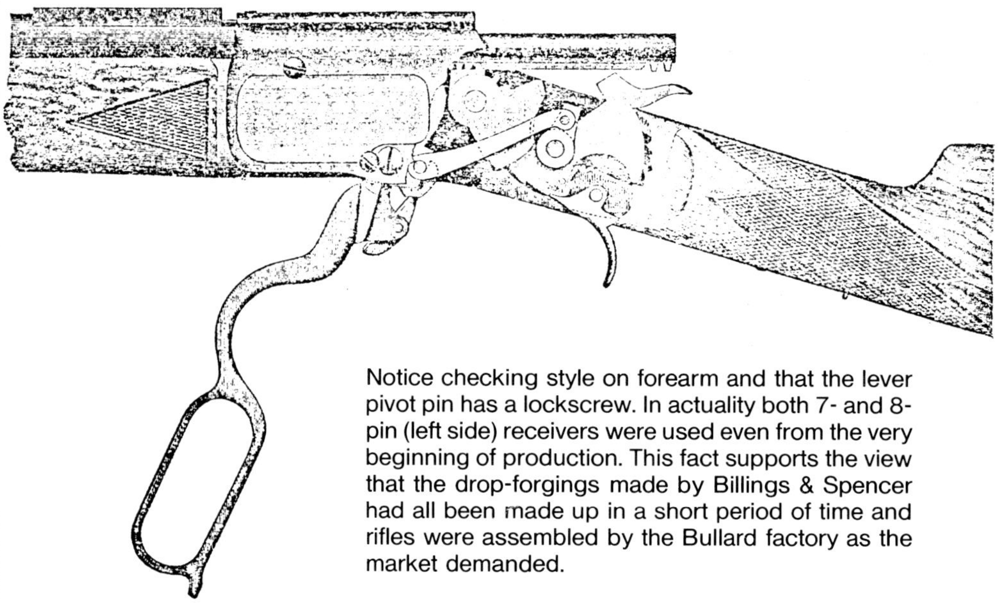

Notice checking style on forearm and that the lever pivot pin has a lockscrew. In actuality both 7- and 8-pin (left side) receivers were used even from the very beginning of production. This fact supports the view that the drop-forgings made by Billings & Spencer had all been made up in a short period of time and rifles were assembled by the Bullard factory as the market demanded.

Fig. 2. Cut of Bullard Rifle showing Bolt withdrawn and Breech Block turned down.

—9—

Bullard Repeating Arms, Springfield, Mass., U. S. A.

Fig. 3. 45—70 Sporting Rifle, Fancy Stock, Globe and Peep Sights.

	Round Barrel.	Half Octagon Barrel.	Octagon Barrel.
Price, Plain,	$33.00	$35.00	$35 00
Length of Barrel,	28 inches.	28 inches.	28 inches.
Number of Shots,	11	11	11
Weight about	9½ lbs.	9¾ lbs.	9¾ lbs.

45—75, 45—60, 40—60 } Same length Barrel, Style and Finish, same price.

Fancy Walnut Butt and Tip, "Checked," with Globe and Peep Sights, $50.00.

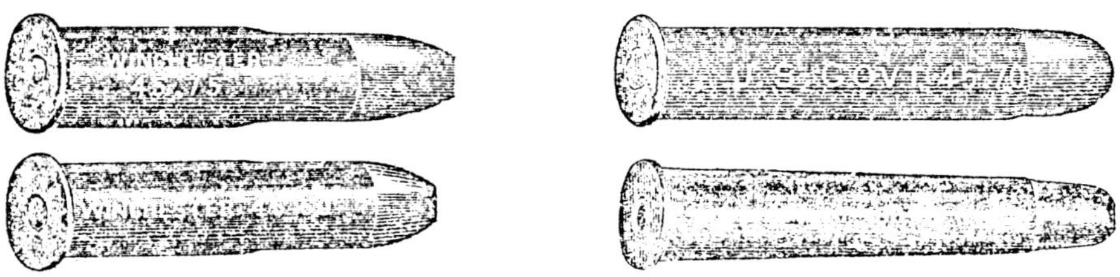

Cartridges for all of these Arms are manufactured by the Winchester Arms Co.

—10—

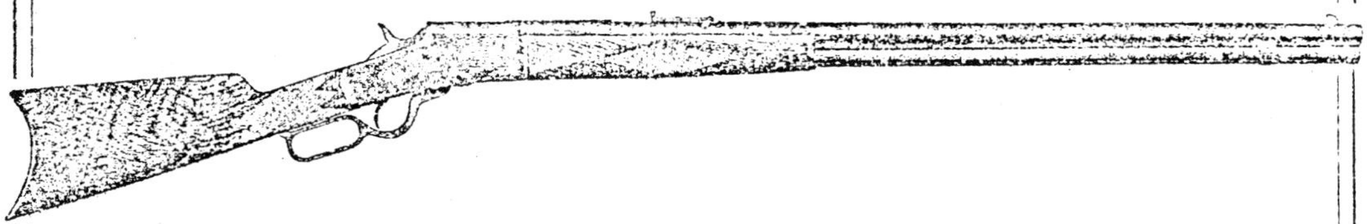

50—95 Express.

PRICES:

Round Barrel, Plain, 27 inches long or under, with Full or Short Magazine, $40 00
Half Octagon Barrel, " " " " " " " " " " 43.00

Extras as per page 14.

We recommend the Winchester Cartridges for our Arms.

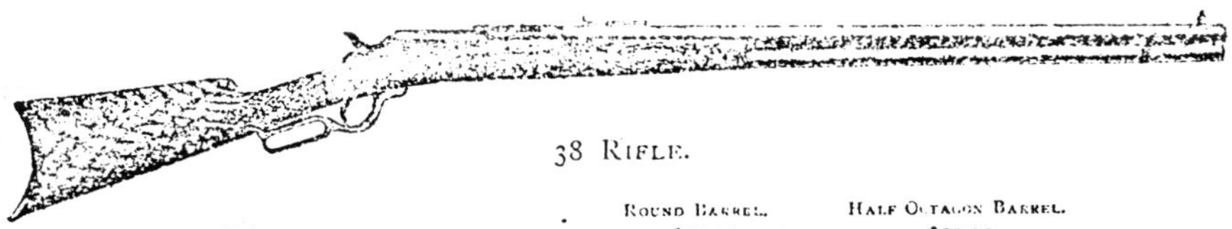

38 Rifle.

	Round Barrel.	Half Octagon Barrel.
Price, ...	$27.00	$29.00
Length of Barrel, . . .	24 inches.	24 inches.
Number of Shots, . . .	16	16

38 Special Rifle, Open.

	Round Barrel.	Half Octagon Barrel.
Price, ...	$30.00	$32.00
Length of Barrel, . . .	26 inches.	26 inches.
Number of Shots, . . .	11	11

Short Magazine and Full, same price. Average Weight, 8¼ lbs.
For Extras, see page 14.
Cartridges for these Arms are manufactured by Winchester Arms Co.

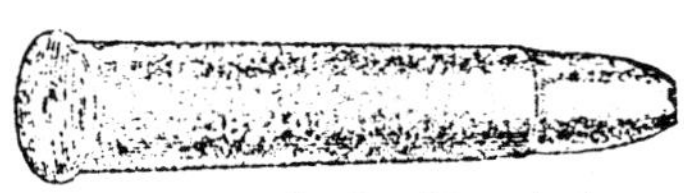

40 gr. powder, Special 245 lead.

Bullard Repeating Arms, Springfield, Mass., U. S. A.

TO DISMOUNT THE GUN.

1. Take out tang screw.
2. Half cock hammer and take out lock-frame and hammer screws.
3. Pull out lock frame down and backwards.
4. Disconnect links from brace.
5. Take out side plate screws and remove side plate.
6. Remove carrier lever spring.
7. Take out guard lever bolt.
8. Take out extractor pin and remove extractor; push back pin before removing the bolt.
9. Draw out guard lever with its connections.
10. Take out breech-block bolt.
11. Take out breech-block.
12. Draw out carrier lever.
13. Take out carrier.

REVERSE OPERATIONS IN ASSEMBLING.

—13—

Bullard Repeating Arms, Springfield, Mass., U. S. A.

TO REMOVE BARREL.

Take out magazine plug screw and the two tip screws, pull out the magazine tube and take off the fore-arm. If the bolt has not been previously removed, pull it back so as to prevent the breaking of the extractor while unscrewing the barrel.

EXTRAS.

All deviations from standard styles will be subject to the following charges:

For additional length of barrel and magazine, add to price $1.00 per inch over regular lengths.

Butt stocks of special shape involving change either in length or drop from standard,	$5.00
Engraving, according to style,	from 5.00 to 100.00
Full nickel plating,	5.00
Fancy walnut stock and fore-arm,	5.00
Checking butt and fore-arm,	5.00
Fancy stock and fore-arm checked,	10.00
Pistol grip stock and fore-arm fancy walnut checked,	15.00
Case hardening,	1.00
Swivels and sling straps,	1.50
Canvas covers	1.50
Stiff leather cases,	6.00
Reloading tools, comprising bullet molds, reloader and recapper, per set,	5.00

Fancy wood cases furnished to order.

—14—

Bullard Repeating Arms, Springfield, Mass., U. S. A.

We furnish with each rifle a flexible wiper that can be carried in the pocket and which will thoroughly clean the arm. We do not think it wise to put an iron rod into a barrel to clean it, after we have taken the greatest possible care to have the grooves and lands of the barrel as smooth and accurate as it is possible for skill and machinery to make it. We have seen barrels ruined by using iron rods for cleaning, and the lands at the muzzle entirely worn away. Another serious objection is that it forces all the dirt and foulness into the mechanism of the breech, causing it to work hard and if neglected will surely rust and injure the arm. The wiper mentioned above is inserted at the butt of the barrel and all dirt is carried out at the muzzle, leaving the breech clean and free from dirt. If wiping with a rod is preferred, use a wooden one and clean the gun with an empty shell in the chamber.

TO LOAD THE GUN.

Carry the guard lever forward as far as it will go, and then insert the cartridges in the magazine through the opening under the carrier.

TO USE AS A SINGLE LOADER.

Carry the lever as far forward as is necessary to extract the spent shell, remove the shell and insert another cartridge in the *chamber* of the gun ; or,

Carry the guard lever as far forward as it will go and insert a cartridge in the magazine through the opening under the carrier.

Bullard Repeating Arms, Springfield, Mass., U. S

If the top cover is drawn back, the magazine can be quickly loaded by simply pressing aga[illegible] under side of the carrier, which will force it up, exposing the entrance to the magazine, and also [illegible] the hammer at half cock.

Many sportsmen desire cartridges that can be reloaded, and all shells, as soon as possible after being fired, should be cleaned and washed out carefully with strong soap suds or soda water and thoroughly dried, otherwise the deposit of nitric acid left on them after firing causes them to oxidize rapidly, and they become worthless. The primer should always be pressed down so that it is sure to be lower than the head of the cartridge shell.

Use American Powder Co.'s Dead Shot, F. G.
Hazard Powder Co., Sea Shooting, F. G.
Orange Rifle (Laflin & Rand Powder Co.), F. G.

Use for bullets one part tin to 16 of lead, and keep the molds very hot and the handles of the molds tight together. It often happens that the handles become too hot for comfort and are not held together properly. The result is a bullet that is not round and one that will not fit the shell. See that the grooves of the bullets are filled with beef tallow or Japan wax, wipe off all surplus grease before loading. When pressing the bullet into the shell, see that it is forced into the chamber of the reloading tool as far as it will go.

We recommend that all sportsmen buy their bullets of cartridge manufacturers, as they [illegible] swaged by powerful machinery and are far more perfect than it is possible to cast them.

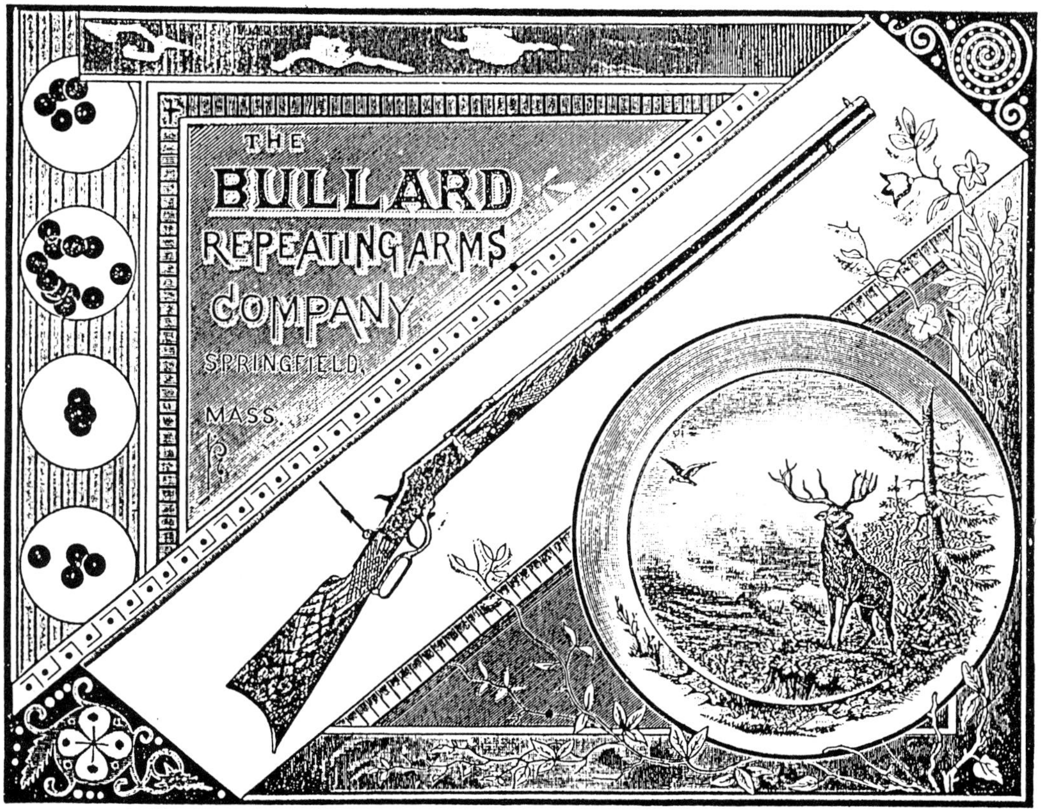

∴ BULLARD ∴

· REPEATING · FIRE · ARMS ·

RIFLED MUSKETS,

CARBINES, HUNTING AND TARGET RIFLES,

FOR MILITARY AND SPORTING USE.

UNEQUALLED FOR
SAFETY, EXCELLENCE OF MATERIAL, WORKMANSHIP, FORCE, ACCURACY OF FIRING, SIMPLICITY OF CONSTRUCTION AND CONVENIENCE OF LOADING.

MANUFACTURED BY THE

BULLARD REPEATING ARMS CO.,

SPRINGFIELD, MASS., U. S. A.

WESTERN HEADQUARTERS:

A. G. SPAULDING & BRO., 108 Madison St., Chicago. LIDDLE & KAEDING, 538 Washington St., San Francisco, Cal.
HENRY C. SQUIRES, 178 Broadway, New York City.

Plate 9-2. Possible second catalog. No single-shots shown. Dates from about October 1883 (date of company name change) to early 1884 (before introduction of single-shots.)

THE statement of Napoleon that "Inventions that are before their time remain useless until the stock of general knowledge comes up to their level," is proved by the fact that although breech-loading fire-arms were invented during the reign of Henry II. of France, previous to 1550, they were never practically used on the battle-field, until the battle of King's Mountain, North Carolina, in 1780, when a breech-loading gun, the invention of Major Patrick Ferguson, of the 71 Highlanders, was used with good effect.

But not until the invention of Sharpes' Rifle in 1857 were the principles of breech-loading of any practical use; and although a repeating breech-loader was invented by S. North, of Middletown, Connecticut, in the year 1825, the principle was not applied in practice until the improvements made by Henry, which resulted in the arm known as the Henry Rifle. The object sought in these and many later inventions was convenience and rapidity in loading, and safety to the user in firing the arm, and these conditions are more fully met in the Bullard Repeater than in any other arm in the market, for the following reasons:

1st. Safety.

The action is positive, not dependent on springs. It is self-locking with a solid breech-block behind the bolt, which must be in place and securely locked before it is possible for the hammer to reach the firing-pin. Before placing this gun on the market it was subjected to the most thorough tests as to its strength and endurance, and has stood the heaviest charges without the least sign of weakness.

2d. Rapidity.

It is possible to fire the Bullard Rifle with greater rapidity than any other repeating rifle, from the fact that works easier and smoother by reason of its more direct leverage on the work to be done, the heaviest work being done with the best leverage, as in extracting the cartridge, which is started when the lever is in position to exert the greatest strain. Cocking the hammer is also done by direct leverage inside the receiver or frame instead of a sliding motion of bolt or firing-pin on and over the top of hammer, which is very often liable to grind and always makes the arm work hard and unpleasantly. (*See Sectional cut, Fig. 2.*) The Bullard has been fired twelve shots in five seconds, the best record of any other gun being eleven shots in seven seconds using U. S. cartridges.

3d. Loading.

The magazine is charged from the under side, and it can be done with equal facility by a right or left-handed person. And as there are no holes or spring covers on the side, it is not possible to have it clogged by passing through brush or laying it on the ground or in trenches, etc. It is also much easier to load on horseback than any other gun, as there is more choice of position than when the opening is on the side. It can be loaded as a single loader, either top or bottom, leaving the magazine full at all times for an emergency.

4th.

The Bullard is the only Lever Repeater that will successfully use the regular United States copper cartridge (45 cal. 70 gr. powder,) owing to its absolute certainty to extract the shell, which being made of copper and with a folded head, does not contract after firing (as is the case of brass shells) but often sticks in the chamber, a difficulty more frequently mentioned in reports of trials of breech-loading fire-arms by government experts than any other.

The Bullard never fails.

Another advantage in using the U. S. cartridge is, that it is the standard amunition and can be obtained anywhere in the country. This point will be appreciated by frontiersmen.

5th. Workmanship.

The quality of work on the Bullard is pronounced by the best judges to be superior to any other repeating rifle in the world, and the fact that Mr. Bullard, the inventor and manager at our manufactory, was for five years master mechanic at the well-known Smith & Wesson works, is a sufficient guarantee that our standard is high and will be kept so. The material used is of the best throughout, no iron or maleable iron being used. All the parts are of forged steel except the carrier, which is composition or gun metal.

6th. Range.

A bullet from the Bullard Rifle using the regular U. S. cartridge (45 cal. 70 gr. powder) has at a distance of *3,000 yards* from muzzle of gun, penetrated a *three-inch spruce plank and eight inches of sand.*

7th. Guarantee.

Our confidence in the points enumerated above is such that, in case of failure of any part by reason of defective material or workmanship, a duplicate will be forwarded free of charge upon receipt of the broken part. All parts are sure to fit.

8th.

To any responsible party we will send one of our guns with the privilege of a three days' trial and examination on receipt of five dollars as guarantee of express charges.

9th.

We challenge the world to produce a repeater that can compare with ours in ease of operation, safety, quality of workmanship, style, balance, finish, accuracy and rapidity of firing.

On the pages following will be found cuts and prices which we trust will be satisfactory, and that you will place your orders with us at your earliest convenience.

Respectfully yours,

BULLARD REPEATING ARMS COMPANY,

SPRINGFIELD, MASS., U. S. A.

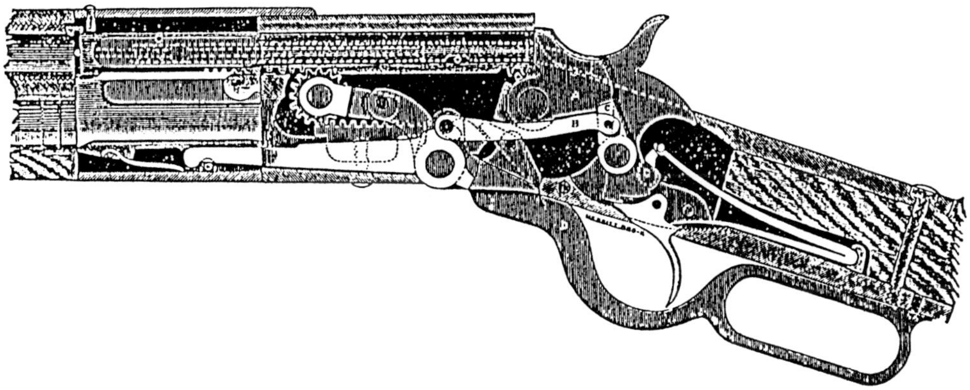

Fig. 1. Section of Gun After Firing.

A, Breech Block held firmly in place by brace C, which is pivoted and cradled on hammer pin and lock frame. C is brought into place and firmly held by links B (only one shown), which are connected by strong pins to guard lever D. It will be seen by the above cut that it is absolutely impossible to get the hammer to strike the firing pin until the brace C is in its proper place. It will also be observed that it is impossible to disarrange the parts, as all are pinned and hinged together, and that there are no sliding surfaces.

BULLARD REPEATING ARMS COMPANY. 9

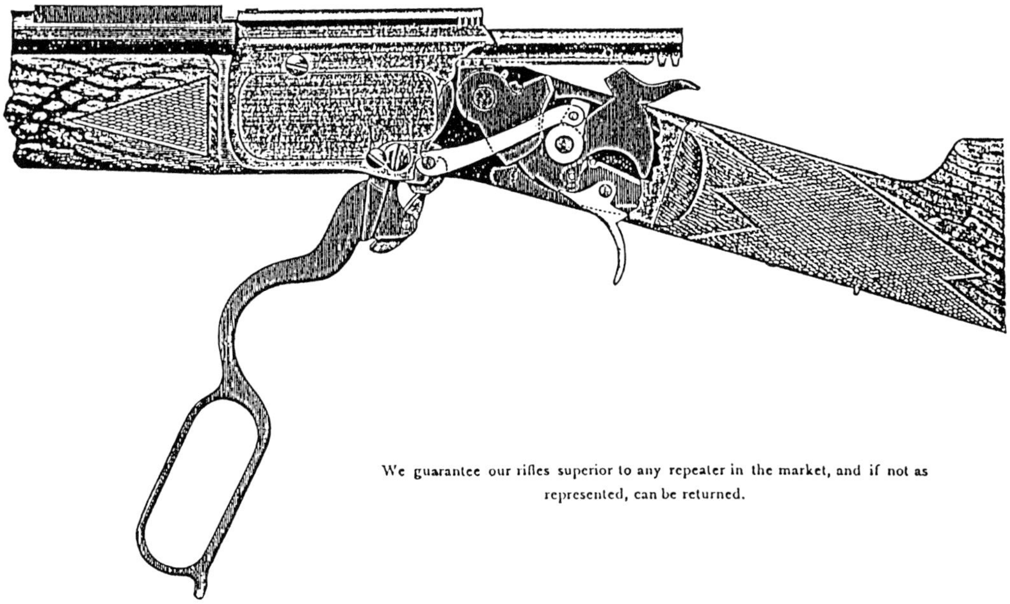

We guarantee our rifles superior to any repeater in the market, and if not as represented, can be returned.

Fig. 2. Cut of Bullard Rifle showing Bolt withdrawn and Breech Block turned down.

SPRINGFIELD, MASS., U. S. A.

10 BULLARD REPEATING ARMS COMPANY.

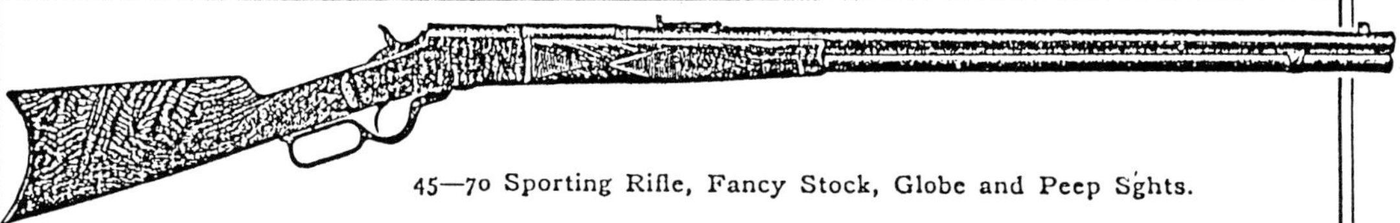

45—70 Sporting Rifle, Fancy Stock, Globe and Peep Sghts.

	Round Barrel.	Half Octagon Barrel.	Octagon Barrel.
Price, Plain, ...	$33.00	$35.00	$35.00
Length of Barrel, . . .	28 inches.	28 inches.	28 inches.
Number of Shots, . . .	11	11	11
Weight about	9½ lbs.	9¼ lbs.	9¼ lbs.

45—60
45—75
45—70
45—85
40—70
40—60 } Same length Barrel, Style and Finish, same price.

Fancy Walnut Butt and Tip, "Checked," with Globe and Peep Sights, $50.00
Fancy Walnut Pistol Grip, "Checked," with open Sights, $48.00 } Either Rifle or Shot Gun butt same price.

4 cal 85 gr. 295.

40 cal. 70 gr.

The height of 200 yd. curve of the 40—75 Cartridge is about 8 inches. 45—85 is but very little more.

Cartridges for all of these Arms are manufactured by the Winchester and Union Metalic Cartridge Co. Cartridges for all of these Arms will be furnished by us at Lowest Prices.

Reject all Cartridges that will not enter the chamber of Rifle with perfect freedom.
Use only solid head shells for re-loading.

SPRINGFIELD, MASS., U. S. A.

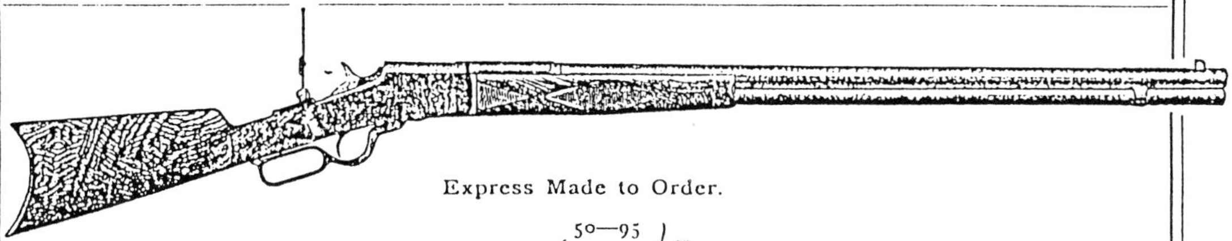

Express Made to Order.

$5.00 Extra for { 50—95, 50—105, 50—115 } Express.

Round Barrel, Plain, 27 inches long or under, with Full or Short Magazine,		$40.00
Half Octagon Barrel, "	" " " " " " " " "	43 00
Full "	" " " " " " " " " "	43.00

Cartridges for these arms will be furnished by us at Lowest Prices.
Ours is the only Repeater in the country that uses 105—115 Express charges.

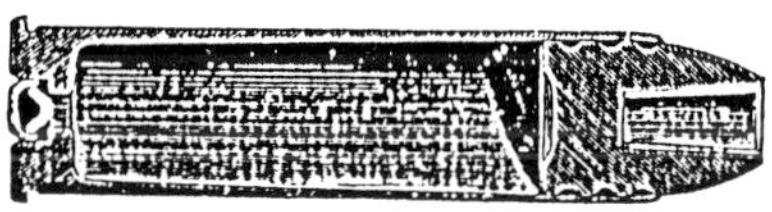

Extras as per page 18.

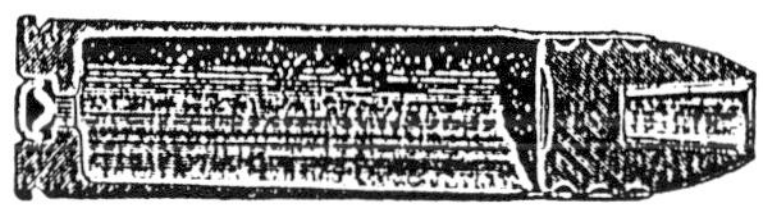

50—105. Bullard Express   50—115. Bullard Express.

Reject all Cartridges that will not enter the chamber of Rifle with perfect freedom.

32, 38, 44 CAL. RIFLES

	Round Barrel.	Half or Full Octagon Barrel.
Price,	$27.00	$29.00
Length of Barrel,	24 inches.	24 inches.
Number of Shots,	16	16
Weight, about	7½	7½

We guarantee our rifles superior to any repeater in the market, and if not as represented can be returned.

Reject all Cartridges that will not enter the chamber of Rifle with perfect freedom.

BULLARD REPEATING ARMS COMPANY. 13

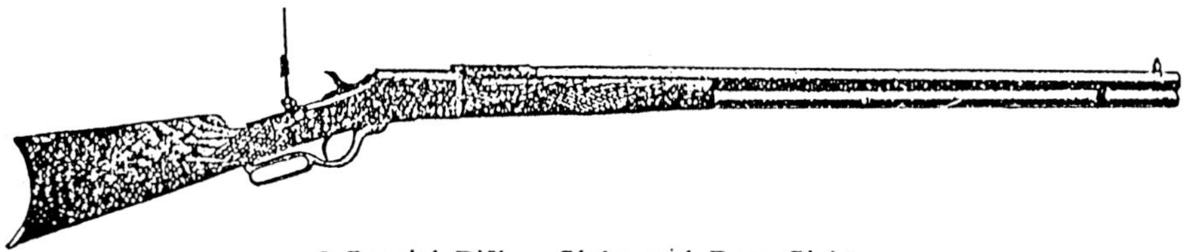

38 Special Rifle. Globe and Peep Sights.

	Round Barrel.	Full Half or Octagon Barrel.
Price, Plain, Open Sights, - - - - -	$30.00	$32.00
Length of Barrel, - - - - - -	26 inches.	26 inches.
Number of Shots, - - - - - -	11	11

Short Magazine and Full, same price. Average Weight, 8¼ lbs.
For Extras, see page 18.
Cartridges for Bullard Arms are manufactured by Winchester and Union Metalic Cartridge Co's.

Special 38 Cal, 50 Gr. Powder, 170 Lead.

We have in preparation a 32 cal., 50 gr. powder, 150 lead. Cartridge same length as above cut.

Reject all Cartridges that will not enter the chamber of Rifle with perfect freedom.

SPRINGFIELD, MASS., U. S. A.

14 BULLARD REPEATING ARMS COMPANY.

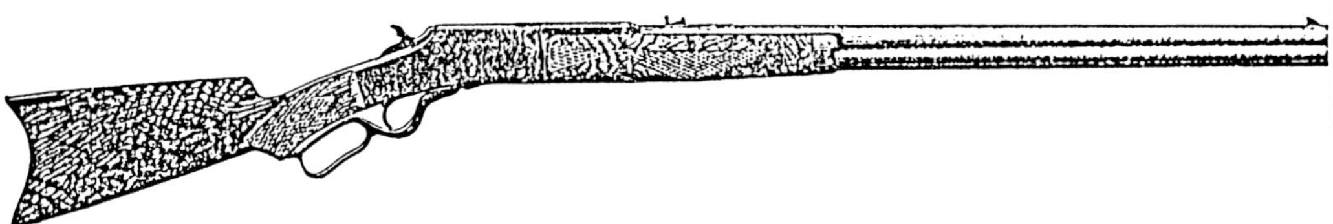

Sporting Rifle, 40—90, Pistol Grip.

	Round Barrel.	Half Octagon Barrel.	Octagon Barrel.
Price, Plain, - - - - -	$35.00	$37.00	$37.00
Length of Barrel, - - -	28 inches.	28 inches.	28 inches.
Number of Shots, - - -	11	11	11
Weight about - - -	10¼ lbs.	10¼ lbs.	10¼

We guarantee our rifles superior to any repeater in the market, and if not as represented can be returned.

40-90.—Bullard.

Cartridges for all our Arms will be furnished by us at Lowest Prices.

Reject all Cartridges that will not enter the chamber of Rifle with perfect freedom.

SPRINGFIELD, MASS., U. S. A.

Manufacture Rifles for the Following Line of Cartridges.

38-50-200 Lead Special.
Bullet rises 14 in. above line of sight going 200 yds.

40-60 Marlin.
Bullet rises 15 in above line of sight going 200 yds.

40-70-225 Bullard.
Bullet rises 8 in. above line of sight going 200 yds.

40-90-300 Bullard
Bullet rises 9 in. above line of sight going 200 yds.

45-70-405 Gov't.

45-85-285 Bullard
Bullet rises 8½ in. above line of sight going 200 yds.

50-105-300 Bullard Ex.

50-115-300 Bullard Ex.

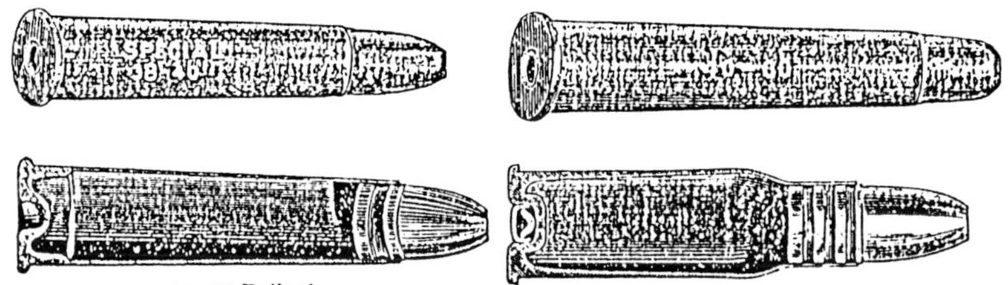

40—70 Bullard.

40—90 Bullard.

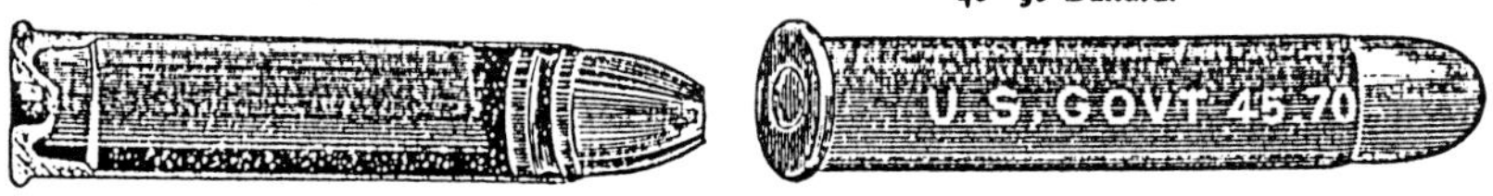

45—85 gr. powder, 285 lead.

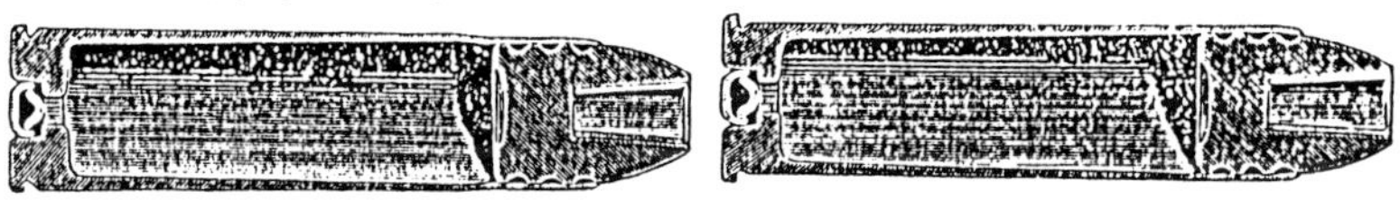

115 gr. powder. Bullard Exp.

105 gr. powder. Bullard Exp.

Bullets carefully Swaged, Grooved and Lubricated for our 40—70, 40—90, 45—85 will be furnished for $1.00 per hundred.

NOTE. 45—70—500 U. S. Cartridge can be used as single loader. The 45 Gov't. and the 45—85 Bullard, both chamber in the same rifle and is strongly recommended for the frontier.

Use only solid head shells for re-loading.

Manufacture Rifles for the Following Line of Cartridges.

32 W. C. F.

45-60 Win.
Bullet rises 16 in. above line of sight going 200 yds.

38 W. C. F.

45-75 Win.
Bullet rises 18 in. above line of sight going 200 yds.

44 W. C. F.

50-95-300 Win. Ex.

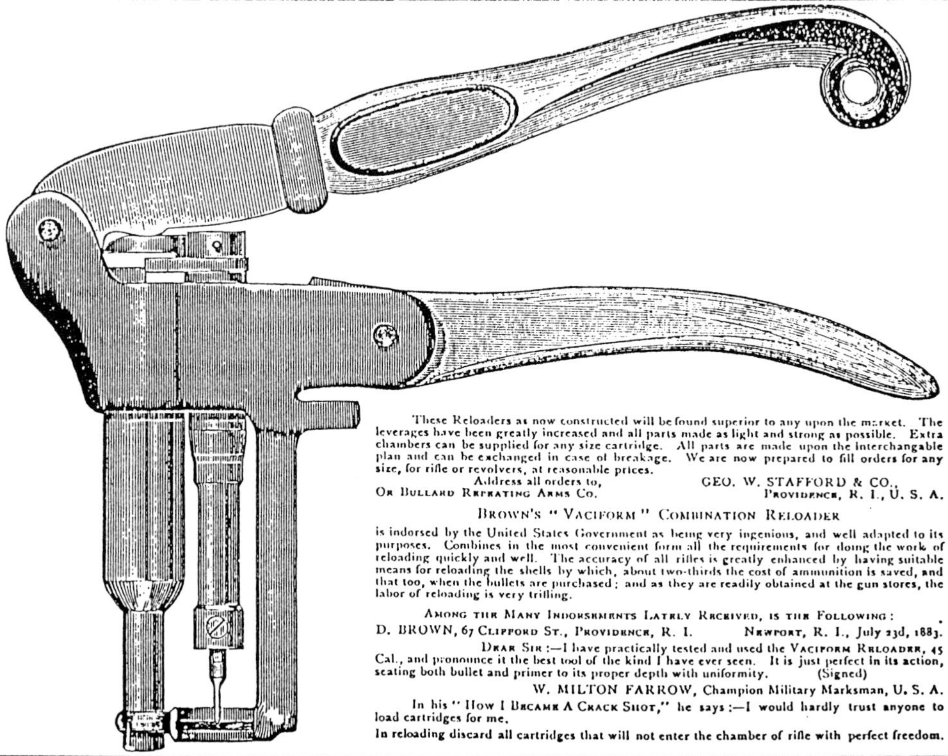

These Reloaders as now constructed will be found superior to any upon the market. The leverages have been greatly increased and all parts made as light and strong as possible. Extra chambers can be supplied for any size cartridge. All parts are made upon the interchangable plan and can be exchanged in case of breakage. We are now prepared to fill orders for any size, for rifle or revolvers, at reasonable prices.

Address all orders to, GEO. W. STAFFORD & CO.,
OR BULLARD REPEATING ARMS CO. PROVIDENCE, R. I., U. S. A.

BROWN'S "VACIFORM" COMBINATION RELOADER

is indorsed by the United States Government as being very ingenious, and well adapted to its purposes. Combines in the most convenient form all the requirements for doing the work of reloading quickly and well. The accuracy of all rifles is greatly enhanced by having suitable means for reloading the shells by which, about two-thirds the cost of ammunition is saved, and that too, when the bullets are purchased; and as they are readily obtained at the gun stores, the labor of reloading is very trifling.

AMONG THE MANY INDORSEMENTS LATELY RECEIVED, IS THE FOLLOWING:

D. BROWN, 67 CLIFFORD ST., PROVIDENCE, R. I. NEWPORT, R. I., July 23d, 1883.

DEAR SIR:—I have practically tested and used the VACIFORM RELOADER, 45 Cal., and pronounce it the best tool of the kind I have ever seen. It is just perfect in its action, seating both bullet and primer to its proper depth with uniformity. (Signed)

W. MILTON FARROW, Champion Military Marksman, U. S. A.

In his "HOW I BECAME A CRACK SHOT," he says:—I would hardly trust anyone to load cartridges for me.

In reloading discard all cartridges that will not enter the chamber of rifle with perfect freedom.

To Dismount the Gun.

1. Take out tang screw.
2. Half cock hammer and take out lock-frame and hammer screws.
3. Pull out lock frame down and backwards.
4. Disconnect links from brace.
5. Take out side plate screws and remove side plate.
6. Remove carrier lever spring.
7. Take out breech-block bolt.
8. Take out breech-block.
9. Take out extractor pin and remove extractor; push back pin before removing the bolt.
10. Take out guard lever bolt.
11. Draw out guard lever with its connections.
12. Draw out carrier lever.
13. Take out carrier.

Reverse Operations in Assembling.

Care being taken that the proper rack connections are made, and the latch lifted, until the lock frame is replaced.

To Remove Barrel.

Take out magazine plug screw and the two tip screws, pull out the magazine tube and take off the fore-arm. If the bolt has not been previously removed, pull it back so as to prevent the breaking of the extractor while unscrewing the barrel.

Extras.

All deviations from standard styles will be subject to the following charges:

For additional length of barrel and magazine, add to price $1.00 per inch over regular lengths.

Butt stocks of special shape involving change either in length or drop from standard,	$5.00
Engraving, according to style,	from $5.00 to 100.00
Full nickel plating,	5.00
Fancy walnut stock and fore-arm,	5.00
Checking grip and fore-arm,	5.00
Fancy stock and fore-arm checked,	10.00
Pistol grip stock and fore-arm fancy walnut checked,	15.00
Case hardening,	1.00
Swivels and sling straps,	1.50
Canvass covers,	1.50
Stiff leather cases,	6.00
Re-loading tools, comprising bullet molds, re-loader and re-capper, per set,	5.00

Peep sights to order.

We furnish with each rifle a flexible wiper that can be carried in the pocket and which will thoroughly clean the arm. We do not think it wise to put an iron rod into a barrel to clean it, after we have taken the greatest possible care to have the grooves and lands of the barrel as smooth and accurate as it is possible for skill and machinery to make it. We have seen barrels ruined by using iron rods for cleaning, and the lands at the muzzle entirely worn away. Another serious objection is that it forces all the dirt and foulness into the mechanism of the breech, causing it to work hard and if neglected will surely rust and injure the arm. The wiper mentioned above is inserted at the butt of the barrel and all dirt is carried out at the muzzle, leaving the breech clean and free from dirt. If wiping with a rod is preferred, use a wooden one and clean the gun with an empty shell in the chamber.

To Load the Gun.

Carry the guard lever forward as far as it will go, and then insert the cartridges in the magazine through the opening under the carrier.

To Use as a Single Loader.

Carry the lever as far forward as is necessary to extract the spent shell, remove the shell and insert another cartridge in the *chamber* of the gun; or,

Carry the guard lever as far forward as it will go and insert a cartridge in the magazine through the opening under the carrier.

If the top cover is drawn back, the magazine can be quickly loaded by simply pressing against the under side of the carrier, which will force it up, exposing the entrance to the magazine, and also place the hammer at half cock.

Many sportsmen desire cartridges that can be re-loaded, and all shells, as soon as possible after being fired, should be cleaned and washed out carefully with strong soap suds or soda water and thoroughly dried, otherwise the deposit of nitric acid left on them after firing causes them to oxidize rapidly, and they become worthless. The primer should always be pressed down so that it is sure to be lower than the head of the cartridge shell.

Use American Powder Co's Dead Shot, F. G.
Hazard Powder Co., Sea Shooting, F. G.
Orange Rifle (Laflin & Rand Powder Co.), F. G.

Use for bullets 1 part tin to 16 of lead, and keep the molds very hot and the handles of the molds tight together. It often happens that the handles become too hot for comfort and are not held together properly. The result is a bullet that is not round and one that will not fit the shell. See that the grooves of the bullets are filled with beef tallow or Japan wax, wipe off all surplus grease before loading. When pressing the bullet into the shell, see that it is forced into the chamber of the re-loading tool as far as it will go.

We recommend that all sportsmen buy their bullets of cartridge manufacturers, as they are all swaged by powerful machinery and are far more perfect than it is possible to cast them.

TERMS

NET CASH in all cases where no special contract is made.

C. O. D. ORDERS from parties not known to us, must be accompanied by thirty per cent. of amount of goods ordered, and the express company's charges for returning the amount collected will be at the expense of the purchaser.

Special terms for large contracts.

Parts can be sent by mail not to exceed 4 pounds weight.

The second catalog has the word Company in its title but shows no single-shot rifles. This catalog probably dates from October 1883 to early 1884 before introduction of the single-shot line and after the name change from Association to Company. It was contemporary with the third piece of literature mentioned below.

The third piece of literature is not really a catalog in the manner that the other four are. According to John Hintlian, who owned an original of this brochure, it contained 14 pages and measured 6x24 inches, with both sides printed and illustrated. Mr. Hintlian supplied the author with a copy of this brochure many years ago. This same brochure was published in *The Canadian Journal of Arms Collecting* (Vol. 21 No. 4) in an article by James C. Tillinghast entitled "Bullard Reloading Tools." This brochure, along with an inserted single-sheet flyer dated April 1885, was sent to H.H. Bigelow on May 14, 1888, in response to a request for information about reloading Bullard cartridges. Why the secretary-treasurer of the Bullard factory would have to send for this information, and such old information at that, is a mystery, although it does support the author's contention that this company's day-to-day affairs were definitely left to the lower management, with Ball and Bigelow maintaining a hands-off approach to the business. Except for one solitary line, no biography of Bigelow that the author has knowledge of contains anything to indicate Bigelow's connection to the Bullard factory.

The author has placed this brochure or smaller catalog third chronologically for several reasons: the company's name change to Bullard Repeating Arms Company; the inclusion of some of the special Bullard calibers and the inclusion of the Brown reloading tool, which only appears in this catalog and the second one which we know was already being superseded by the Bullard reloader, as the "visitor" to the factory had seen in September 1884; and by virtue of the fact that 1,000 copies of the Bullard reloader flyer had been printed for distribution in April 1885. Based on these facts, the author has concluded that this brochure must have been printed sometime in early to mid-1884. No mention of single-shot rifles is made in the brochure either, yet the March 20, 1884 issue of *Forest and Stream* carries an article on the single-shot action. This brochure was available about the same time as the second catalog.

Plate 9-3. Envelope cover that was sent May 14, 1888, to H.H. Bigelow and contained catalog and reload flyer. *From the collection of James C. Tillinghast, copied from The Canadian Journal of Arms Collecting.*

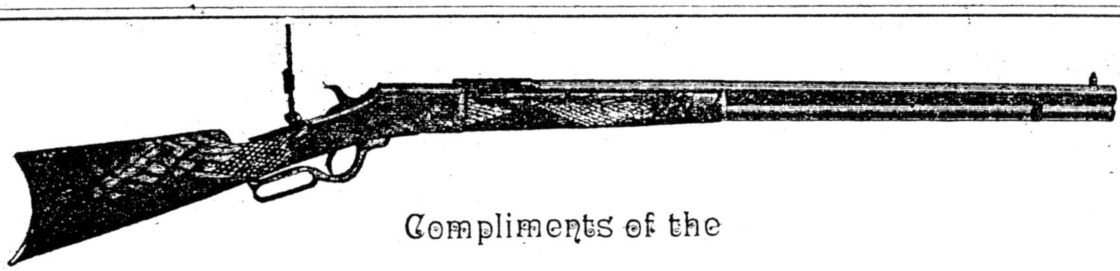

Compliments of the

Bullard Repeating Arms Company,

Springfield, Mass., U. S. A.

40 Cal., 90 Grs. powder.

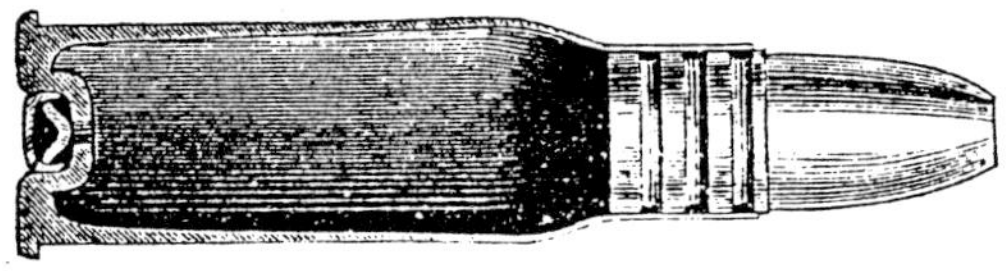

Bullet 300 Grains.

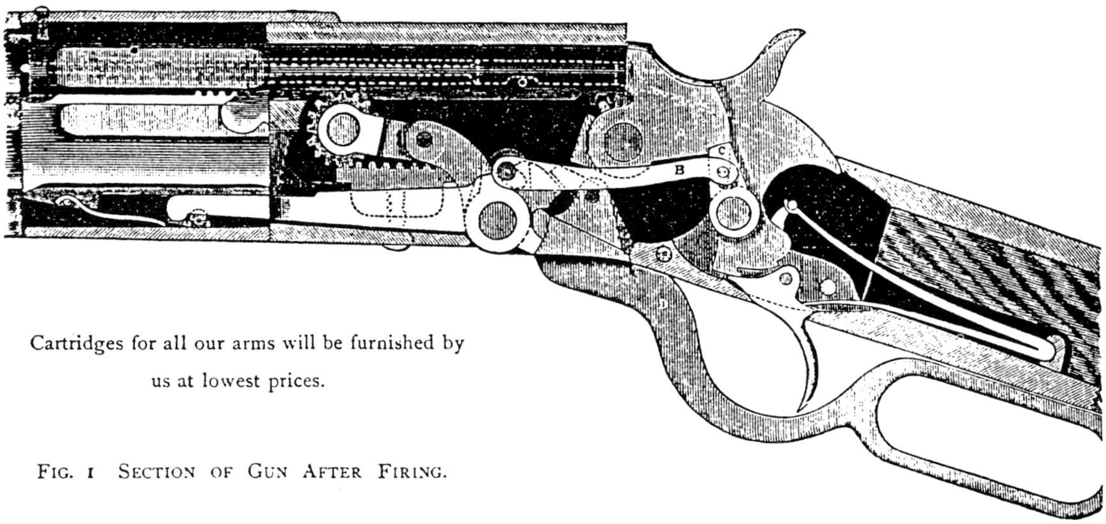

Cartridges for all our arms will be furnished by us at lowest prices.

FIG. I SECTION OF GUN AFTER FIRING.

Plate 9-4. Brochure shown to original size. Printed early to mid-1884 (possibly very late 1883). *Courtesy James C. Tillinghast*

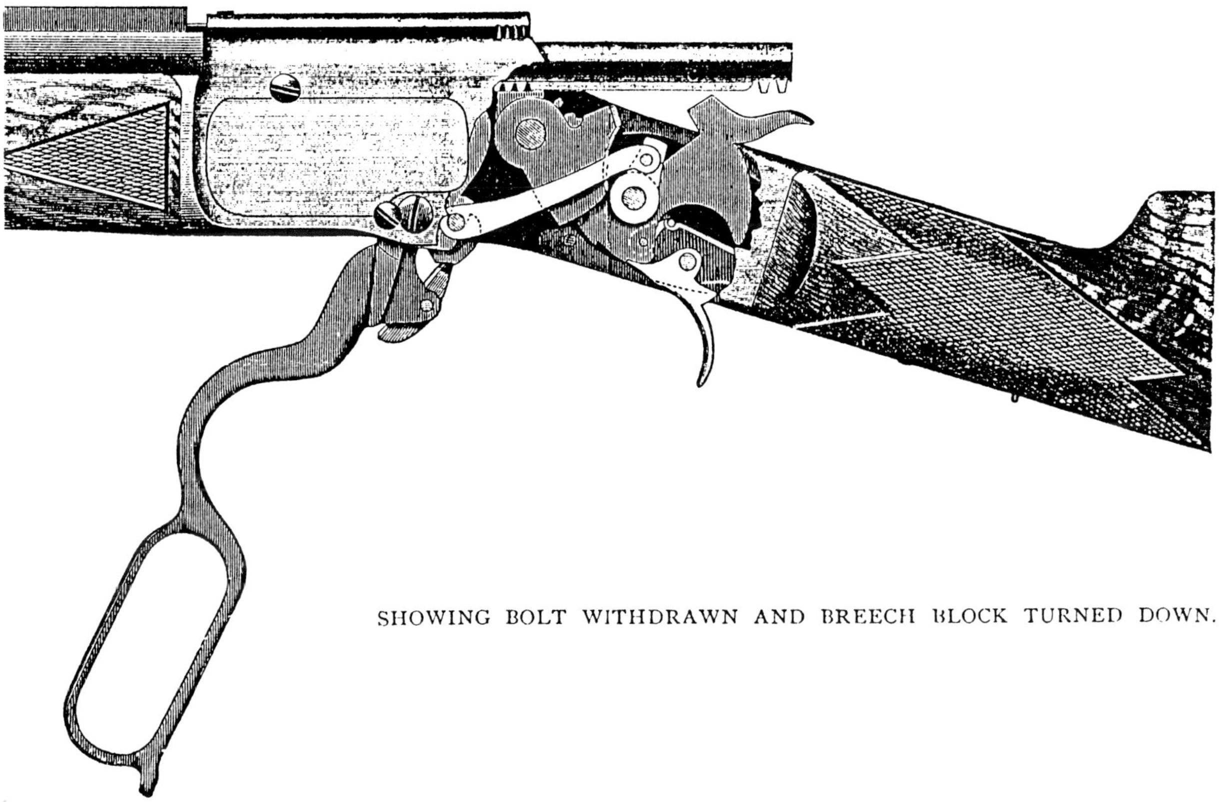

SHOWING BOLT WITHDRAWN AND BREECH BLOCK TURNED DOWN.

1st. Safety.

The action is positive in all its parts, not dependent on springs. It is self locking, with a solid breech-block behind the bolt, which must be in place and securely locked, before it is possible for the hammer to reach the firing pin. Before placing this rifle on the market, it was subjected to the most thorough tests as to its strength and endurance, and has stood the heaviest charges without the least sign of weakness.

2d. Rapidity.

It is possible to fire the Bullard with greater rapidity than any other repeating rifle, as it works easier and smoother; for its more direct leverage on the work to be done, the heaviest work being done with the best leverage, as in extracting the cartridge which is started when the lever is in position to exert the greatest strain. The hammer is brought to full cock by direct leverage inside the frame, instead of a sliding motion of bolt or firing-pin, on and over the top of hammer, which is very often liable to grind and always makes the arm work hard and unpleasantly. (*See sectional cut, Fig. 2.*) The Bullard has been fired twelve shots in five seconds, the best record of any other gun being eleven shots in seven seconds, using U. S. cartridges.

Bullard Repeating Arms Company, Springfield, Mass., U. S. A. 6

3d. Loading.

The magazine is charged from the under side, and it can be done with equal facility by a right or left handed person and as there are no holes or Spring-Covers on side, it is not possible to have it clogged by passing through brush or laying it on the ground or in trenches, etc. It is much easier to load on horseback than any other gun, as there is more choice of position than when the opening is on the side. It can be loaded as a single loader, either top or bottom, leaving the magazine full at all times for an emergency.

4th.

The Bullard is the only Lever Repeater that will successfully use the regular United States copper cartridge (*45 Cal., 70 Grs. Powder,*) owing to its absolute certainty to extract the shell, which being made of copper and with a folded head, does not contract after firing (as is the case of brass shells) but often sticks in the chamber, a difficulty more frequently mentioned in reports of trials of breech-loading fire arms by Government experts, than any other. The Bullard never fails.

Bullard Repeating Arms Company, Springfield, Mass., U. S. A. 7

We Manufacture Repeating Rifles for the following Line of Cartridges.

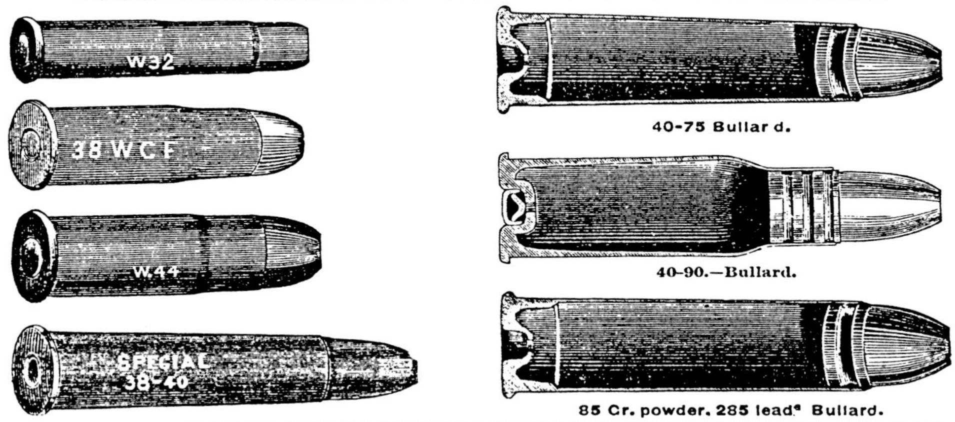

Note the special Bullard cartridges of .40-75, .40-90 and .45-85 have appeared. *Courtesy James C. Tillinghast*

Bullard Repeating Arms Company, Springfield Mass., U, S. A. 8

We Manufacture Repeating Rifles for the Following Line of Cartridges.

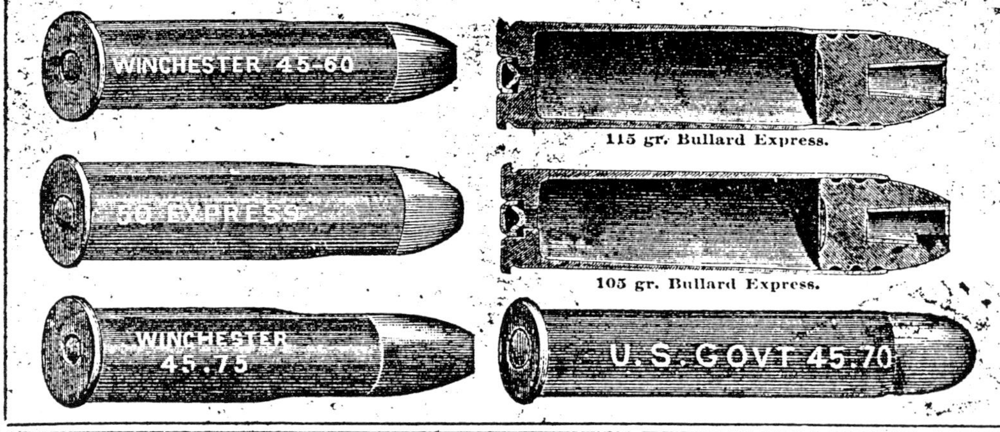

Note the picture of a .50-105 Bullard express, a round offered in only one other catalog seen by the author. Unfortunately since no .50-115 cartridges have been seen to be head-stamped, we will have to wait for someone to discover a box of ammunition so marked before we will know if it was ever marketed with this powder load. *Courtesy James C. Tillinghast*

BULLARD REPEATING ARMS COMPANY, SPRINGFIELD, MASS., U. S. A.

We furnish with each rifle a flexible wiper that can be carried in the pocket and which will thoroughly clean the arm. We do not think it wise to put an iron rod into a barrel to clean it, after we have taken the greatest possible care to have the grooves and lands of the barrel as smooth and accurate as it is possible for skill and machinery to make it. We have seen barrels ruined by using iron rods for cleaning, and the lands at the muzzle entirely worn away. Another serious objection is that it forces all the dirt and foulness into the mechanism of the breech, causing it to work hard and if neglected will surely rust and injure the arm. The wiper mentioned above is inserted at the chamber of the barrel and all dirt is carried out at the muzzle, leaving the breech clean and free from dirt. If wiping with a rod is preferred, use a wooden one and clean the gun with an empty shell in the chamber.

TO LOAD THE GUN.

Carry the guard lever forward as far as it will go, and then insert the cartridges in the magazine through the opening under the carrier.

TO USE AS A SINGLE LOADER.

Carry the lever as far forward as is necessary to extract the spent shell, remove the shell and insert another cartridge in the *chamber* of the gun; or,

Carry the guard lever as far forward as it will go and insert the cartridge in the magazine through the opening under the carrier.

BULLARD REPEATING ARMS COMPANY, SPRINGFIELD, MASS., U. S. A. 10

40-90—SPORTING RIFLE, PISTOL GRIP,

FANCY STOCK, CHECKED.

	Round Barrel.	*Half Octagon.*	*Full Octagon.*
Price, Plain,	$35.00.	$37.00.	$37.00
Length of Barrel,	28 Inches.	28 Inches,	28 Inches.
Number of Shots,	11,	11,	11.
Weight, about	10 1-2 lbs..	10 3-4 lbs.	10 3-4 lbs.

The 40-90 Bullard cartridge is loaded with great care. A highy compressed, grooved and lubricated bullet is used, weighing 300 Grains. It has a flatter trajectory than any other 40 Cal. cartridge in the market.

BULLARD REPEATING ARMS COMPANY, SPRINGFIELD, MASS., U. S. A. 11

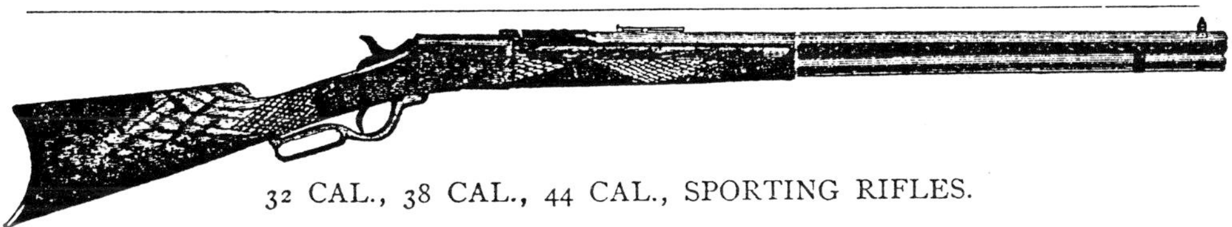

32 CAL., 38 CAL., 44 CAL., SPORTING RIFLES.

	Round Barrel.	*Full or Half Octagon.*
Price, Plain	$27.00.	$29.00.
Length of Barrel,	24 Inches.	24 Inches.
Number of Shots,	16.	16.
Weight, about	7 1-2 lbs.	7 1-2 lbs.

PRICE OF RIFLES FOR THE BULLARD CARTRIDGES, 40-75, 45-85 AND 45 GOV'T.

	Round Barrel.	*Half Octagon.*	*Full Octagon.*
Price, Plain	$33.00.	$35.00.	$35.00
Length of Barrel,	28 Inches.	28 Inches.	28 Inches.
Number of Shots,	11.	11.	11.
Weight, about	9 1-2 lbs.	9 3-4 lbs.	9 3-4 lbs.

Cartridges for all our arms will be furnished by us at lowest rates.

Page 11 shows a small-frame offered in .44 caliber, a model the author has never seen. *Courtesy James C. Tillinghast*

BULLARD REPEATING ARMS COMPANY, SPRINGFIELD, MASS., U. S. A. 12

50 CAL. EXPRESS RIFLES MADE TO ORDER ONLY.

50-95 Grs, Hollow Point Bullet, Price, Plain Round Barrel,	27 Inches long or under,	$40.00
50-105 Grs., 50-115 Grs, Hollow Point Bullet, same style and finish $5 extra. Half or Full Octagon Barrel,	" "	$42.00

For list of Extras on all Rifles, see Page 14.

OUR SPECIAL 38 CAL. 50 GRS. POWDER, 170 LEAD

is made with reference to the wants of riflemen who practice at two hundred yds. off-hand. The heavy patched bullet can be used as a single loader.

Round Barrel,	26 Inches,	Half or Full Magazine,	$30.00
Half or Full Octagon Barrel,		Full Magazine,	$32.00

We consider this to be the most accurate of the small Calibres.

BULLARD REPEATING ARMS COMPANY, SPRINGFIELD, MASS., U. S. A. 13

These Reloaders as now constructed will be found superior to any upon the market. The leverages have been greatly increased and all parts made as light and strong as possible. Extra chambers can be supplied for any size cartridge. All parts are made upon the interchangable plan and can be exchanged in case of breakage. We are now prepared to fill orders for any size, for rifle or revolvers, at reasonable prices. Address all orders to

GEO. W. STAFFORD & CO.,
OR BULLARD REPEATING ARMS CO. PROVIDENCE, R. I., U. S. A.

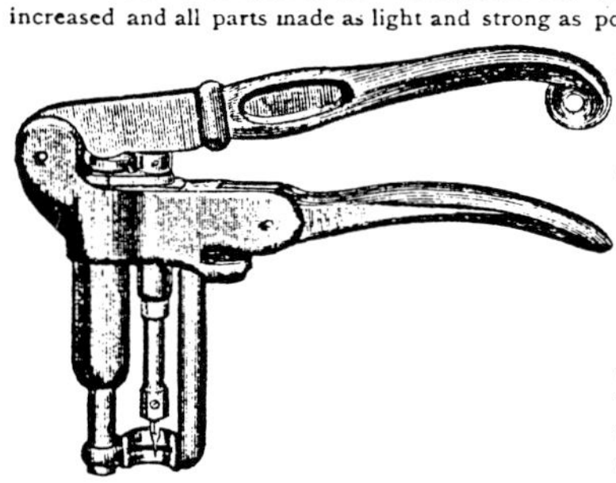

BROWN'S "VACIFORM" COMBINATION RELOADER

Is indorsed by the United States Government as being very ingenious, and well adapted to its purposes. Combines in the most convenient form all the requirements for doing the work of reloading quickly and wel . The accuracy of all rifles is greatly enhanced by having suitable means for reloading the shells by which, about two-thirds the cost of ammunition is saved, and that too when the bullets are purchased; and as they are readily obtained at the gun stores, the labor of reloading is very trifling.

AMONG THE MANY INDORSEMENTS LATELY RECEIVED, IS THE FOLLOWING:

D. BROWN, 67 CLIFFORD STREET, PROVIDENCE, R. I. NEWPORT, R. I., July 23d, 1883.

DEAR SIR:—I have practically tested and used the VACIFORM RELOADER, 45 Cal., and pronounce it the best tool of the kind I have ever seen. It is just perfect in its action, seating both bullet and primer to its proper depth with uniformity.

(Signed) W. MILTON FARROW, Champion Military Marksman, U. S. A.

In his "HOW I BECAME A CRACK SHOT," he says:—" I would hardly trust anyone to load cartridges for me.

In reloading discard all cartridges that will not enter the chamber of rifle with perfect freedom.

Note on page 13 that they call it a "Vaciform" instead of the correct "Variform." *Courtesy James C. Tillinghast*

BULLARD REPEATING ARMS COMPANY, SPRINGFIELD MASS., U. S. A. 14

EXTRAS.

All deviations from standard styles subject to the following.

For additional length of barrel,	$1. per inch
Butt stocks of special shape, drop or length,	5.00
Engraving, according to style,	5.00 to 100.00
Full nickel plating,	5.00
Fancy walnut stock and forearm,	5.00
Checking grip and forearm,	5.00
Fancy stock and forearm checked,	10.00
Pistol grip stock and forearm, fancy walnut, checked,	15.00
Case hardening,	1.00
Swivels and sling straps,	1.50
Canvas covers,	1.50
Stiff leather cases,	6.00
Reloading tools, bullet mould re-loader, re- and de-capper	5.00
Peep Sights, Lyman, Vernier and Graduated to order,	3.50 to 5.00

TO DISMOUNT THE GUN.

1. Take out tang screw.
2. Half cock hammer and take out lock-frame and hammer screws.
3. Pull out lock-frame, down and backwards.
4. Disconnect links from brace.
5. Take out side plate screws and remove side plate.
6. Remove carrier lever spring.
7. Take out breech-bolt.
8. Take out breech-block.
9. Take out extractor pin and remove extractor; push back pin before removing bolt.
10. Take out guard lever bolt.
11. Draw out guard lever and its connections.
12. Draw out carrier lever.
13. Take out carrier.

Reverse Operations in Assembling.

Press of Louis H. Orr & Co., Springfield.

1 m. 4. '85.

DIRECTIONS

FOR RELOADING CARTRIDGES

WITH THE

BULLARD RELOADER

To prepare shell. Place the fired shell in sizing die. Place sizing die and shell in anvil with the shallow cup upward. Strike the sizing die until the shell is forced completely in. Turn over anvil, put plunger into the shell, which will extract shell from the sizing die and expel the exploded primer.

Now, put this reduced shell into the ball-seater with the head end up. ~~Place over it the anvil with the long cup upward~~, and seat the primer by driving ~~the primer-seater through the small hole in the anvil as far as it will go.~~

Now the shell is in its original shape and size, and the primer properly seated. Fill the shell full of powder well shaken down, preferably loaded through a tube so that the powder shall be even full in the shell; if no tube is used, shake the powder well down and fill again even full.

Place the ball evenly on the cartridge with the shell in the anvil, shallow cup upward. Place over the ball seater, and drive down to the shoulder, preferably with a wooden mallet. Use occasionally a drop of oil in the reloader, which will prevent the cartridge and ball sticking.

Great care should be used in preparing cartridges of an even length, and in every case the plunger should be driven down to the shoulder. Shells should be washed in soda water and dried before using. To make lubrication for bullets, use equal parts sperm-oil and beeswax.

We will guarantee our Rifles in every case when the cartridges are re-loaded with the above tools, according to these directions.

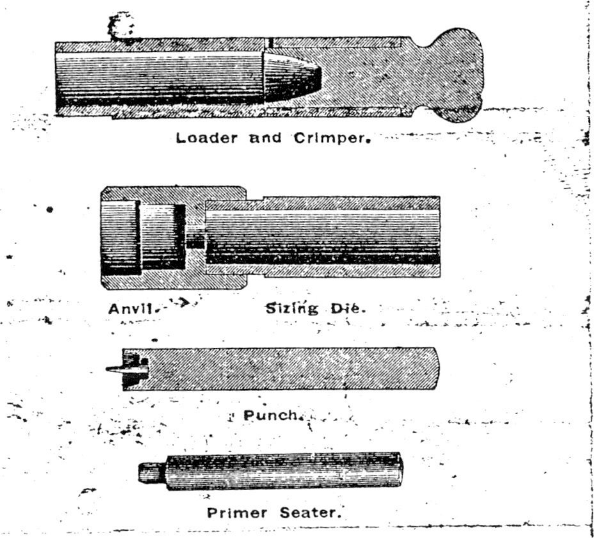

TO DISMOUNT THE GUN.

1. Take out the tang screw.
2. Half cock hammer, and take out lock-frame and hammer screws.
3. Pull out lock frame, down and backwards.
4. Disconnect links from brace.
5. Take out side plate screws, and remove side plate.
6. Remove carrier lever spring.
7. Take out breech-bolt.
8. Take out breech-block.
9. Take out extractor pin and remove extractor; push back pin before removing bolt.
10. Take out guard lever bolt.
11. Draw out guard lever and its connections.
12. Draw out carrier lever.
13. Take out carrier.

Reverse operations in Assembling.

Plate 9-5. The one-page flyer inserted for Bigelow's letter. Note the upper left-hand corner of the side marked "directions," there you can see the 1 m. 4. '85, meaning 1,000 copies, April 1885. *Courtesy James C. Tillinghast*

The third catalog reproduced here introduces, probably for the first time in catalog form, the solid-frame single-shot civilian and military rifle. The first paragraph on page 3 makes quite clear that they wish the reader to note their new single-shot rifle, and yet articles in the March 20, 1884 issue of *Forest and Stream* carried an article on the single-shot action. An advert in the May 16, 1885 *American Field* shows the single-shot. Page 3 (in this catalog) is a most important page in that it also mentions various dates of tests and articles of Bullard arms, thus making it somewhat easier to date this catalog. January 7, 1886, is the last date mentioned, therefore this catalog was probably issued in the first half of 1886. This catalog consisted of 29 pages and its format was 8x6 inches. It had a light pastel green cover.

The catalog now covers their complete line of proprietary cartridges, and pages 16 and 17 show the solid-frame single-shots in all their versions. The last two pages of the catalog show the Bullard reloading tool and sights for target rifles.

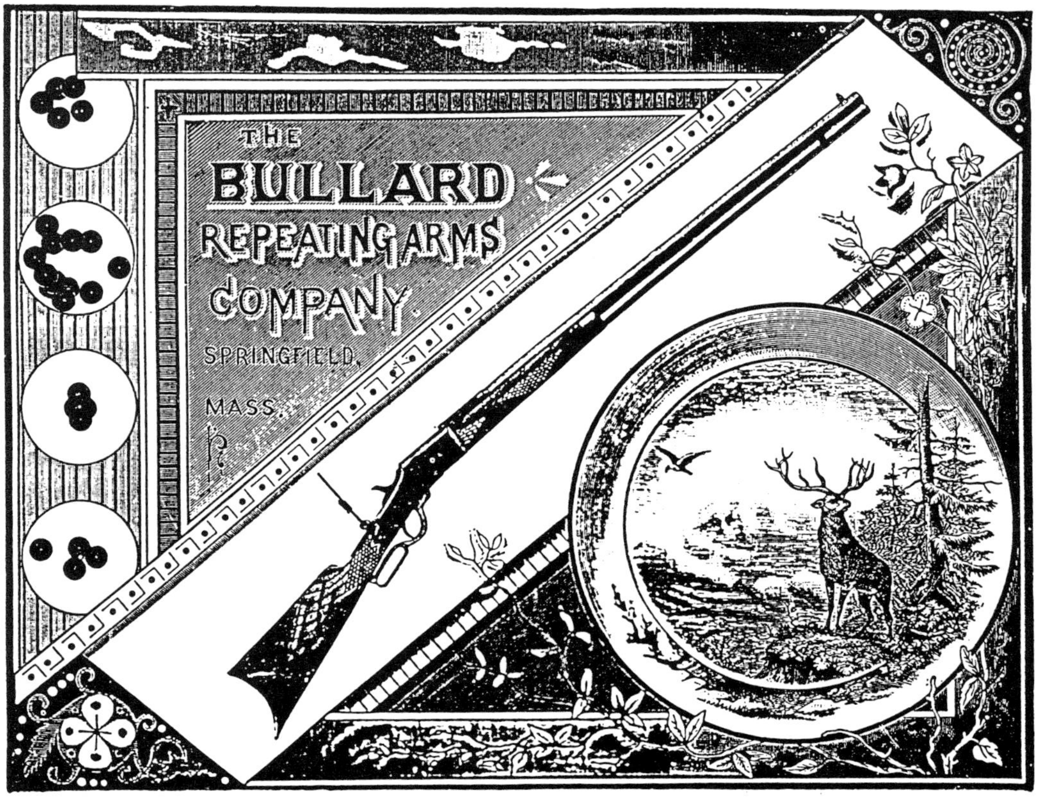

1

BULLARD

REPEATING FIRE ARMS.

RIFLED MUSKETS,

CARBINES, HUNTING AND TARGET RIFLES,

FOR MILITARY AND SPORTING USE.

UNEQUALLED FOR

SAFETY, EXCELLENCE OF MATERIAL, WORKMANSHIP, FORCE, ACCURACY OF FIRING, SIMPLICITY OF CONSTRUCTION AND CONVENIENCE OF LOADING.

MANUFACTURED BY THE

BULLARD REPEATING ARMS CO.,

SPRINGFIELD, MASS., U. S. A.

F. S. BLANCHARD & CO., Printers, 154 Front St., Worcester, Mass.

Plate 9-6. Possible third catalog. Reprinted per contents *Gun Digest*, 8th edition, 1954.

The Bullard Repeating Arms Company catalog reproduced here was published about 1887. Among their proprietary and interesting cartridges, the Bullard 50-115-300 Express, the first semi-rimmed case commercially produced, will be seen on page 15 of this catalog. Operating on a rack-and-pinion principle, the Bullard is a remarkably smooth-working action. Notwithstanding an extremely narrow sectioned receiver, the Bullard was strong and safe with the then existing cartridges.

Bullard single-shot rifles will also be found with detachable and interchangeable barrels, a type not shown in the catalog pages pictured here.

The company is thought to have started in business about 1883, and ceased operations sometime in the early 1890's.

For a more extensive treatment of the Bullard rifles, see *Single Shot Rifles*, by James Grant, and *The Rifle in America*, by Phil Sharpe.

The Gun Digest Company is pleased to announce an authorized reprint of L. D. Satterlee's *Fourteen Old Gun Catalogs*, a most valuable and instructive reference work that has been out of print for many years. Containing all of the known Sharps Rifle Company catalogs, as well as several others, the new reprint will be limited to 1,000 copies only, price $7.50. Orders may be sent to The Gun Digest Company, 1255 S. Wabash Ave., Chicago 5, Ill.

TERMS:

NET CASH in all cases where no special contract is made.

C. O. D. ORDERS from parties not known to us, must be accompanied by thirty per cent. of amount of goods ordered, and the express company's charges for returning the amount collected will be at the expense of the purchaser.

Special terms for large contracts.

Parts can be sent by mail not to exceed 4 pounds weight.

IN ISSUING this Catalogue, we desire to call especial attention to the new single-shot rifle, recently brought out, for military, hunting and target purposes. The system adopted is the same as in repeaters, so far as necessary. The same breech block, brace, lever, links, hammer, trigger and springs are used. By reference to the cuts, showing the system opened and closed, it will be seen that we have the strongest system made and consequently, the safest arm. The parts are few, made of the best material, of sufficient strength and pinned together so as to be positive in their movements.

The ammunition for the military arm, is the 45-70-405 and 45-70-500, the standard U. S. Government Service.

The standard fixed ammunition for the repeater is especially prepared for hunting purposes, where a flat trajectory is required, and on reference to the recent tests made in Boston, in August, 1885, and the "*Forest and Stream*" test, made at Creedmoor, in September and October, 1885, it will be seen that the Bullard stands first. Complete report of the said tests of "*Forest and Stream*," can be found in issue of January 7, 1886, which show the merits of our repeaters in these trials.

For target purposes, the repeaters are unsurpassed, by using special ammunition as given under the various calibres. It is admitted by all riflemen, and it is a fact that a rifle using a large amount of powder and a light bullet is not the most accurate at the target, yet, for hunting purposes, where a flat trajectory is required, the heavy powder charge and light bullet is of vital importance.

The single shot rifles are made for hunting and target shooting. The same ammunition can be used in the single shot rifles as in the repeater. The hunting and target rifles are made in all calibres, from 22 short, to 50 cal. and chambered for any of the standard shells.

We guarantee our arms to be first-class in every particular and superior to all others, for the following reasons:

1st. Safety.

The action is positive, not dependent on springs. It is self-locking, with a solid breech-block behind the bolt, which must be in place and securely locked before it is possible for the hammer to reach the firing-pin. Before placing this gun on the market, it was subjected to the most thorough tests as to its strength and endurance and has stood the heaviest charges without the least sign of weakness.

2nd. Rapidity.

It is possible to fire the Bullard Rifle with greater rapidity than any other repeating rifle, from the fact that it works easier and smoother, by reason of its more direct leverage on the work to be done, the heaviest work being done with the best leverage, as in extracting the cartridge, which is started when the lever is in position to exert the greatest strain. Cocking the hammer is also done by direct leverage inside the receiver or frame, instead of a sliding motion of bolt or firing-pin on and over the top of hammer, which is very often liable to grind and always makes the arm work hard and unpleasantly. (*See Sectional Cut, Fig. 2.*) The Bullard has been fired twelve shots in five seconds, the best record of any other gun being eleven shots in seven seconds, using U. S. cartridges.

3d. Loading.

The magazine is charged from the under side, and it can be done with equal facility by a right or left-handed person. And as there are no holes or spring covers on the side, it is not possible to have it clogged by passing through brush or laying it on the ground or in trenches, etc. It is also much easier to load on horseback than any other gun, as there is more choice of position than when the opening is on the side. It can be loaded as a single loader, either top or bottom, leaving the magazine full at all times for an emergency.

4th.

The Bullard is the only Lever Repeater that will successfully use the regular United States copper cartridge (45 cal. 70 gr. powder), owing to its absolute certainty to extract the shell, which being made of copper and with a folded head, does not contract after firing (as is the case of brass shells) but often sticks in the chamber, a difficulty more frequently mentioned in reports of trials of breech-loading fire-arms by government experts than any other.

Another advantage in using the U. S. cartridge is, that it is the standard ammunition and can be obtained anywhere in the country. This point will be appreciated by frontiersmen.

5th. Workmanship.

The quality of work on the Bullard is pronounced by the best judges to be superior to any other repeating rifle in the world. The material used is of the best throughout, no iron or malleable iron being used. All the parts are of forged steel except the carrier, which is composition or gun metal. Special care is taken in rifting and chambering the barrels.

6th. Range.

A bullet from the Bullard Rifle using the regular U. S. cartridge (45 cal. 70 gr. powder) has at a distance of *3,000 yards* from muzzle of gun, penetrated a *three-inch spruce plank and eight inches of sand.*

7th. Guarantee.

Our confidence in the points enumerated above is such that, in case of failure of any part by reason of defective material or workmanship, a duplicate will be forwarded free of charge upon receipt of the broken part. All parts are sure to fit.

8th.

To any responsible party we will send one of our guns with the privilege of a three days' trial and examination on receipt of five dollars as guarantee of express charges.

9th.

We challenge the world to produce a repeater that can compare with ours in ease of operation, safety, quality of workmanship, style, balance, finish, accuracy and rapidity of firing.

8 BULLARD REPEATING ARMS COMPANY.

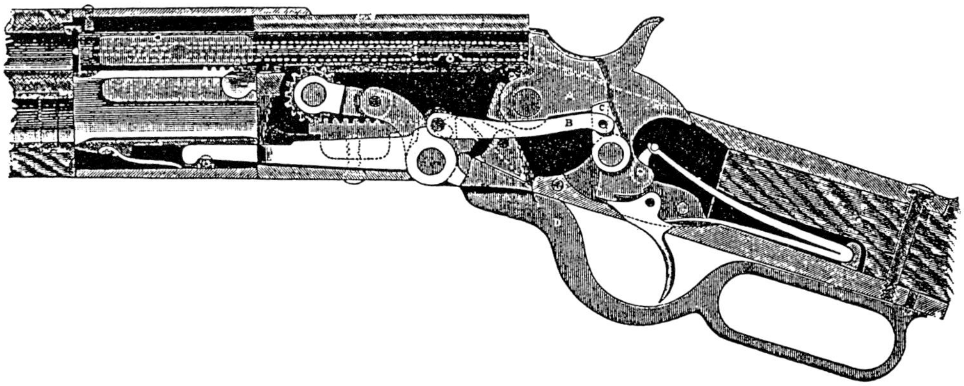

Fig. 1. Section of Gun After Firing.

A, Breech Block held firmly in place by brace C, which is pivoted and cradled on hammer pin and lock frame. C is brought into place and firmly held by links B (only one shown), which are connected by strong pins to guard lever D. It will be seen by the above cut that it is absolutely impossible to get the hammer to strike the firing pin until the brace C is in its proper place. It will also be observed that it is impossible to disarrange the parts, as all are pinned and hinged together, and that there are no sliding surfaces.

SPRINGFIELD, MASS., U. S. A.

BULLARD REPEATING ARMS COMPANY. 9

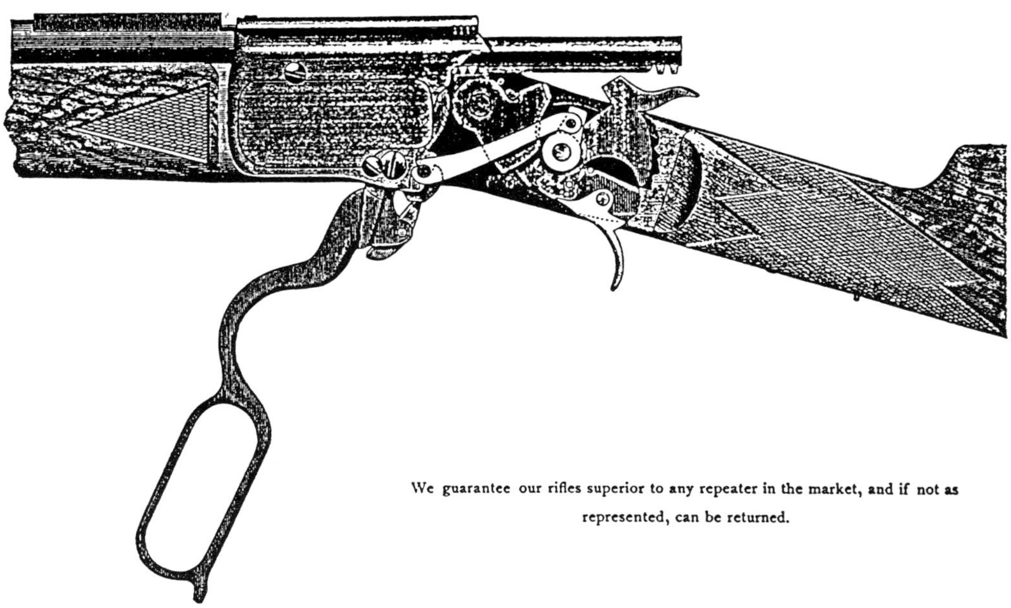

We guarantee our rifles superior to any repeater in the market, and if not as represented, can be returned.

Fig. 2. Cut of Bullard Rifle showing Bolt withdrawn and Breech Block turned down.

SPRINGFIELD, MASS., U. S. A.

10 BULLARD REPEATING ARMS COMPANY.

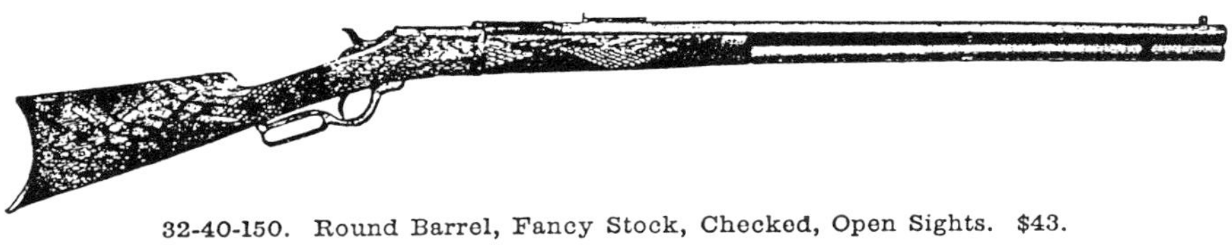

32-40-150. Round Barrel, Fancy Stock, Checked, Open Sights. $43.

	ROUND BARREL.	HALF OR FULL OCTAGON BARREL.
Price, Plain Stock	$33.00	$35.00
Length of Barrel	26 inches.	26 inches.
Number of Shots	11	11
Weight, about	8 lbs.	8½ lbs.

Full or Half Magazine, Rifle or Shot, Rubber Butt, same price.

32 Special Bullard.

Bullet rises 11 inches above line of sight going 200 yards. Penetration 15 inches solid pine, 50 yards.

Reject all Cartridges that will not enter the chamber of the Rifle with perfect freedom.

To dismount the Rifle, see page 19.

SPRINGFIELD, MASS., U. S. A.

BULLARD REPEATING ARMS COMPANY. 11

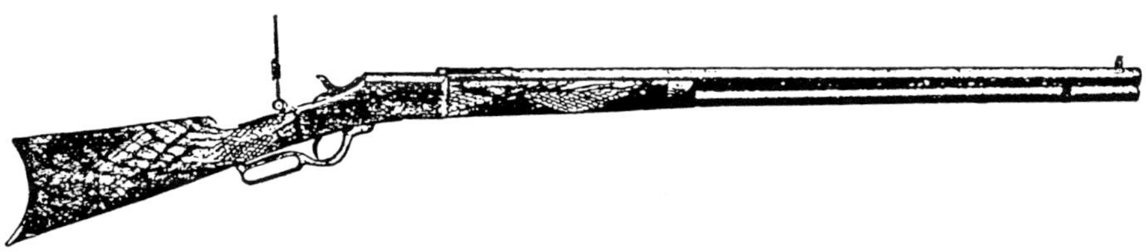

38 Special Rifle. Round Barrel, Fancy Stock, Checked, Globe and Peep Sight, $47.25.

	ROUND BARREL.	HALF OR FULL OCTAGON BARREL.
Price, Plain Stock, Open Sights	$33.00	$35.00
Length of Barrel	26 inches.	26 inches.
Number of Shots	11	11
Weight, about	8 lbs.	8¼ lbs.

Full or Half Magazine, Rifle or Shot, Rubber Butt, Same Price.

For Target practice, 200 yards, this Rifle is unexcelled, by using 35 grains of American F. G. Powder and the regular 190 grain bullet.

Special 38 Cal. 45 Gr. Powder, 190 Lead.

Bullet rises about 12 inches going 200 yards. Penetration 14 inches solid pine, 50 yards.

Reject all Cartridges that will not enter the chamber of Rifle with perfect freedom.

To dismount the Rifle, see page 19.

SPRINGFIELD, MASS., U. S. A.

12 BULLARD REPEATING ARMS COMPANY.

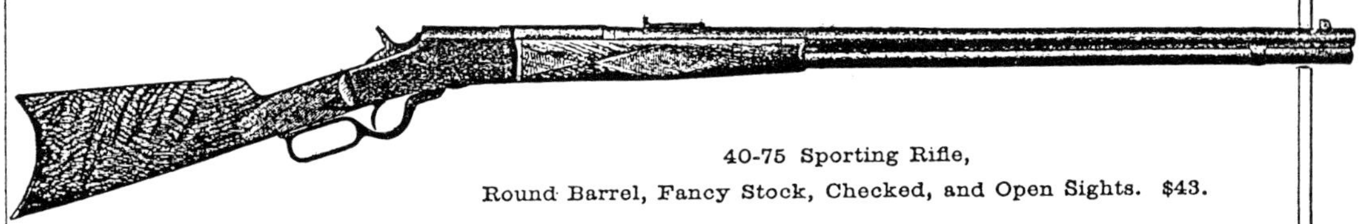

40-75 Sporting Rifle,
Round Barrel, Fancy Stock, Checked, and Open Sights. $43.

	ROUND BARREL.	HALF OR FULL OCTAGON BARREL.
Price, Plain Stock	$33.00	$35.00
Length of Barrel	28 inches.	28 inches.
Number of Shots	11	11
Weight, about	10 lbs.	10¼ lbs.

For target practice, 200 yards, this Rifle is first-class, and by using 55 grains American F. G. powder and a 300 grain bullet the most satisfactory results are obtained. The Marlin 40-60-260 Cartridges can be used in this caliber rifle, being of the same length as 40-75-258 Bullard.

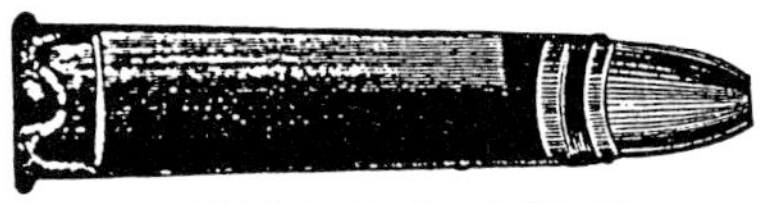

40 Cal. 75 Gr. Powder. 258 Lead.

Bullet rises about 10 inches going 200 yards. Penetration 19 inches solid pine, 50 yards.

Reject all Cartridges that will not enter the chamber of Rifle with perfect freedom.

To dismount the Rifle, see page 19.

SPRINGFIELD, MASS., U. S. A.

BULLARD REPEATING ARMS COMPANY. 13

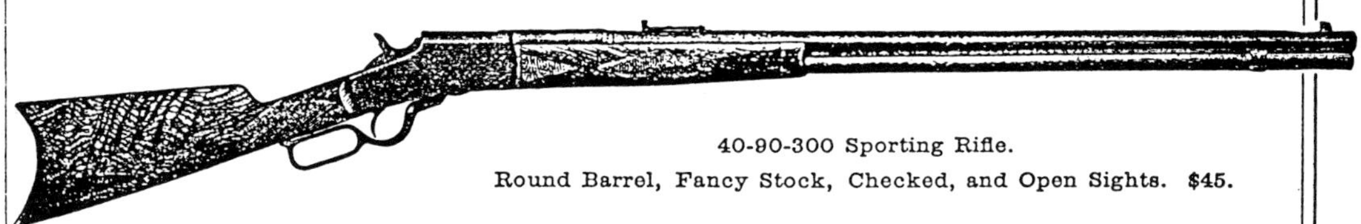

40-90-300 Sporting Rifle.
Round Barrel, Fancy Stock, Checked, and Open Sights. $45.

	ROUND BARREL.	HALF OR FULL OCTAGON BARREL.
Price, Plain Stock	$35.00	$37.00
Length of Barrel	28 inches.	28 inches.
Number of Shots	11	11
Weight, about	10 lbs.	10 lbs.

Full or Half Magazine, Rifle or Shot, Rubber Butt, same price.

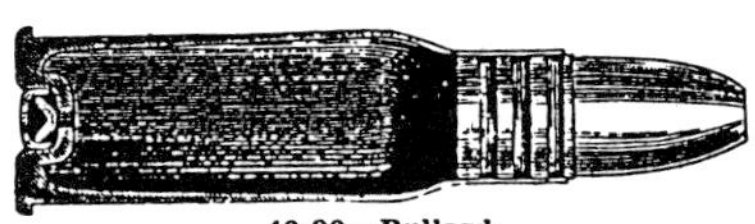

40-90.—Bullard.

Bullet rises 10 inches above line of sight, going 200 yards. Penetration 16 inches solid pine, 50 yards.

Reject all Cartridges that will not enter the chamber of Rifle with perfect freedom.

To dismount the Rifle, see page 19.

SPRINGFIELD, MASS., U. S. A.

14 BULLARD REPEATING ARMS COMPANY.

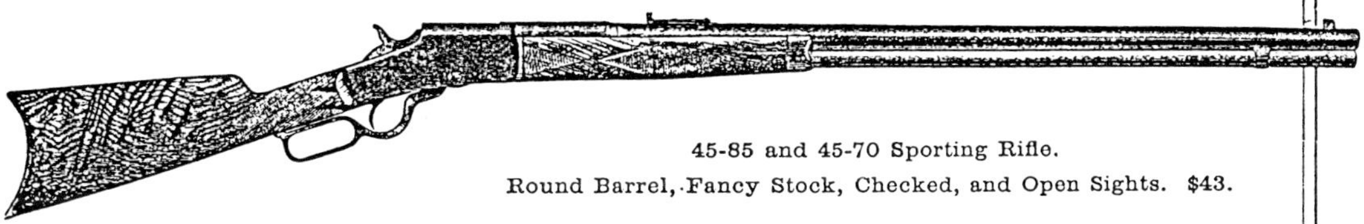

45-85 and 45-70 Sporting Rifle.
Round Barrel, Fancy Stock, Checked, and Open Sights. $43.

	Round Barrel.	Half or Full Octagon Barrel.
Price, Plain Stock	$33.00	$35.00
Length of Barrel	28 inches.	28 inches.
Number of Shots	11	11
Weight, about	10 lbs.	10½ lbs.

Full or Half Magazine, Rifle or Shot, Rubber Butt, same price.

45-85-290 Bullard.
Bullet rises about 10 inches, 200 yards.

45-70-405 and 45-70-500.
Bullet rises about 13 inches, 200 yards. Penetration 21 in. solid pine, 50 yds.

Reject all Cartridges that will not enter the chamber of Rifle with perfect freedom.

To dismount the rifle, see page 19.

SPRINGFIELD, MASS., U. S. A.

BULLARD REPEATING ARMS COMPANY. 15

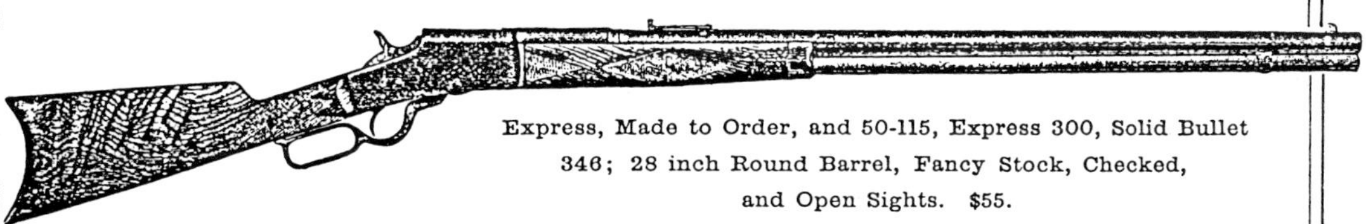

Express, Made to Order, and 50-115, Express 300, Solid Bullet 346; 28 inch Round Barrel, Fancy Stock, Checked, and Open Sights. $55.

	Round Barrel.	Half or Full Octagon Barrel.
Price, plain	$45.00	$47.00
Length of Barrel	28 inches.	28 inches.
Number of Shots	11	11
Weight, about	10 lbs.	10½ lbs.

Price for 50-95 Express, same style as above, $5 less.

The measured trajectory for 200 yards, about 10 inches for 50-115 Express. The measured trajectory for 100 yards is less than 2 inches for 50-115 Express.

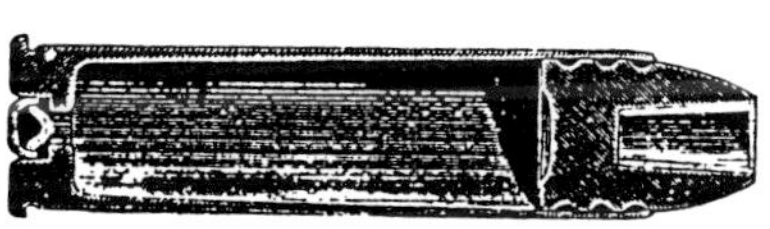

50-115-300 Bullard Express.

50-95-300 Win. Ex.

Reject all Cartridges that will not enter the chamber of Rifle with perfect freedom.

To dismount the Rifle, see page 19.

SPRINGFIELD, MASS., U. S. A.

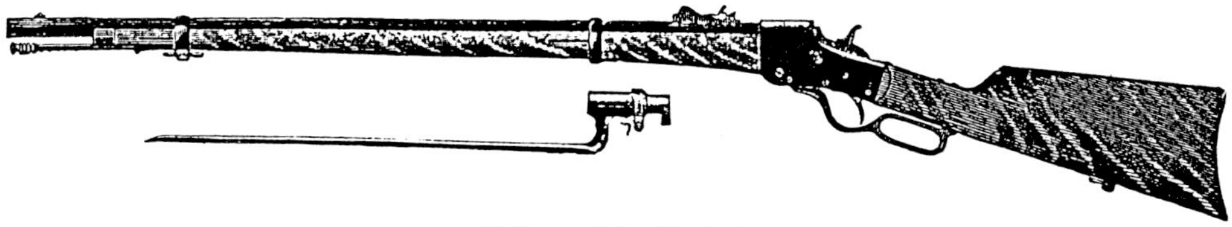

Military Rifle Musket.

Adapted to the U. S. Government service ammunition.

Weight, about . . . . 8 1-2 lbs.

Length of barrel . . . . 32 inches.

Weight of carbine, 26 in. barrel, about . . . . 8 lbs.

TERMS: Subject to special contract.

This rifle is one of the most rapid firing single loaders made.

45-70-405 and 45-70-500.

Extreme range of bullet, 3,680 yards.

Target Rifle. 28 in. Barrel, Fancy Stock, Checked, Swiss Butt. Vernier and Wind Gauge Sights. $47.

PRICE LIST—HUNTING RIFLES.

Cal. 32, 38, 40, 45 and 50. Octagon barrels, 28 in. or less, pistol grip, rifle butt, open sights, $25.
Cal. 22. " " 26 in. " " " rubber butt, open sights, $25.

REGULAR AMMUNITION FOR SPORTING RIFLES.

22, short or long.
32, 40, 150, 38, 45, 190, 40, 75, 258.
45, 85, 290, 45, 70, 405, 45, 70, 500.
50, 115, 346, solid bullet, 50, 115, 300, express bullet.

TARGET RIFLES.

The 32, 35, 38, 40 and 45 Cal. Target Rifles, chambered for any shell in general use, and for lubricated or patched bullets Weight, from 8 to 10 lbs. Without Sights, $25; Swiss butt, half octagon barrel, $27.

The 22 Cal. Gallery or Hunting, Octagon barrel, 26 inches, with open sights and rubber shot butt, $25. Weight 6½ to 8 lbs

To dismount the Rifle, see page 18.

Single Target and Hunting Rifle.

The same breech block, lever, brace and brace links are used in this Rifle as in Repeater. The same forward movement of the lever operates the hammer and cocks it. The system is the strongest and safest that is made, and reference to Fig. 1, showing the system open, and Fig. 2, showing the system closed, and the parts in position after firing, clearly indicate the safety of the Arm.

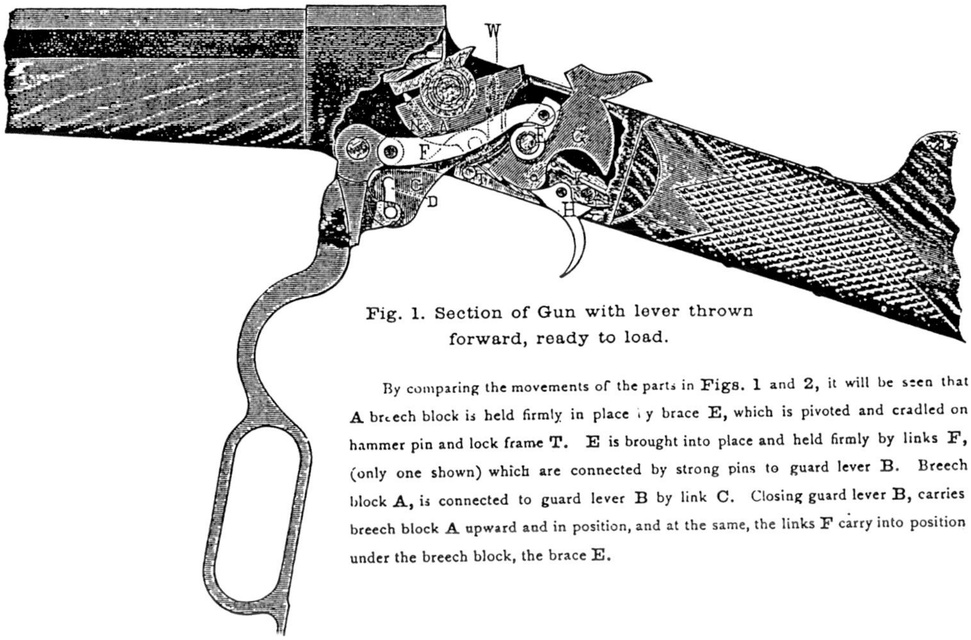

Fig. 1. Section of Gun with lever thrown forward, ready to load.

By comparing the movements of the parts in Figs. 1 and 2, it will be seen that A breech block is held firmly in place by brace E, which is pivoted and cradled on hammer pin and lock frame T. E is brought into place and held firmly by links F, (only one shown) which are connected by strong pins to guard lever B. Breech block A, is connected to guard lever B by link C. Closing guard lever B, carries breech block A upward and in position, and at the same, the links F carry into position under the breech block, the brace E.

It will be seen by these cuts that it is impossible to get the hammer to strike the firing pin until the brace E is in its proper position under the breech block. It will also be observed that it is impossible to disarrange the parts, as they are pinned together, have no sliding surfaces and have positive motions.

To Dismount the Gun.

1. Take out tang screws and remove butt stock.
2. Half cock the hammer and take out lock frame and hammer screw.
3. Pull out lock frame — down and backwards.
4. Disconnect links from brace.
5. Take out breech block and guard lever screws.
6. Remove guard lever and breech block.

To Assemble the Gun.

Reverse operations as given to dismount the gun.

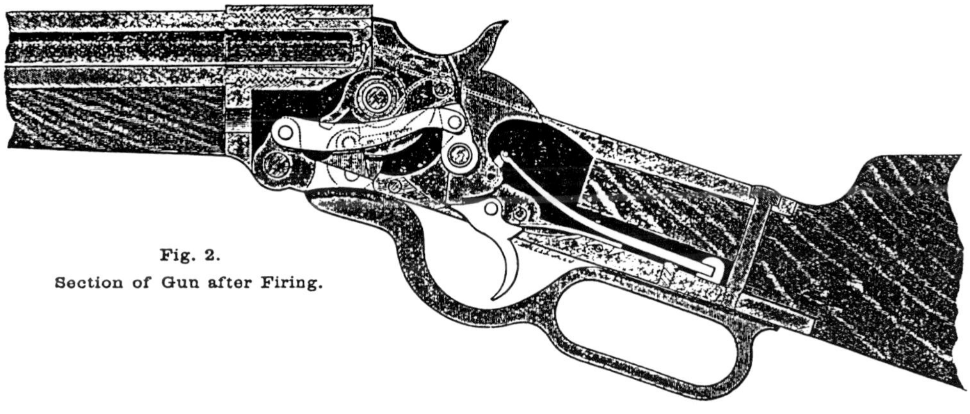

Fig. 2.
Section of Gun after Firing.

To Dismount the Magazine Rifle.

1. Take out tang screw.
2. Half cock hammer and take out lock-frame and hammer screws.
3. Pull out lock frame down and backwards.
4. Disconnect links from brace.
5. Take out side plate screws and remove side plate.
6. Remove carrier lever spring.
7. Take out breech-block bolt.
8. Take out breech-block.
9. Take out extractor pin and remove extractor; push back pin before removing bolt.
10. Take out guard lever bolt.
11. Draw out guard lever with its connections.
12. Draw out carrier lever.
13. Take out carrier.

Reverse Operations in Assembling.

Care being taken that the proper rack connections are made, and the latch lifted, until the lock frame is replaced.

To Remove Barrel.

Take out magazine plug screw and the two tip screws, pull out the magazine tube and take off the fore-arm. If the bolt has not been previously removed, pull it back so as to prevent the breaking of the extractor while unscrewing the barrel.

Extras.

All deviations from standard styles will be subject to the following charges:

For additional length of barrel, add to price $1 00 per inch over regular lengths.

Butt stocks of special shape involving change either in length or drop from standard,	$5.00
Engraving, according to style, - - - - - from $5.00 to	100.00
Full nickel plating,	5.00
Fancy walnut stock and fore-arm,	5.00
Checking grip and fore-arm,	5.00
Fancy stock and fore-arm checked,	10.00
Pistol grip stock and fore-arm fancy walnut checked,	15.00
Case hardening,	1.00
Swivels and sling straps,	1.50
Canvas Covers,	2.25
Stiff leather cases,	6.00
Re-loading tools, comprising bullet molds, re-loader and re-capper, for repeaters, per set,	5.00
Combination molds for solid and hollow point bullets,	4.50
Re- and de-capper for target rifles	1.50
Wiping rods,	.25
Brushes,	.25
Wad cutter,	.50

We furnish with each magazine rifle a flexible wiper that can be carried in the pocket and which will thoroughly clean the arm. We do not think it wise to put an iron rod into a barrel to clean it, after we have taken the greatest possible care to have the grooves and lands of the barrel as smooth and accurate as it is possible for skill and machinery to make it. We have seen barrels ruined by using iron rods for cleaning, and the lands at the muzzle entirely worn away. Another serious objection is that it forces all the dirt and foulness into the mechanism of the breech, causing it to work hard, and if neglected will surely rust and injure the arm. The wiper mentioned above is inserted at the butt of the barrel and all dirt is carried out at the muzzle, leaving the breech clean and free from dirt. If wiping with a rod is preferred, use a wooden one and clean the gun with an empty shell in the chamber. Target rifles we furnish rod and brush wiper.

To Load the Magazine Rifle.

Carry the guard lever forward as far as it will go, and then insert the cartridges in the magazine through the opening under the carrier.

To Use a Single Loader Magazine Rifle.

Carry the lever as far forward as is necessary to extract the spent shell, remove the shell and insert another cartridge in the *chamber* of the gun; or,

Carry the guard lever as far forward as it will go and insert a cartridge in the magazine through the opening under the carrier.

If the top cover is drawn back, the magazine can be quickly loaded by simply pressing against the under side of the carrier, which will force it up, exposing the entrance to the magazine, and also place the hammer at half cock.

Many sportsmen desire cartridges that can be re-loaded, and all shells, as soon as possible after being fired, should be cleaned and washed out carefully with strong soap suds or soda water and thoroughly dried, otherwise the deposit of nitric acid left on them after firing causes them to oxidize rapidly, and they become worthless. The primer should always be pressed down so that it is sure to be lower than the head of the cartridge shell.

Use American Powder Co.'s Dead Shot,	F. G.
Hazard Powder Co., Sea Shooting,	F. G.
Orange Rifle (Laflin & Rand Powder Co.),	F. G.

Use for bullets 1 part tin to 16 of lead, and keep the molds very hot and handles of the molds tight together. It often happens that the handles become too hot for comfort and are not held together properly. The result is a bullet that is not round and one that will not fit the shell. See that the grooves of the bullets are filled with beef tallow or Japan wax; wipe off all surplus grease before loading. When pressing the bullet into the shell, see that it is forced into the chamber of the re-loading tool as far as it will go.

We recommend that all sportsmen buy their bullets of cartridge manufacturers, as they are all swaged by powerful machinery and are far more perfect than it is possible to cast them.

24 BULLARD REPEATING ARMS COMPANY.

Parts of Bullard Repeating Rifle.

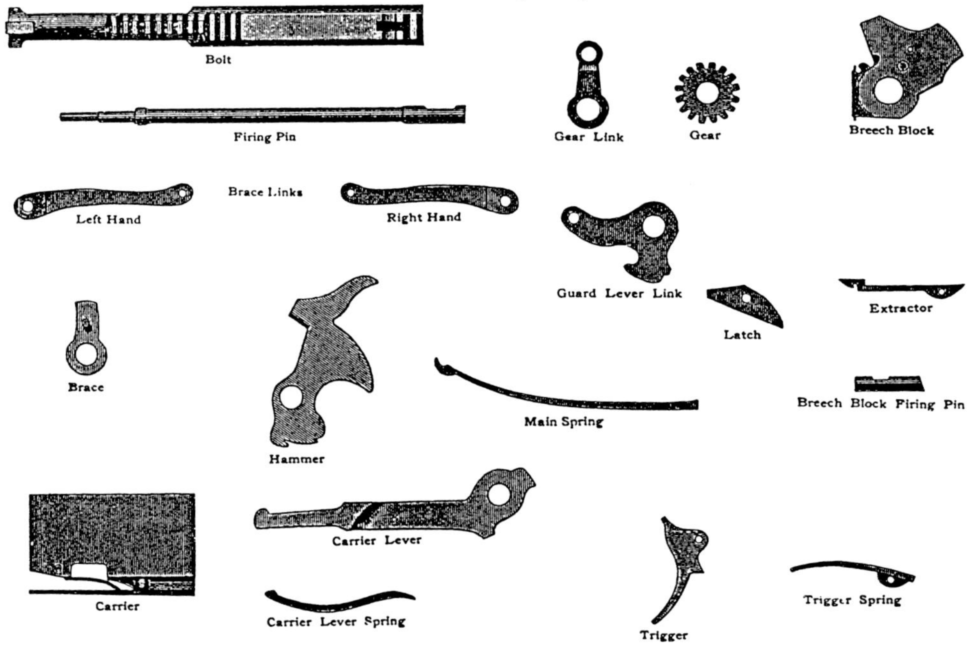

SPRINGFIELD, MASS., U. S. A.

BULLARD REPEATING ARMS COMPANY. 25

Nomenclature and Price List of the Component Parts of the Bullard Repeating Fire Arms.

Barrel, round, half otagon or octagon,	$10.00
Bolt	2.00
Bolt firing pin	.30
Bolt firing pin stop pin	.02
Bolt firing pin spring	.05
Bottom cover	.35
Bottom cover spring	.25
Bottom cover spring pin	.02
Bottom cover pin	.02
Butt plate, rifle	1.00
Butt plate, shot	1.00
Butt plate screws, 2	.12
Butt stock, standard rifle	1.75
Brace link, right or left	.25
Brace	.50
Brace link pin, large or small	.05
Breech block	1.50
Breech block firing pin	.10
Breech block axis pin	.25
Breech block axis firing pin screw	.05
Breech block firing pin stop pin	.02
Breech block spring	.05
Carrier	2 00
Carrier lever	.50
Carrier lever spring	.25
Extractor	.30
Extractor pin	$ .02
Forestock	1.25
Gear	.15
Gear pin	.05
Gear link	30
Gear link pin	.02
Guard lever	1.50
Guard lever, pistol grip	1.75
Guard lever latch	.35
Guard lever latch pin	.02
Guard lever latch spring	.02
Guard lever latch plunger	.05
Guard lever bolt	.10
Guard lever bolt screws	.05
Guard lever steady pin	.02
Guard lever link	.75
Hammer	.90
Hammer bushing	.10
Hammer screw	.08
Lock frame	3 00
Lock frame screws	.06
Lyman sight screws, 2	.10
Magazine tube	2 25
Magazine muzzle cap	.10
Magazine muzzle cap screw	.05
Magazine spring	.30
Magazine follower	$ .25
Magazine ring	.50
Magazine tenon	.10
Main spring	.50
Main spring, pistol grip	.60
Main spring strain screw	.05
Rack	.30
Rack screw	.06
Receiver	10.00
Side plate	1.25
Side plate screw	.05
Sling strap	.50
Stirrup	.20
Stirrup pin	.02
Swivel, for sling strap	.50
Tang screw	.10
Tip for forestock	1.00
Tip screws, 2	.10
Top cover	.50
Top cover screw	.05
Trigger	.50
Trigger, pistol grip	.60
Trigger pin	.02
Trigger spring	.40
Trigger spring, pistol grip	.50
Wiper, complete	.30

SPRINGFIELD, MASS., U. S. A.

26 BULLARD REPEATING ARMS COMPANY.

Parts of Single Shot Rifle.

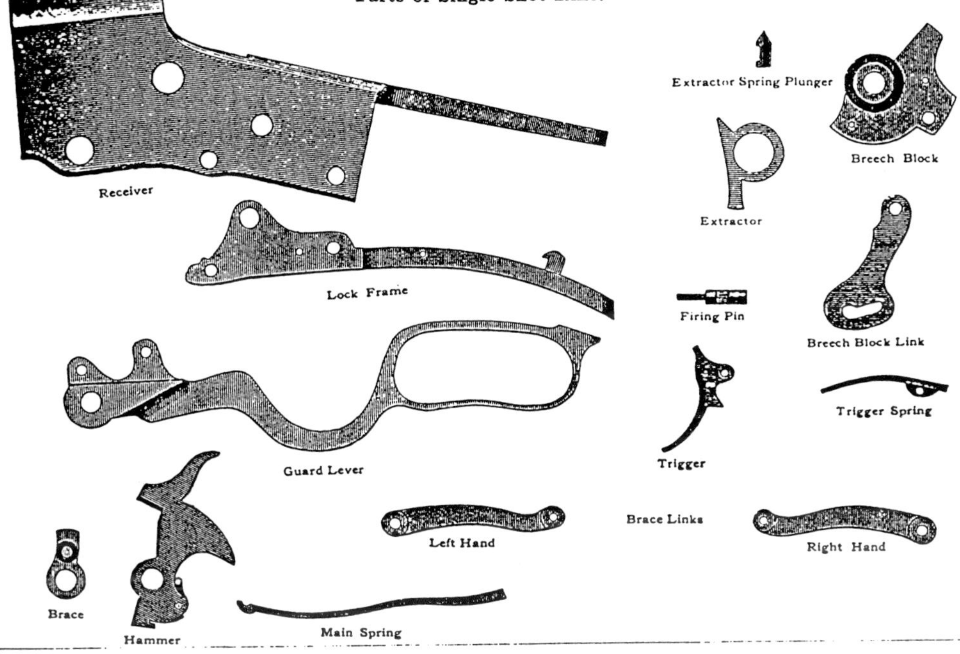

SPRINGFIELD, MASS., U. S. A.

BULLARD REPEATING ARMS COMPANY. 27

Nomenclature and Price List of the Component Parts of the Single Shot Rifle.

Part	Price	Part	Price
Barrel, 28 inches or under,	$10 00	Hammer,	$ .90
Brace,	.50	Hammer bushing,	.10
Brace links, right and left, each,	.25	Hammer screw,	.08
Breech block,	1.50	Lock frame,	3.00
Breech block bolt,	.25	Lock frame screw,	.08
Breech block link,	.35	Main spring,	.40
Buckhorn sight,	.80	Receiver,	5 00
Butt stock,	3.50	Rifle butt plate (steel) and screws,	1.00
Extractor,	1.00	Rifle butt plate, Swiss, nickel plated,	2 00
Extractor spring and plunger,	.10	Rubber butt plate and screws,	1.00
Firing pin,	.15	Sight base screws, each,	.05
Firing pin pin,	.05	Strain screw,	.05
Firing pin spring,	.05	Tang screw,	.10
Forestock,	1.25	Trigger,	.50
Forestock screw,	.10	Trigger pin,	.05
Front sight,	.50	Trigger spring,	.40
Guard lever,	1.50		

SPRINGFIELD, MASS., U. S. A.

Price List of Cartridges for Hunting Rifles.

Cartridges for all our arms are manufactured by the U. M. C. Co., and are guaranteed to be of first quality. Solid head and re-loadable shells.

32 Cal	40	grains Powder,	150	Grains lead,	$31.50	per M.	Shells,	$20	Bullets,	$7.25	per M.
38 Cal	45	"	190	"	31.50	"	"	20	"	8.00	"
40 Cal	75	"	258	"	33.00	"	"	18	"	9.00	"
40 Cal	90	"	300	"	40.00	"	"	22	"	10.50	"
45 Cal	60	"	300	"	33.00	"	"	20	"	9.00	"
45 Cal	75	"	350	"	35.00	"	"	20	"	9.50	"
45 Cal	70	"	405	"	35.00	"	"	22	"	12.00	"
45 Cal	85	"	290	"	37.00	"	"	22	"	9.50	"
50 Cal	95	"	300	"	40.00	"	"	23	"	11.00	"
50 Cal.	115	"	300	" Exp.	90.00	"	"	50	"	11.00	"
50 Cal.	115	"	346	"	90.00	"	"	50	"	9.00	"

Primers $2 per M.

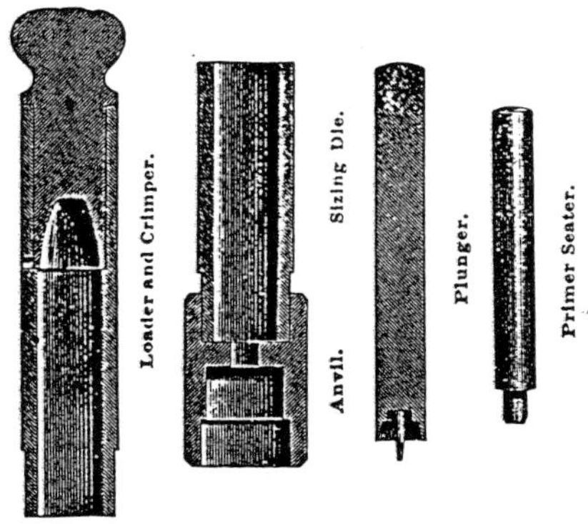

Directions for Re-loading Cartridges with the Bullard Re-loader.

To prepare the shell. Wash the fired shells in strong soap suds or soda water, and dry thoroughly. Place the fired shell in sizing die. Place sizing die and shell in anvil with the shallow cup upward. Strike the sizing die until the shell is forced completely in. Turn over anvil, put plunger into the shell, which will extract shell from the sizing die and expel the exploded primer.

Now put this reduced shell into the loader with the head end up. Put on the primers and seat with the primer seater. Remove the shell, fill with powder, well shaken down, preferably loaded through a tube, so that the powder shall be even full in the shell. Place the ball evenly on the shell, with the shell in the anvil, shallow cup upward. Place over the crimper and drive down to the shoulder, preferably with a wooden mallet.

Great care should be used in preparing cartridges of an even length and in every case the crimper should be driven down to the shoulder.

We will guarantee our rifles in every case when the cartridges are re-loaded with the above tools, according to these directions.

Vernier Peep Sight. Price, $8.00.

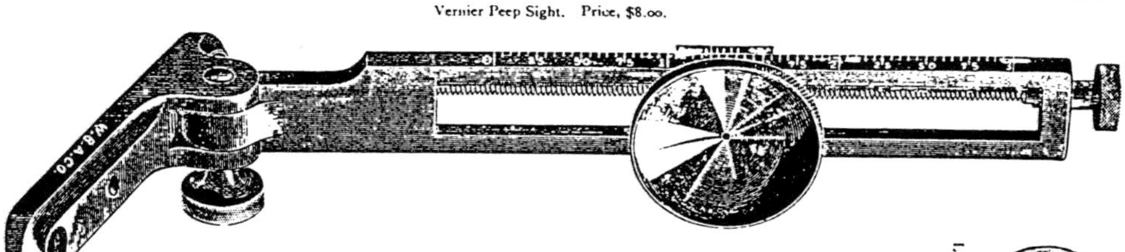

Improved Vernier Mid-Range Peep Sight. $6.

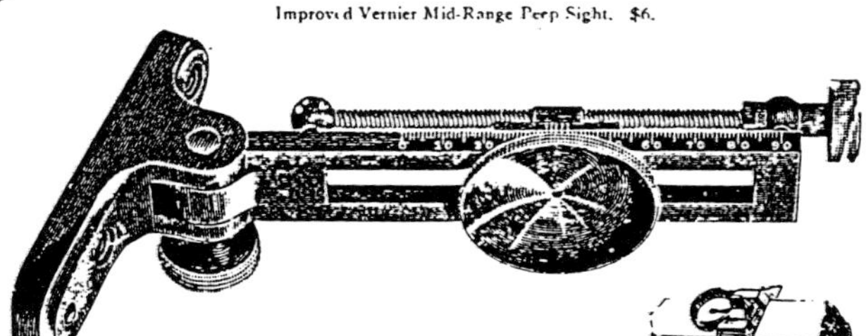

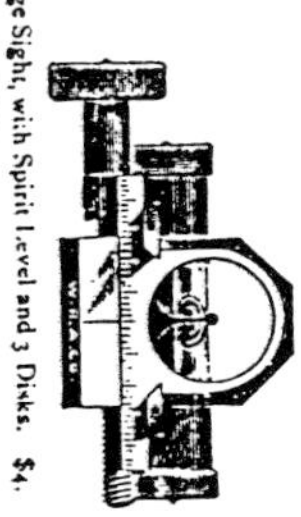

Wind Gauge Sight, with Spirit Level and 3 Disks. $4.

Beach Front Sight. $2.

Lyman Ivory Front. $1.

Lyman Rear Sight. $5.

Graduated Peep Sight. $3.

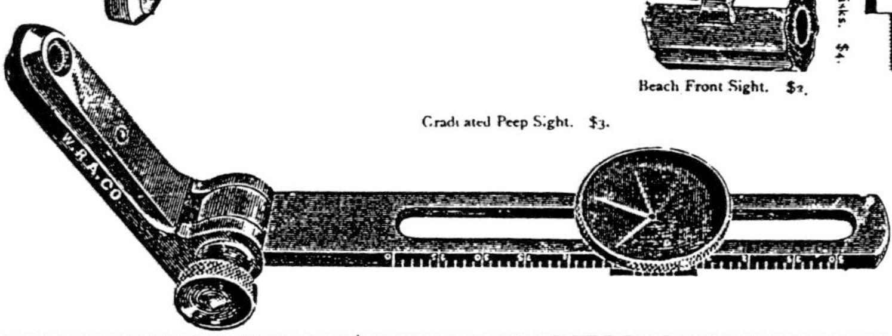

BULLARD REPEATING ARMS CO.

SPRINGFIELD, MASS., U. S. A.

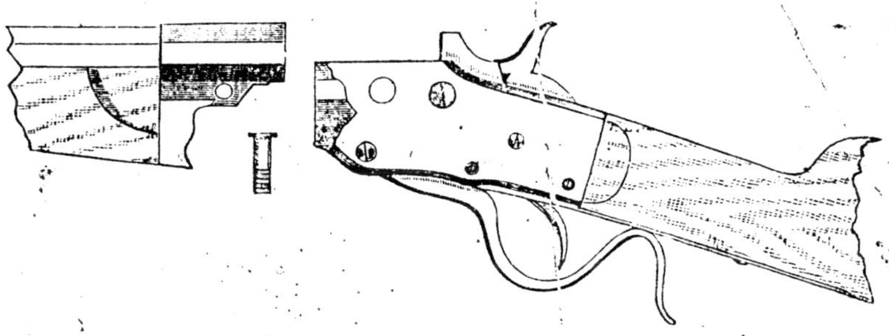

NEW SINGLE-SHOT DETACHABLE BARREL RIFLE

We take pleasure to announce to riflemen for both hunting and target purposes, a new Detachable Barrel Rifle, simple in construction, strong in its parts and system and easily taken apart. It will be seen by reference to the above cut that the barrel is screwed into a sleeve on lug where it permanently remains The tip-stock is screwed to lug and barrel where it remains. The frame is cut open and dovetailed to receive the lug and barrel The frame and lug are fastened together by a screw. To take off barrel, remove the screw, and draw the barrel, lug and tip stock from the frame It will be seen by sliding the lug into the frame on a dovetail that the barrel is securely held and is immovable The objection to Detachable Barrel Rifles, heretofore has been the looseness on the bolts or screws rendering them not reliable for fine shooting. We are prepared to make these rifles in cals. 22, 32 and 38. The 22 cal. to take the rim or center fire. The 32 for 32, 40, 185 patched or 165 grooved bullets The 38, for 55, 255 patched and grooved bullets, and 38, 45, 190 grooved bullets. These rifles will be made with interchangeable barrels when desired We shall soon be prepared to receive orders for the 40, 45 and 50 cals. The 22 cal. standard gun will weigh about 7 lbs., 24 in. octagon barrel, rubber butt and open sights.

The 32 and 38 cal. standard gun will weigh 8 to 8½ lbs., ½ oct. barrel 28 in., rubber butt and open sights.

Pistol grip, fancy stocks and checked Case Hardening and Nickel Plating when ordered.

The following recent scores have been made with the Bullard Single Shot Rifle:

E. F. Richardson, at Lawrence, Mass., December 18, 1886, 200 yds. off-hand, Hinman Target.

10- 9-10-10- 8-10- 9-10-10-10- 9—95, possible 100.

A. C. White, January 1, 1887, at Worcester, Mass., in 100 shot match.

8- 6- 7- 6- 8- 9-10- 7- 8- 8—77
8-10-10- 8-10- 7- 9- 7- 9-10—88
9- 8- 8- 9- 7-10- 9-10-10-10—90
10- 7- 6-10- 7- 7- 8- 9- 7- 8—79
6-10- 6- 8- 8- 8-10-10- 8-10—84
9- 6- 8- 8- 9- 6- 8-10-10- 6—80
8- 9-10- 8- 6-10- 6- 9- 7-10—83
7- 6- 7- 9-10- 7- 7- 8- 6- 7—74
9- 7-10- 9- 9-10- 9- 7- 7- 6—83
7- 8- 7- 9- 9- 9- 9- 5- 7- 7—77—**815**, possible 1000.

E. F. Richardson, Lawrence, Mass., November 25, 1886, in a 50 shot match, 200 yds. off-hand, Hinman Target.

8-10- 9- 8- 9- 7- 6- 9-10-10—86
9- 8- 8- 6- 8-10- 8-10- 8-10—85
9-10- 9- 7- 8-10-10- 8-10- 9—90
8-10- 9- 9-10- 6- 7- 9- 8- 9—85
10- 6- 9-10- 9- 8- 9- 9- 7- 9—86—**432**, possible 500.

These scores so far as known are the best on record. We guarantee all our rifles to be first class in every particular or may be returned. Send for price list and catalog.

Springfield, Mass., January 15, 1887.

Plate 9-7. This one-page flyer printed on both sides was dated January 15, 1887, and predates the rifle advertisement of February 1887 for this type of rifle. This advertisement would also predate the 1887 catalog reproduced next in this chapter, which is the first known catalog to show the detachable-interchangeable barrel model.

As can be seen in this flyer the Bullard Company was prepared to bore its barrels for either paper-patched or grooved bullets. When a barrel was rifled for a grooved (greased) bullet, the rifling started right at the end of the chamber, whereas when chambered for a paper-patched bullet, there was an unrifled gap between the end of the cartridge chamber and the start of the rifling. *Courtesy Ed Curtis*

The fourth catalog is dated 1887 and is reproduced complete. Its format too, was 8x6 inches with a green cover. This fourth catalog covered the complete Bullard line of rifles with the exception of the "Hindley" and Schuetzen model with the quickly detachable triggers.

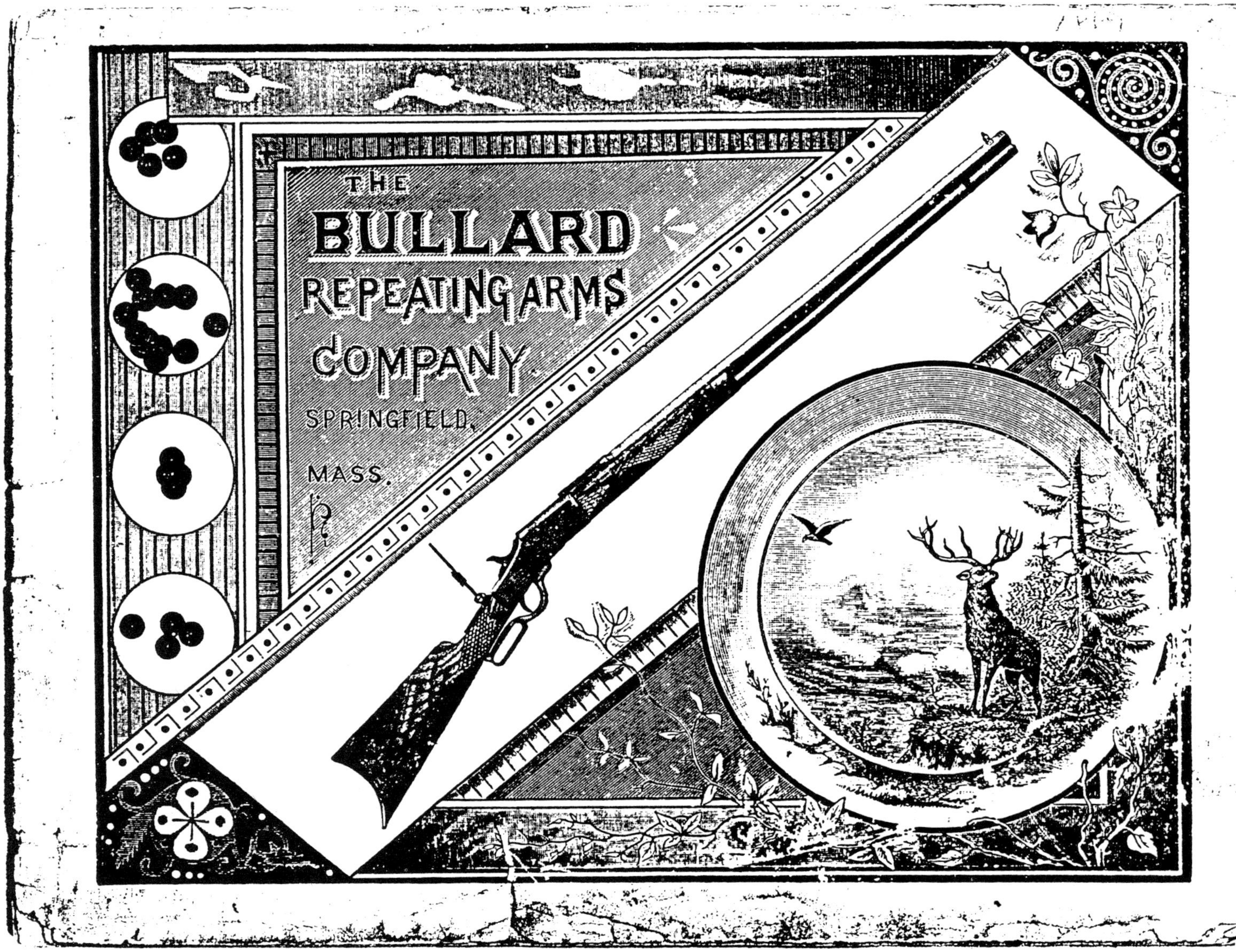

Plate 9-8. The fourth and final known Bullard catalog, dated 1887. This catalog represents the complete Bullard line of firearms with the exception of the detachable-trigger model Schuetzen rifle. *Courtesy G. Scott Jamieson*

INDEX.

BULLARD
REPEATING FIRE ARMS.

RIFLED MUSKETS,

CARBINES, HUNTING AND TARGET RIFLES,

FOR MILITARY AND SPORTING USE.

UNEQUALLED FOR

SAFETY, EXCELLENCE OF MATERIAL, WORKMANSHIP, FORCE, ACCURACY OF FIRING, SIMPLICITY OF CONSTRUCTION AND CONVENIENCE OF LOADING.

MANUFACTURED BY THE

BULLARD REPEATING ARMS CO.,

SPRINGFIELD, MASS., U. S. A.

1887.

2 BULLARD REPEATING ARMS COMPANY.

✦TERMS:✦

NET CASH in all cases where no special contract is made.

C. O. D. ORDERS from parties not known to us, must be accompanied by thirty per cent. of amount of goods ordered, and the express company's charges for returning the amount collected will be at the expense of the purchaser.

Special terms for large contracts.

Parts can be sent by mail not to exceed 4 pounds weight.

SPRINGFIELD, MASS., U. S. A.

BULLARD REPEATING ARMS COMPANY. 3

IN issuing this Catalogue we desire to call especial attention to our recent improvements in Magazine Rifles, also to the new Detachable and Interchangeable Barrel single shot Rifles. By reference to figure 1, (Section of Magazine Gun after Firing,) it will be seen that the breech block A is held firmly in place by brace C, which is pivoted and cradled on hammer-pin and lock-frame; C is brought into place and firmly held by links B, (only one shown) which are connected by strong pins to guard lever D. The guard-lever D engages the breech-block A at its lowest point, which cams on the said breech-block when closing the guard-lever, which forces the bolt forward and carries the cartridge into the chamber, the breech-block and guard-lever serving as an additional strength to the system. The power applied to the breech-block by the guard-lever is sufficient to force into the chamber a cartridge that may be a trifle large, and which could not be forced into the chamber by ordinary repeaters. The power thus applied we find to be sufficient, so that shells require no re-sizing, and may be used a greater number of times than when re-sizing is practiced. It will also be observed that it is impossible to disarrange the parts, as all are pinned together, and that there are no sliding surfaces.

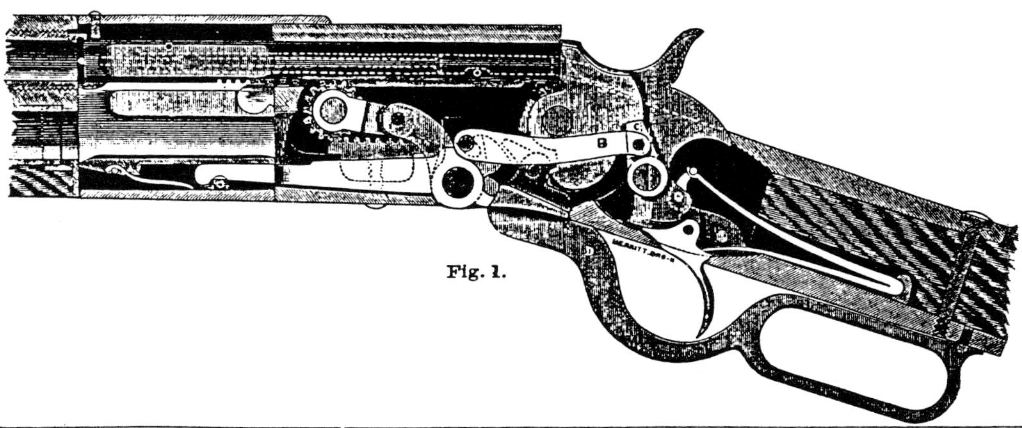

Fig. 1.

SPRINGFIELD, MASS., U. S. A.

For target purposes, the repeaters are unsurpassed, by using special ammunition as given under the various calibres. It is admitted by all riflemen, and it is a fact that a rifle using a large amount of powder and a light bullet is not the most accurate at the target, yet for hunting purposes, where a flat trajectory is required, the heavy powder charge and light bullet is of vital importance.

The single shot rifles are made for hunting and target shooting. The same ammunition can be used in the single shot rifles as in the repeater. The hunting and target rifles are made in all calibres, from 22 short to 50 cal., and chambered for any of the standard shells.

We guarantee our arms to be first-class in every particular, and superior to all others, for the following reasons:

1st. Safety.

The action is positive, not dependent on springs. It is self-locking, with a solid breech-block behind the bolt, which must be in place and securely locked before it is possible for the hammer to reach the firing-pin. Before placing this gun on the market, it was subjected to the most thorough tests as to its strength and endurance, and has stood the heaviest charges without the least sign of weakness.

2nd. Rapidity.

It is possible to fire the Bullard Rifle with greater rapidity than any other repeating rifle, from the fact that it works easier and smoother, by reason of its more direct leverage on the work to be done, the heaviest work being done with the best leverage, as in extracting the cartridge, which is started when the lever is in position to exert the greatest strain. Cocking the hammer is also done by direct leverage inside the receiver or frame, instead of a sliding motion of bolt or firing-pin on and over the top of hammer, which is very often liable to grind, and always makes the arm work hard and unpleasantly. The Bullard has been fired ten shots in five seconds, the best record of any other gun being eleven shots in seven seconds, using U. S. cartridges.

3rd. Loading.

The magazine is charged from the under side, and it can be done with equal facility by a right or left-handed person. And as there are no holes or spring covers on the side, it is not possible to have it clogged by passing through brush or laying it on the ground or in trenches, etc. It is also much easier to load on horseback than any other gun, as there is more choice of position than when the opening is on the side. It can be loaded as a single loader, either top or bottom, leaving the magazine full at all times for an emergency.

4th.

The Bullard is the only Lever Repeater that will successfully use the regular United States copper Cartridge (45 cal. 70 gr. powder), owing to its absolute certainty to extract the shell, which being made of copper, and with a folded head, does not contract after firing (as is the case of brass shells,) but often sticks in the chamber, a difficulty more frequently mentioned in reports of trials of breech-loading firearms by government experts than any other.

Another advantage in using the U. S. Cartridge is, that it is the standard ammunition, and can be obtained anywhere in the country. This point will be appreciated by frontiersmen.

5th. Workmanship.

The quality of work on the Bullard is pronounced by the best judges to be superior to any other repeating rifle in the world. The material used is the best throughout, no iron or malleable iron being used. All the parts are of forged steel except the carrier, which is composition or gun metal. Special care is taken in rifling and chambering the barrels.

6th. Range.

A bullet from the Bullard Rifle using the regular U. S. cartridge (45 cal. 70 gr. powder,) has at a distance of *3,000 yards* from muzzle of gun, penetrated a *three-inch spruce plank and eight inches of sand.*

7th. Guarantee.

Our confidence in the points enumerated above is such that, in case of failure ot any part by reason of defective material or workmanship, a duplicate will be forwarded free of charge upon receipt of the broken part. All parts are sure to fit.

3d.

To any responsible party we will send one of our guns with the privilege of a three days' trial and examination, on receipt of five dollars as guarantee of express charges.

9th.

We challenge the world to produce a repeater that can compare with ours in ease ot operation, safety, quality of workmanship, style, balance, finish, accuracy and rapidity of firing.

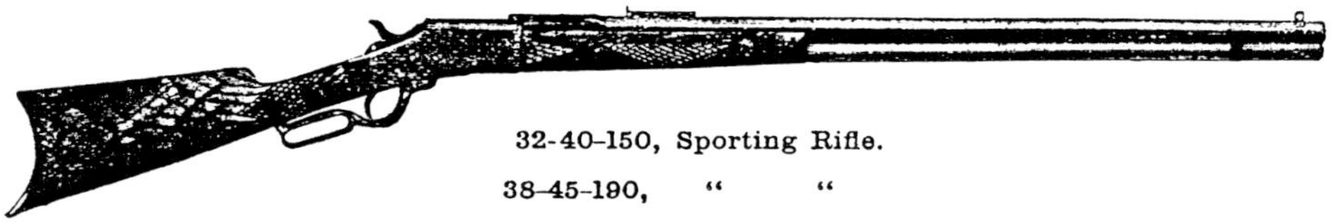

32-40-150, Sporting Rifle.

38-45-190, " "

Round barrel, 26 in. standard stock, rifle or shot butt, full or half magazine and open sights . . $33 00 .
Octagon or half-octagon barrel 35 00
Number of shots, 11; weight, 8 to 8½ lbs. Extras on page 16.

For Target practice, 200 yards, the 38 calibre Rifle is unexcelled, by using 35 grains of American F. G. Powder and the regular 190 grain bullet.

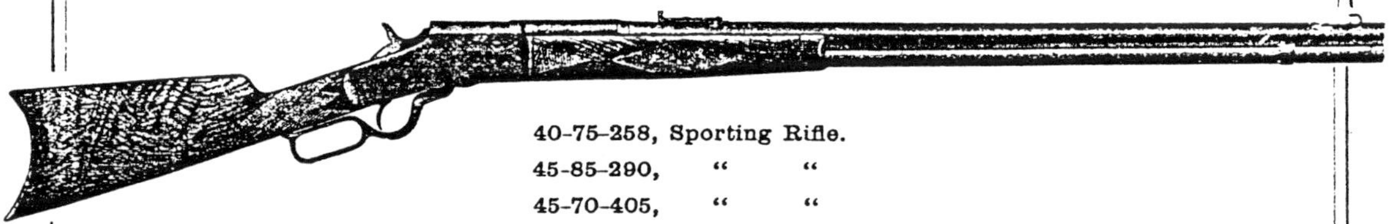

40-75-258, Sporting Rifle.

45-85-290, " "

45-70-405, " "

Round barrel, 28 in. standard stock, rifle or shot butt, full or half magazine and open sight . . . $33 00
Octagon or half-octagon barrel 35 00
Number of shots, 11; weight, 10 to 10½ lbs. Extras on page 16.

For Target practice, 200 yards, the 40 calibre Rifle is first-class, and by using 55 grains American F. G. powder and a 300 grain bullet the most satisfactory results are obtained. The Marlin 40-60-260 Cartridges can be used in this calibre rifle, being of the same length as 40-75-258 Bullard.

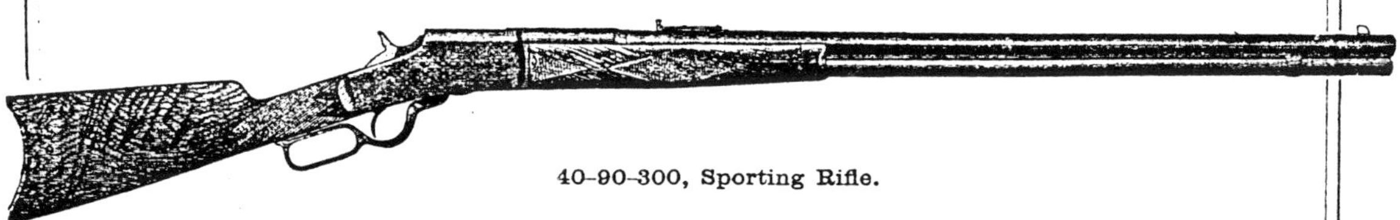

40-90-300, Sporting Rifle.

Round barrel, 28 in. standard stock, rifle or shot butt, full or half magazine and open sights . . $35 00
Octagon or half-octagon barrel 37 00
Number of shots, 11; weight, 10 to 10½ lbs. Extras on page 16.

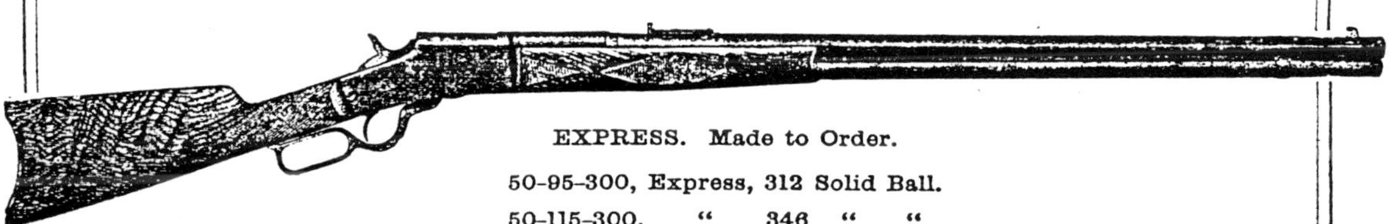

EXPRESS. Made to Order.

50-95-300, Express, 312 Solid Ball.

50-115-300, " 346 " "

50-115 Express, 28 in. round barrel, full or half magazine, rifle or shot butt and open sights . . . $45 00
50-95 Express, 28 in. round barrel, full or half magazine, rifle or shot butt and open sights . . . 40 00
Extras on page 16.

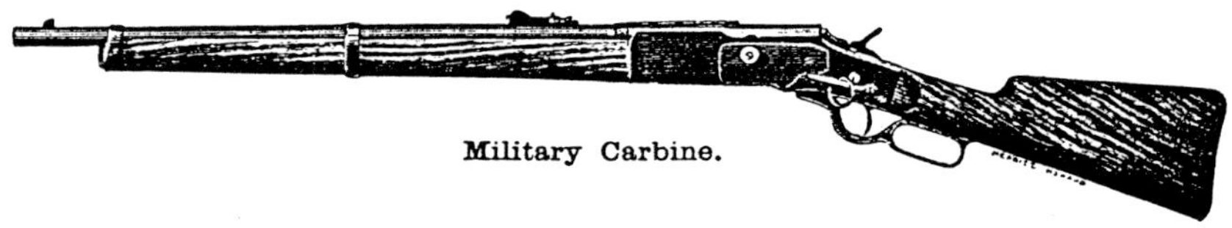

Military Carbine.

Military Magazine Musket and Carbine.

Musket 28 in. barrel, full magazine, weight 10½ lbs.
Carbine 22 in. " " " " 9½ "

Adapted to the Government 45-70-405 cartridge.

The musket and carbine has a device on the left side of the frame for changing the arm from a magazine to a single loader. The change is quickly made by the sliding of a button. The magazine being filled can be held in reserve for an emergency, and the rifle used as a single loader.

As a single loader the rifle has been fired 30 times in a minute, and as a repeater ten shots in five seconds.

TERMS: Subject to special contract.

The military carbine and musket make their first appearance in catalog form.

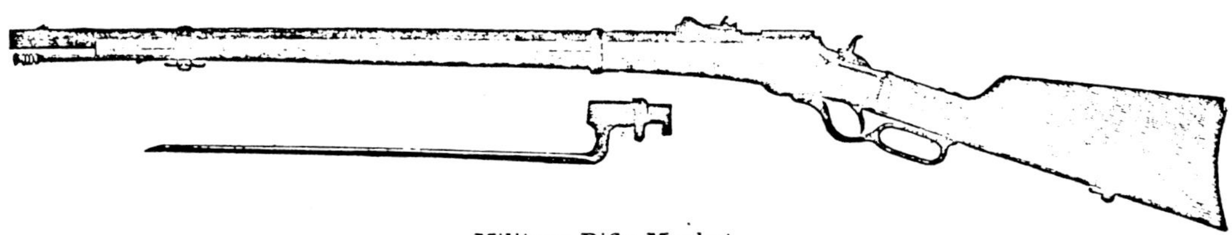

Military Rifle Musket.

Adapted to the U. S. Government service ammunition.

Weight, about . . . . 8½ lbs.
Length of barrel . . . . 32 inches.
Weight of carbine, 26 in. barrel, about . . . . 8 lbs.

TERMS: Subject to Special contract.

This rifle is the most rapid firing single loader made.

The same breech block, lever, hammer and trigger as used in repeater.

45-70-405 and 45-70-500.

Extreme range of bullet, 3,680 yards.

First mention in catalog form of the detachable and interchangeable barrel rifle model. Note that at issuance of this catalog this model was not available in the .40 or .,50 calibers.

12 BULLARD REPEATING ARMS COMPANY.

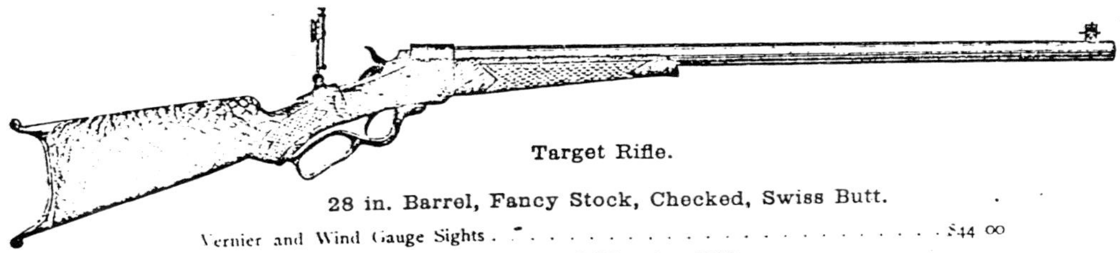

Target Rifle.

28 in. Barrel, Fancy Stock, Checked, Swiss Butt.

Vernier and Wind Gauge Sights $44 00

Target, Gallery and Hunting Rifles.

Cal. 32, 38, 40, 45 and 50. Ro., ½ oct. or oct. barrels, 28 in. or less, pistol grip, rifle butt, open sights, 8 to 10 lbs., . . $25.
Cal. 22, " " 26 in. " " " " rubber butt, open sights, 7 to 8 lbs., $25.

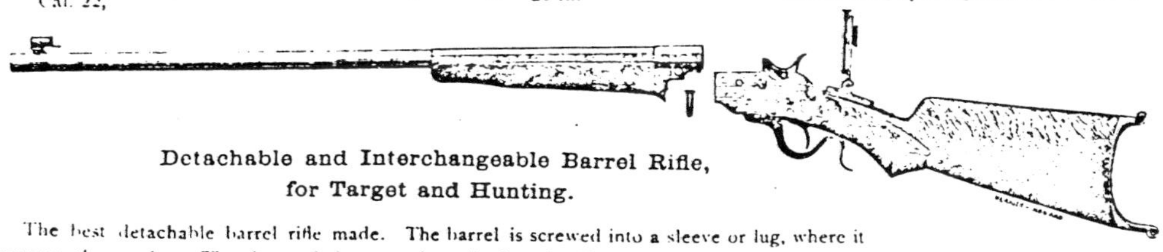

Detachable and Interchangeable Barrel Rifle, for Target and Hunting.

The best detachable barrel rifle made. The barrel is screwed into a sleeve or lug, where it permanently remains. The tip stock is secured to the barrel. The frame and lug are fastened together by a screw. To take off barrel, remove the screw and draw the barrel lug and tip stock from the frame. The lug on barrel enters the frame on a dovetail, and is held immovable by the screw. Calibres from 22 to 38 now ready. Calibres from 40 to 50 soon. Price, $20.00, without pistol grip, open sights, shot butt.

Calibre 22, 26 in., Octagon barrel . . . . 7 lbs. 32-38 calibre, 28 in., Octagon barrel . . . . . . 8 lbs.

Rim or center fire, pistol grip . . . . $25.

Extra barrel, complete . . . . 12.

22 Rim fire barrel will interchange with 32 and 38 calibre center fire. Extras on page 16.

SPRINGFIELD, MASS., U. S. A.

SINGLE TARGET AND HUNTING RIFLE.

The same-breech block, lever, brace and brace links are used in this Rifle as in the Repeater. The same forward movement of the lever operates the hammer and cocks it. The system is the strongest and safest that is made, and reference to Fig. 1, showing the system open, and Fig. 2, showing the system closed and the parts in position after firing, clearly indicate the safety of the Arm.

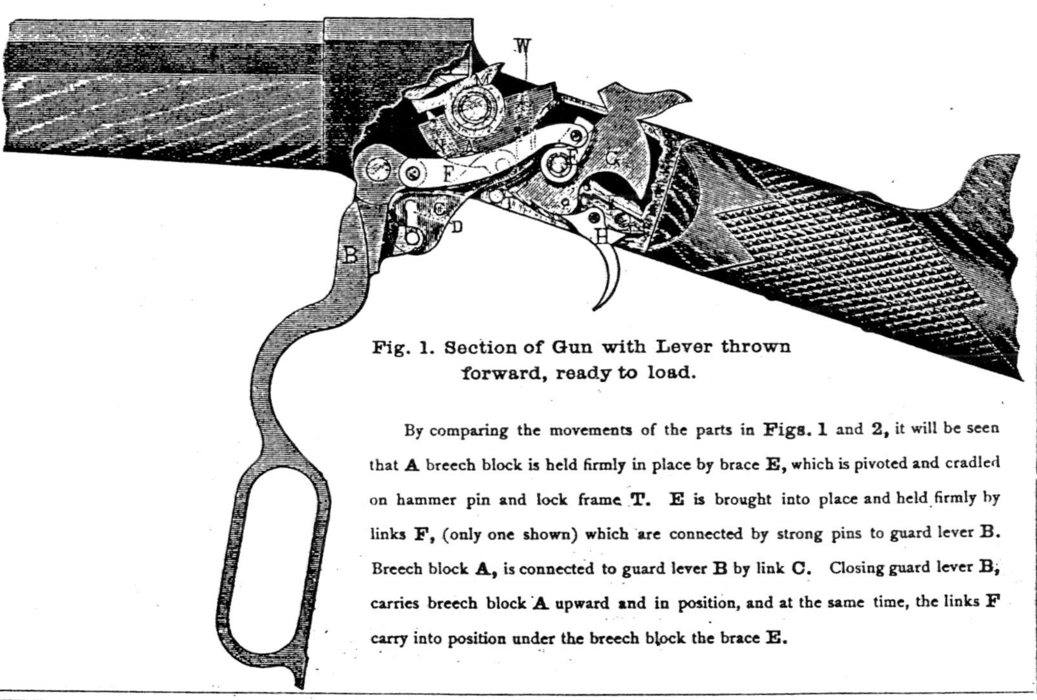

Fig. 1. Section of Gun with Lever thrown forward, ready to load.

By comparing the movements of the parts in **Figs. 1** and **2**, it will be seen that **A** breech block is held firmly in place by brace **E**, which is pivoted and cradled on hammer pin and lock frame **T**. **E** is brought into place and held firmly by links **F**, (only one shown) which are connected by strong pins to guard lever **B**. Breech block **A**, is connected to guard lever **B** by link **C**. Closing guard lever **B**, carries breech block **A** upward and in position, and at the same time, the links **F** carry into position under the breech block the brace **E**.

It will be seen by these cuts that it is impossible to get the hammer to strike the firing pin until the brace **E** is in its proper position under the breech block. It will also be observed that it is impossible to disarrange the parts, as they are pinned together, have no sliding surfaces and have positive motions.

To Dismount the Gun.

1. Take out tang screws and remove butt stock.
2. Half cock the hammer and take out lock frame and hammer screw.
3. Pull out lock frame — down and backwards.
4. Disconnect links from brace.
5. Take out breech block and guard lever screws.
6. Remove guard lever and breech block.

To Assemble the Gun.

Reverse operations as given to dismount the gun.

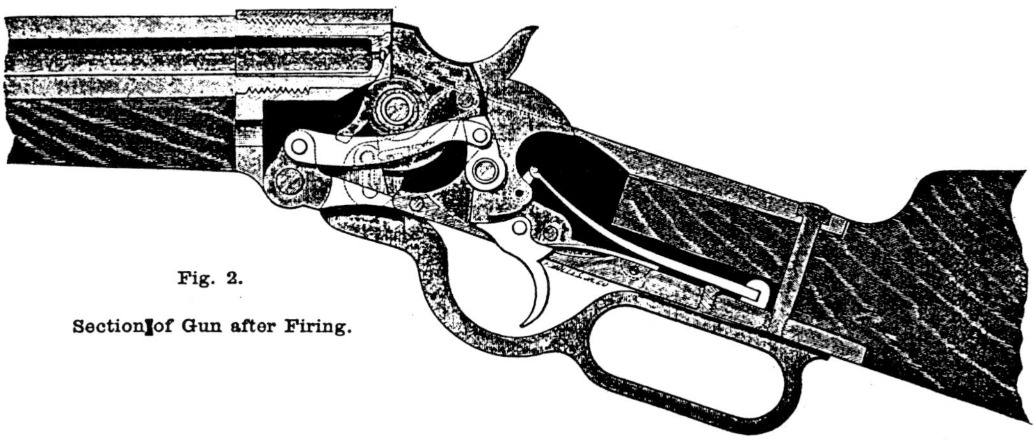

Fig. 2.

Section of Gun after Firing.

To Dismount the Magazine Rifle.

1. Take out tang screw.
2. Half cock hammer and take out lock-frame and hammer screws.
3. Pull out lock-frame down and backwards.
4. Disconnect links from brace.
5. Take out side plate screws and remove side plate.
6. Remove carrier lever spring.
7. Take out breech-block bolt.
8. Take out breech-block.
9. Take out extractor pin, and remove extractor; push back pin before removing bolt.
10. Take out guard lever bolt.
11. Draw out guard lever with its connections.
12. Draw out carrier lever.
13. Take out carrier.

Reverse Operations in Assembling,

Care being taken that the proper rack connections are made, and the latch lifted, until the lock-frame is replaced.

To Remove Barrel.

Take out magazine plug screw and the two tip screws, pull out the magazine tube and take off the fore-arm. If the bolt has not been previously removed, pull it back so as to prevent the breaking of the extractor while unscrewing the barrel.

Extras.

All deviations from standard styles will be subject to the following charges:

For additional length of barrel, add to price $1.00 per inch over regular lengths.

Butt stocks of special shape involving change either in length or drop from standard	$5 00
Engraving, according to style	from $5.00 to 100 00
Full nickel plating	5 00
Pistol grip stock	5 00
Fancy walnut stock and fore-arm	5 00
Checking grip and fore-arm	5 00
Fancy stock and fore-arm checked	10 00
Pistol grip stock and fore-arm fancy walnut checked	15 00
Case hardening	1 00
Swivels and sling straps	1 50
Canvas Covers	2 25
Stiff leather cases	6 00
Re-loading tools, comprising bullet moulds, re-loader and re-capper, for repeaters, per set,	5 00
Combination moulds for solid and hollow point bullets	3 50
Re- and de-capper for target rifles	1 25
Bullet moulds	1 50
Wiping rods	25
Brushes	25
Wad cutter	50
Ball seater	1 35
Powder Measure	15

We furnish with each magazine rifle a flexible wiper that can be carried in the pocket and which will thoroughly clean the arm. We do not think it wise to put an iron rod into a barrel to clean it, after we have taken the greatest possible care to have the grooves and lands of the barrel as smooth and accurate as it is possible for skill and machinery to make it. We have seen barrels ruined by using iron rods for cleaning, and the lands at the muzzle entirely worn away. Another serious objection is that it forces all the dirt and foulness into the mechanism of the breech, causing it to work hard, and if neglected will surely rust and injure the arm. The wiper mentioned above is inserted at the butt of the barrel, and all dirt is carried out at the muzzle, leaving the breech clean and free from dirt. If wiping with a rod is preferred, use a wooden one and clean the gun with an empty shell in the chamber. Target rifles we furnish rod and brush wiper.

To Load the Magazine Rifle.

Carry the guard lever forward as far as it will go, and then insert the cartridges in the magazine through the opening under the carrier.

To Use as a Single Loader Magazine Rifle.

Carry the lever as far forward as is necessary to extract the spent shell, remove the shell and insert another cartridge in the *chamber* of the gun; or,

Carry the guard lever as far forward as it will go, and insert a cartridge in the magazine through the opening under the carrier.

If the top cover is drawn back, the magazine can be quickly loaded by simply pressing against the under side of the carrier, which will force it up, exposing the entrance to the magazine, and also place the hammer at half cock.

Many sportsmen desire cartridges that can be re-loaded, and all shells, as soon as possible after being fired, should be cleaned and washed out carefully with strong soap suds or soda water and thoroughly dried, otherwise the deposit of nitric acid left on them after firing causes them to oxidize rapidly, and they become worthless. The primer should always be pressed down, so that it is sure to be lower than the head of the cartridge shell.

Use American Powder Co.'s Dead Shot,	F. G.
Hazard Powder Co. Sea Shooting,	F. G.
Orange Rifle (Laflin & Rand Powder Co.),	F. G.
American. Powder Mills Rifle Ctg. Powder,	$3.00

Use for bullets 1 part tin to 16 of lead, and keep the moulds very hot and handles of the moulds tight together. It often happens that the handles become too hot for comfort and are not held together properly. The result is a bullet that is not round and one that will not fit the shell. See that the grooves of the bullets are filled with beef tallow or Japan wax; wipe off all surplus grease before loading. When pressing the bullet into the shell, see that it is forced into the chamber of the re-loading tool as far as it will go.

We recommend that all sportsmen buy their bullets of cartridge manufacturers, as they are all swaged by powerful machinery, and are far more perfect than it is possible to cast them.

Parts of Bullard Repeating Rifle.

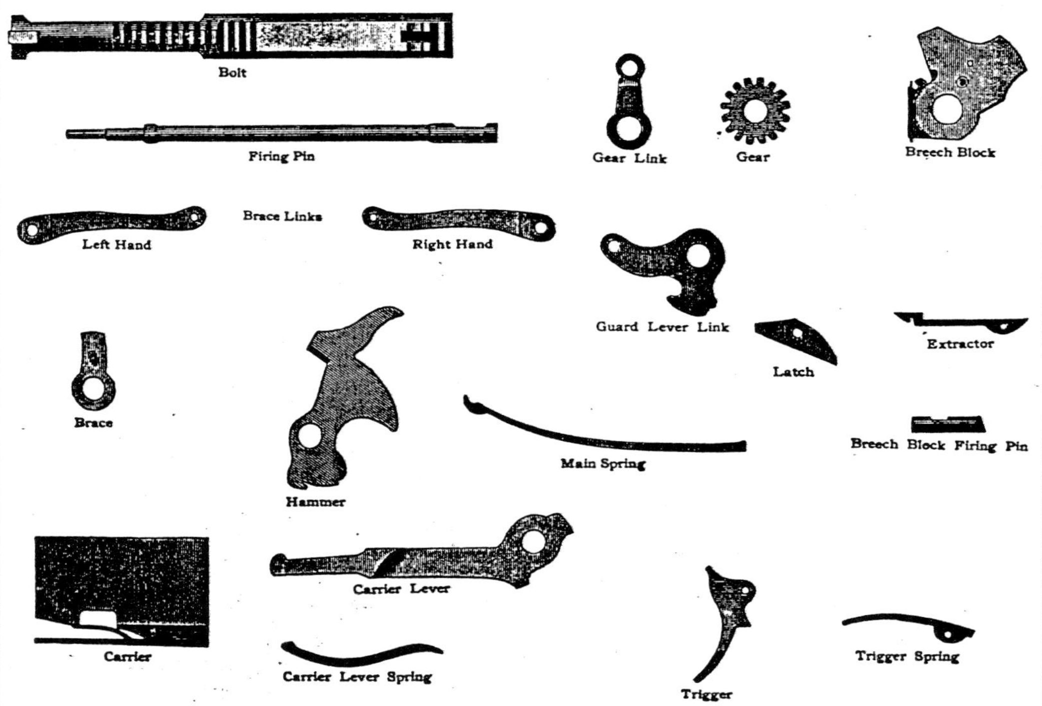

20 BULLARD REPEATING ARMS COMPANY.

Nomenclature and Price List of the Component Parts of the Bullard Repeating Fire Arms.

Barrel, round, half oct. or octagon	$10.00
Bolt	2.00
Bolt firing pin	.30
Bolt firing pin stop pin	.02
Bolt firing pin Spring	.05
Bottom cover	.35
Bottom cover spring	.25
Bottom cover spring pin	.02
Bottom cover pin	.02
Butt plate, rifle	1.00
Butt plate, shot	1.00
Butt plate screws, 2	.12
Butt stock, standard rifle	1.75
Brace link, right or left	.25
Brace	.50
Brace link pin, large or small	.05
Breech Block	1.50
Breech Block firing pin	.10
Breech block axis pin	.25
Breech block axis firing pin screw	.05
Breech block firing pin stop pin	.02
Breech block spring	.05
Carrier	2.00
Carrier lever	.50
Carrier lever spring	.25
Extractor	.30
Extractor pin	$ .02
Forestock	1.25
Gear	.15
Gear pin	.05
Gear link	.30
Gear link pin	.02
Guard lever	1.50
Guard lever, pistol grip	1.75
Guard lever latch	.35
Guard lever latch pin	.02
Guard lever latch spring	.02
Guard lever latch plunger	.05
Guard lever bolt	.10
Guard lever bolt screws	.05
Guard lever steady pin	.02
Guard lever link	.75
Hammer	.90
Hammer bushing	.10
Hammer screw	.08
Lock frame	3.00
Lock frame screws	.06
Lyman sight screws, 2	.10
Magazine tube	2.25
Magazine muzzle cap	.10
Magazine muzzle cap screw	.05
Magazine spring	.30
Magazine follower	$ .25
Magazine ring	.50
Magazine tenon	.10
Main spring	.50
Main spring, pistol grip	.60
Main spring strain screw	.05
Rack	.30
Rack screw	.06
Receiver	10.00
Side plate	1.25
Side plate screw	.05
Sling strap	.50
Stirrup	.20
Stirrup pin	.02
Swivel, for sling strap	.50
Tang screw	.10
Tip for forestock	1.00
Tip screws, 2	.10
Top cover	.50
Top cover screw	.05
Trigger	.50
Trigger, pistol grip	.60
Trigger pin	.02
Trigger spring	.40
Trigger spring, pistol grip	.50
Wiper, complete	.30

SPRINGFIELD, MASS., U. S. A.

Parts of Single Shot Rifle.

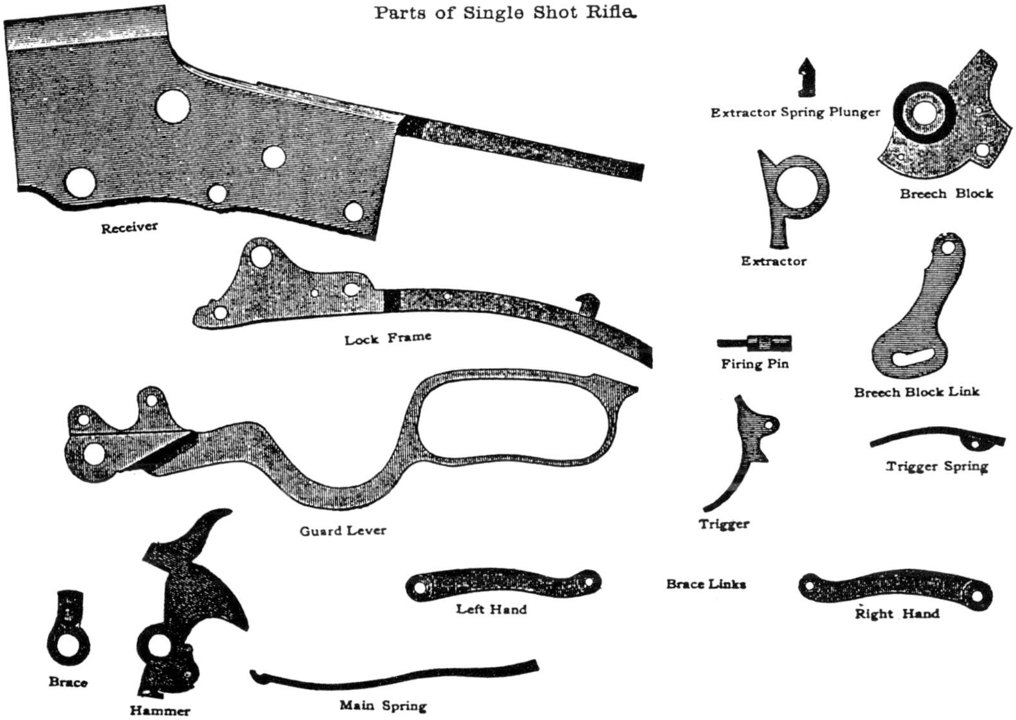

Nomenclature and Price List of the Component Parts of the Single Shot Rifle.

Barrel, 28 inches or under,	$10.00	Hammer,	$ .90
Brace,	.50	Hammer bushing,	.10
Brace links, right and left,	.25	Hammer screw,	.08
Breech block,	1.50	Lock frame,	3.00
Breech block bolt,	.25	Lock frame screw,	.08
Breech block link,	.35	Main spring,	.60
Buckhorn sight,	.80	Receiver,	5.00
Butt stock,	3.50	Rifle butt plate (steel) and screw,	1.00
Extractor,	1.00	Rifle butt plate, Swiss, nickel plated,	2.00
Extractor spring and plunger,	.10	Rubber butt plate and screws,	1.00
Firing-pin,	.15	Sight base screws, each,	.05
Firing-pin pin,	.05	Strain screw,	.05
Firing-pin spring,	.05	Tang screw,	.10
Forestock,	1.25	Trigger,	.60
Forestock screw,	.10	Trigger pin,	.05
Front sight,	.50	Trigger spring,	.40
Guard lever,	1.75		

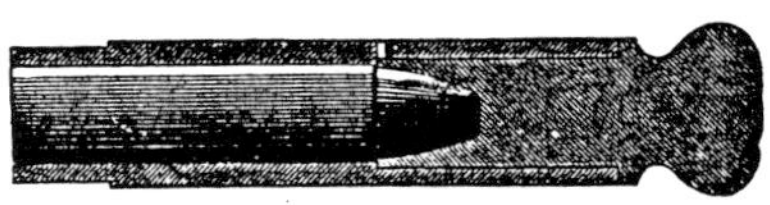

Loader and Crimper.

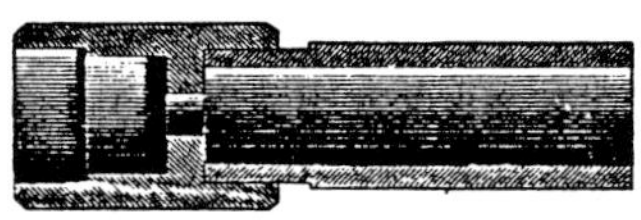

Anvil. Sizing Die.

Plunger.

Primer Seater.

Directions for Re-loading Cartridges with the Bullard Re-loader.

To prepare the shell. Wash the fired shells in strong soap suds or soda water, and dry thoroughly. Place the fired shell in sizing die. Place sizing die and shell in anvil with the shallow cup upward. Strike the sizing die until the shell is forced completely in. Turn over anvil, put plunger into the shell, which will extract shell from the sizing die and expel the exploded primer.

Now put this reduced shell into the loader with the head end up. Put on the primers and seat with the primer seater. Remove the shell, fill with powder, well shaken down, preferably loaded through a tube, so that the powder shall be even full in the shell. Place the ball evenly on the shell, with the shell in the anvil, shallow cup upward. Place over the crimper and drive down to the shoulder, preferably with a wooden mallet.

Great care should be used in preparing cartridges of an even length, and in every case the crimper should be driven down to the shoulder.

We will guarantee our rifles in every case when the cartridges are re-loaded with the above tools, according to these directions.

Bullet moulds	$1 50
Re-loading tools	3 50
Combination bullet moulds for solid and hollow point bullets	3 50

Cartridges for Bullard Repeating Rifles.

32-40-150

Price per 1000 . . . . $31 50
Shells 20 00
Bullets 7 25

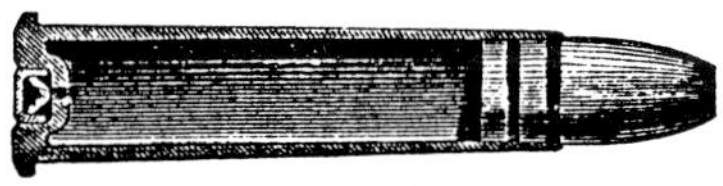

38-45-190, lead.

Price per 1000 . . . . $31 50
Shells 20 00
Bullets 8 00

40-75-258. 40-60-260.

Price per 1000 . . . . $33 00
Shells 18 00
Bullets 9 00

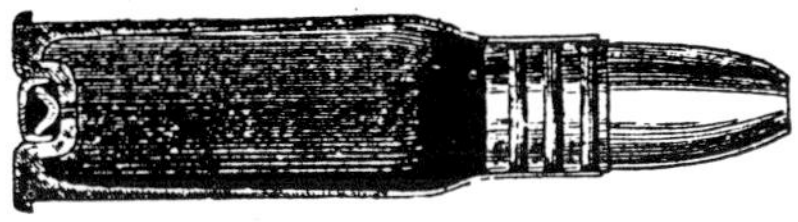

40-90-300.

Price per 1000 . . . . $40 00
Shells 22 00
Bullets 10 50

Cartridges for Bullard Repeating Rifles.

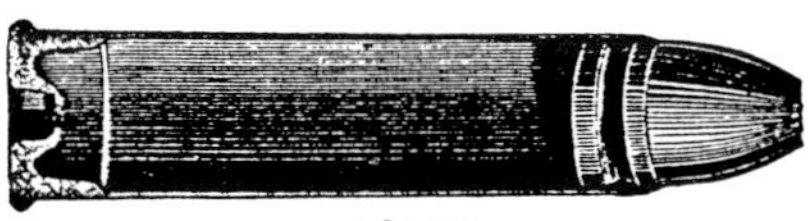

45-85-290.

Price per 1000 . . . . $37 00
Shells 22 00
Bullets 9 50

45-75-350.

Price per 1000 . . . . $35 00
Shells 20 00
Bullets 9 50

45-60-300.

Price per 1000 . . . . $33 00
Shells 20 00
Bullets 9 00

45-70-405.

Price per 1000 . . . . $35 00
Shells 22 00
Bullets 12 00

50-95-312 solid ball. 50-95-300 Express.

Price per 1000 . . . . $40 00
Shells 23 00
Solid ball 9 00
Express ball 11 00

50-115-346 solid ball. 50-115-300 Express.

Price per 1000 . . . . $90 00
Shells 50 00
Solid ball 9 00
Express ball 11 00

Cartridges for Bullard Single Shot Rifles.

RIM FIRE.

Price per 1000, $5.00.

Price per 1000, $6.00.

Price per 1000, $9.00.

Price per 1000, $10.00.

Price per 1000, $11.50.

Price per 1000, $16.50.

Price per 1000, $16.00.

Price per 1000, $18.00.

Price per 1000, $25.00.

CENTER FIRE.

Price per 1000, $11.00.
Primed Shells per 1000, $6.00.

Price per 1000, $12.00.
Primed Shells per 1000, $6.50.

Price per 1000, $18.00.
Primed Shells per 1000, $8.00.

Price per 1000, $13.50.
Primed Shells per 1000, $7.00.

Price per 1000, $14.50.
Primed Shells per 1000, $7.50.

Price per 1000, $23.00.
Primed Shells per 1000, $11.00.

Price per 1000, $13.00.

Price per 1000, $16.00.
Primed Shells per 1000, $9.00.

Price per 1000, $27.00.
Primed Shells per 1000, $15.00.

Price per 1000, $33.00.
Primed Shells per 1000, $18.00.

Everlasting Shells, 32-40 . . . . 6 cts. each.
" " 38-55 . . . . 7 cts. each.

The single shot Rifles are chambered also for the same cartridges as Repeaters.

Velocities.

32-40-150,	1492 feet per second.
38-45-190,	1388 " " "
40-75-258,	1513 " " "
45-85-290,	1548 " " "
40-90-300,	1569 " " "
50-115-300,	1539 " " "
50-115-346,	1539 " " "

Trajectory.

	Highest Point 100 Yards.	200 Yards Highest Point.
32-40-150,	2.443 in.	11.342 in.
38-45-190,		12.691 in.
40-70-230,	2.157 in.	10.503 in.
40-90-300,	2.4 in.	10.3 in.
45-85-290,	2.3 in.	10.4 in.
50-115-346,	1.817 in.	10.289 in.
50-115-300, Express,	2.047 in.	10.493 in.

Penetration of Bullets.

32-40-150,	15 inches solid pine, 50 yards.
38-45-190,	14 " " " 50 "
40-75-258,	19 " " " 50 "
40-90-300,	16 " " " 50 "
45-70-500,	21 " " " 50 "

SIGHTS FOR TARGET RIFLES.

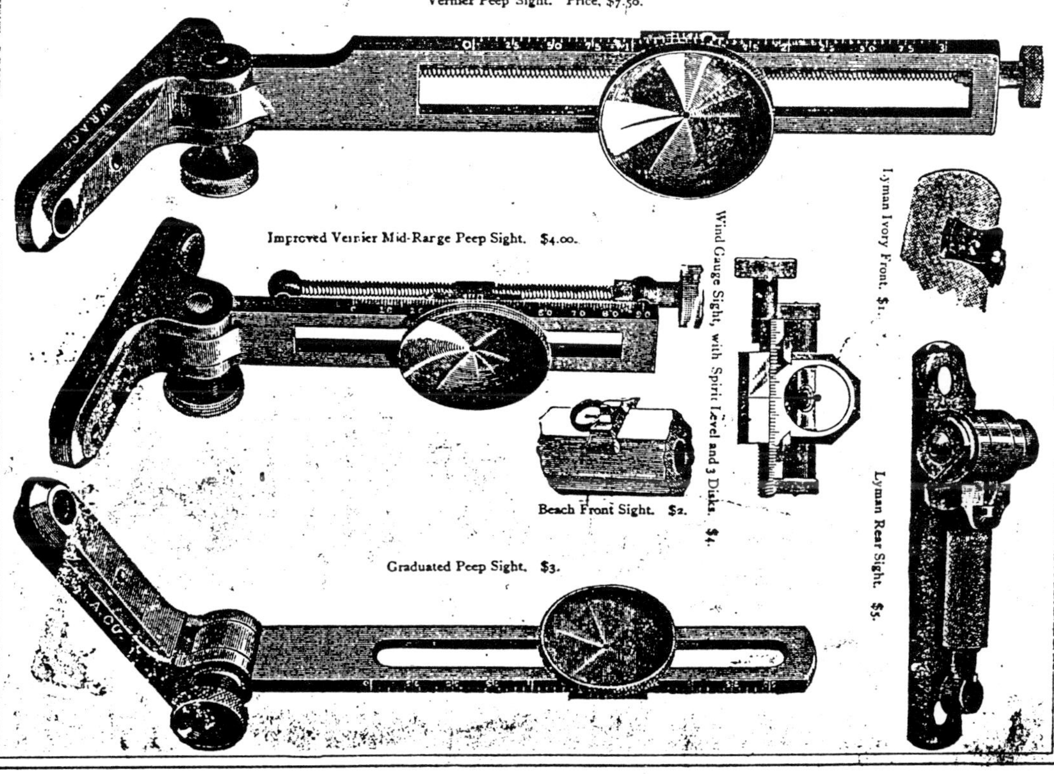

The following four pages were contained in one brochure 8½x11 inches in format. This brochure lists the complete Bullard line of firearms and cartridges then available. It is noteworthy for showing all the military designs with the exception of the "Hindley" rifle and for showing the Schuetzen rifle with quickly detachable triggers, which is first mentioned in the May 1887 issue of *The Rifle*, page 8. This is chronologically the last documentable piece of factory literature uncovered, with the exception of magazine advertisements. This piece was reproduced through the courtesy of John Hintlian.

The Schuetzen rifle in this advertisement is of the type shown in Plate 9-18.

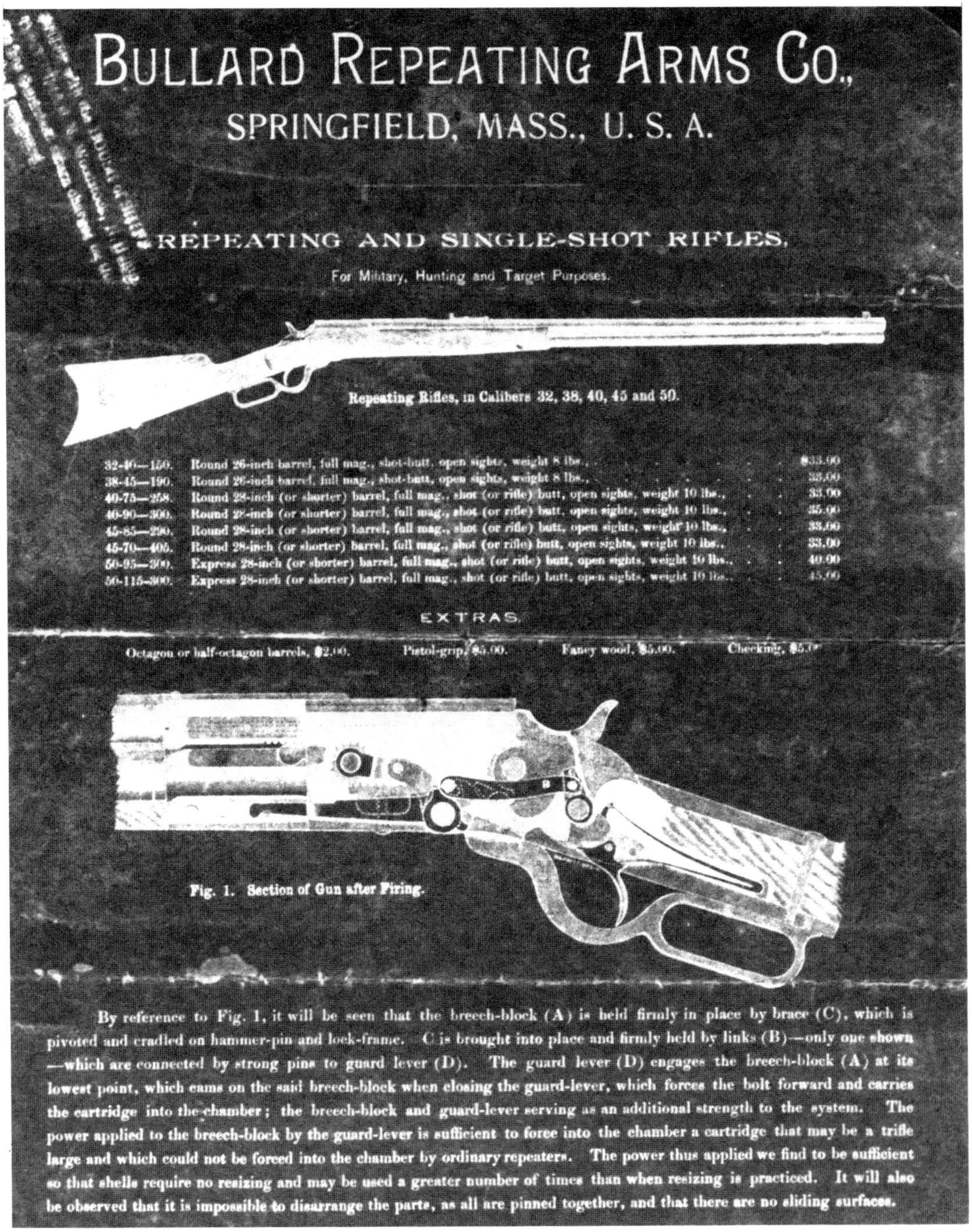

BULLARD REPEATING ARMS CO.,

SPRINGFIELD, MASS., U. S. A.

REPEATING AND SINGLE-SHOT RIFLES,

For Military, Hunting and Target Purposes.

Repeating Rifles, in Calibers 32, 38, 40, 45 and 50.

32-40—150.	Round 26-inch barrel, full mag., shot-butt, open sights, weight 8 lbs.,	$33.00
38-45—190.	Round 26-inch barrel, full mag., shot-butt, open sights, weight 8 lbs.,	33.00
40-75—258.	Round 28-inch (or shorter) barrel, full mag., shot (or rifle) butt, open sights, weight 10 lbs.,	33.00
40-90—300.	Round 28-inch (or shorter) barrel, full mag., shot (or rifle) butt, open sights, weight 10 lbs.,	35.00
45-85—290.	Round 28-inch (or shorter) barrel, full mag., shot (or rifle) butt, open sights, weight 10 lbs.,	33.00
45-70—405.	Round 28-inch (or shorter) barrel, full mag., shot (or rifle) butt, open sights, weight 10 lbs.,	33.00
50-95—300.	Express 28-inch (or shorter) barrel, full mag., shot (or rifle) butt, open sights, weight 10 lbs.,	40.00
50-115—300.	Express 28-inch (or shorter) barrel, full mag., shot (or rifle) butt, open sights, weight 10 lbs.,	45.00

EXTRAS.

Octagon or half-octagon barrels, $2.00. Pistol-grip, $5.00. Fancy wood, $5.00. Checking, $5.00

Fig. 1. Section of Gun after Firing.

By reference to Fig. 1, it will be seen that the breech-block (A) is held firmly in place by brace (C), which is pivoted and cradled on hammer-pin and lock-frame. C is brought into place and firmly held by links (B)—only one shown—which are connected by strong pins to guard lever (D). The guard lever (D) engages the breech-block (A) at its lowest point, which cams on the said breech-block when closing the guard-lever, which forces the bolt forward and carries the cartridge into the chamber; the breech-block and guard-lever serving as an additional strength to the system. The power applied to the breech-block by the guard-lever is sufficient to force into the chamber a cartridge that may be a trifle large and which could not be forced into the chamber by ordinary repeaters. The power thus applied we find to be sufficient so that shells require no resizing and may be used a greater number of times than when resizing is practiced. It will also be observed that it is impossible to disarrange the parts, as all are pinned together, and that there are no sliding surfaces.

Plate 9-9.

We guarantee our Arms to be First-class in every particular, and superior to all others, for the following reasons:

1st, Safety. The action is positive, not dependent on springs. It is self-locking with a solid breech-block behind the bolt, which must be in place and securely locked before it is possible for the hammer to reach the firing-pin. Before placing this gun on the market it was subjected to the most thorough tests as to its strength and endurance, and has stood the heaviest charges without the least sign of weakness.

2d, Rapidity. It is possible to fire the Bullard Rifle with greater rapidity than any other repeating rifle, from the fact that it works easier and smoother by reason of its more direct leverage on the work to be done, the heaviest work being done with the best leverage, as in extracting the cartridge, which is started when the lever is in position to exert the greatest strain. Cocking the hammer is also done by direct leverage inside the receiver or frame instead of a sliding motion of bolt or firing-pin on and over the top of hammer, which is very often liable to grind and always makes the arm work hard and unpleasantly. The Bullard has been fired twelve shots in five seconds, the best record of any other gun being eleven shots in seven seconds, using U. S. cartridges.

3d, Loading. The magazine is charged from the under side, and it can be done with equal facility by a right or left-handed person. And as there are no holes or spring covers on the side, it is not possible to have it clogged by passing through brush or laying it on the ground or in trenches, etc. It is also much easier to load on horseback than any other gun, as there is more choice of position than when the opening is on the side. It can be loaded as a single loader, either top or bottom, leaving the magazine full at all times for an emergency.

4th. The Bullard is the only Lever Repeater that will successfully use the regular United States copper cartridge (45 cal. 70 gr. powder), owing to its absolute certainty to extract the shell, which, being made of copper and with a folded head, does not contract after firing (as is the case of brass shells, which often stick in the chamber), a difficulty more frequently mentioned in reports of trials of breech-loading fire-arms by government experts than any other.

The Bullard never fails.

Another advantage in using the U. S. cartridge is, that it is the standard ammunition and can be obtained anywhere in the country. This point will be appreciated by frontiersmen.

5th. Workmanship. The quality of work on the Bullard is pronounced by the best judges to be superior to any other repeating rifle in the world, and the fact that Mr. Bullard, the inventor and manager at our manufactory, was for five years master mechanic at the well-known Smith & Wesson works, is a sufficient guarantee that our standard is high and will be kept so. The material used is of the best throughout, no iron or malleable iron being used. All the parts are of forged steel except the carrier, which is composition or gun metal.

6th, Range. A bullet from the Bullard Rifle, using the regular U. S. cartridge (45 cal. 70 gr. powder) has, at a distance of *3,000 yards* from muzzle of gun, penetrated a *three-inch spruce plank and eight inches of sand.*

7th, Guarantee. Our confidence in the points enumerated above is such that, in case of failure of any part by reason of defective material or workmanship, a duplicate will be forwarded free of charge upon receipt of the broken part. All parts are sure to fit.

8th. To any responsible party we will send one of our guns with the privilege of a three days' trial and examination, on receipt of five dollars as guarantee of express charges.

9th. We challenge the world to produce a repeater that can compare with ours in ease of operation, safety, quality of workmanship, style, balance, finish, accuracy and rapidity of firing.

Military Magazine Musket.

Carbine.

Musket, 28-inch Round Barrels—weight, 10½ lbs. Carbine, 22-inch Round Barrels—weight, 9¼ lbs.

Adapted to the U. S. 45-70—405 cartridge.

Military Rifle Musket.

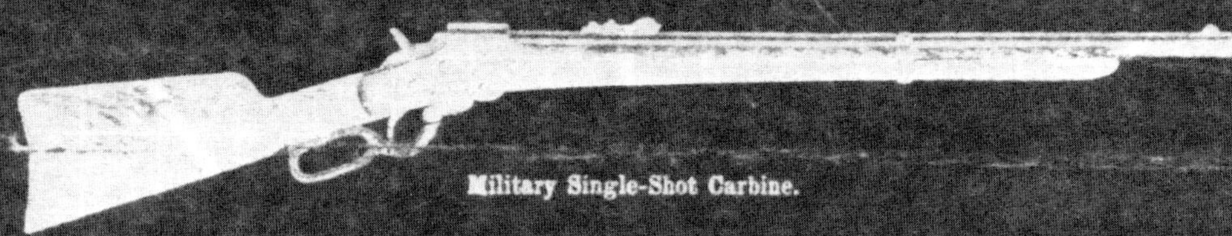

Military Single-Shot Carbine.

Adapted to the U. S. Government service, ammunition 45-70—405, and 45-70—500.

Musket, 32-inch round barrel—weight, 8½ lbs. Carbine, 26-inch round barrel—weight, 8 lbs.

The same breech-block, guard-lever, hammer and trigger, as used in Repeater.

Target Rifle—Solid Receiver

22 caliber, 26-inch one-half octagon barrel, pistol-grip, shot-butt and open sight. . . . . . . $25.00.

32 to 50 caliber, 28-inch one-half octagon barrel, pistol-grip, shot-butt and open sight. . . . . . . 25.00.

Detachable and Interchangeable Barrel Rifle for Target and Hunting.

The best detachable barrel rifle made. The barrel is screwed into a sleeve or lug, where it permanently remains. The tip stock is secured to the barrel. The frame and lug are fastened together by a screw. To take off the barrel, remove the screw and draw the barrel, lug and tip stock from the frame. The lug on barrel enters the frame on a dove-tail, and is held immovable by the screw. Calibers from 22 to 50.

22 caliber, one-half octagon 26-inch barrel, chambered for short, long or extra long rim-fire, center-fire, 22—9.45, and 22—15.45, pistol-grip, shot-butt, open sights, weight about 7 lbs, $25.00.

32, 38, 40, 45 and 50 caliber, 28 inch one-half octagon barrel, pistol-grip, shot-butt, open sights, $25.00. 32 and 38 caliber, rim or center-fire; 40, 45 and 50 caliber, center-fire. Eight to ten lbs., as desired.

EXTRAS.

Extra length of barrel, per inch,	$1.00	Cast hardening,	$1.00	Bullet moulds,	$1.50
Rifles over 10 lbs., per lb.,	1.00	Re and de-capper,	1.25	Combination moulds,	4.50
Fancy walnut stock and fore-arm,	5.00	Reloader,	1.35	Wiping rod,	.25
Checking,	5.00	Wad cutter,	.50	Brush,	.25
Swiss butt plate, nickel plated,	2.00	Powder measure,	.15	Reloading tools for Repeater,	3.50

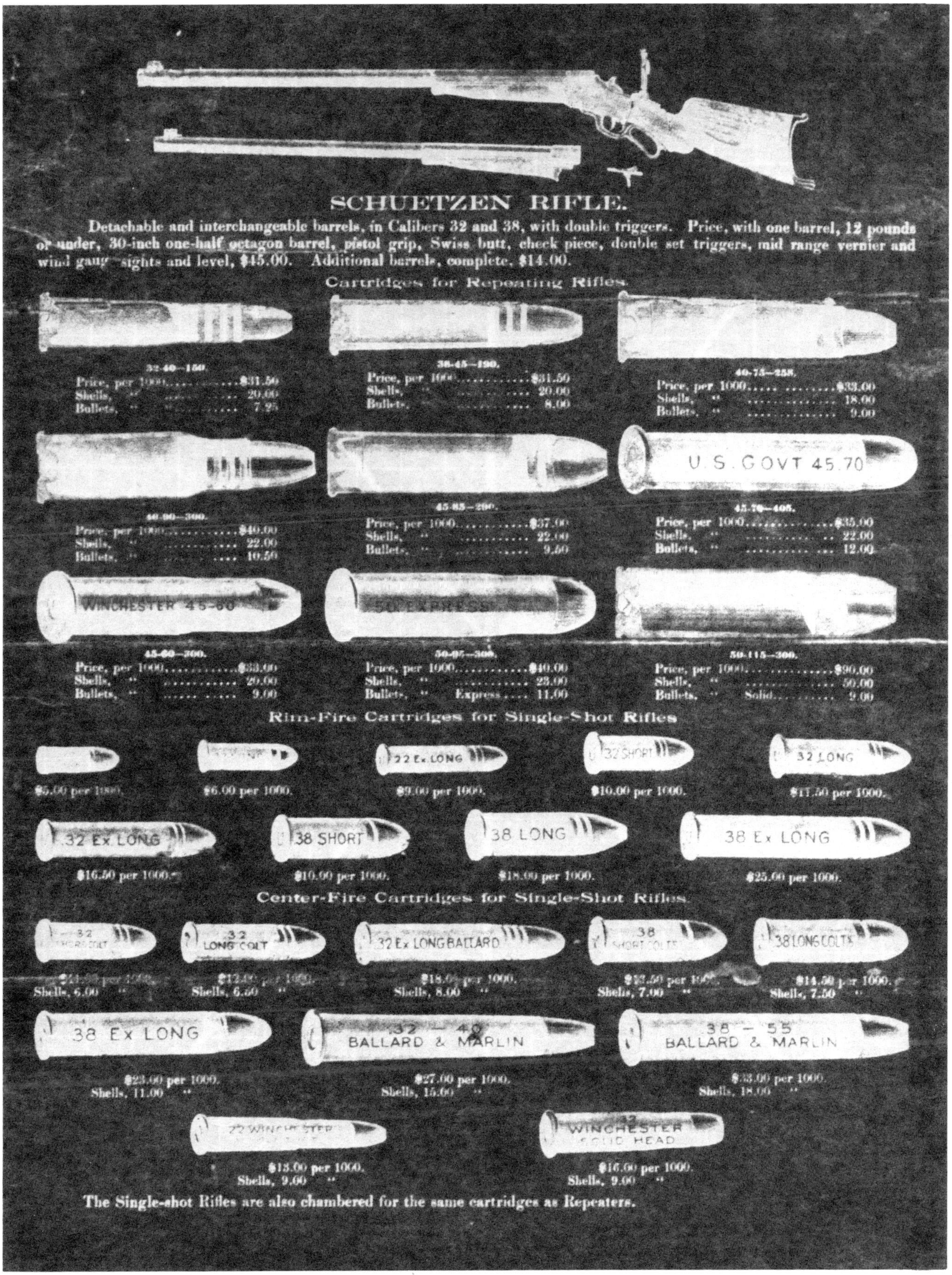

SCHUETZEN RIFLE.

Detachable and interchangeable barrels, in Calibers 32 and 38, with double triggers. Price, with one barrel, 12 pounds or under, 30-inch one-half octagon barrel, pistol grip, Swiss butt, cheek piece, double set triggers, mid range vernier and wind gauge sights and level, $45.00. Additional barrels, complete, $14.00.

Cartridges for Repeating Rifles.

32-40—150.
Price, per 1000..........$31.50
Shells, " 20.00
Bullets, " 7.25

38-45—190.
Price, per 1000..........$31.50
Shells, " 20.00
Bullets, " 8.00

40-75—258.
Price, per 1000..........$33.00
Shells, " 18.00
Bullets, " 9.00

40-90—300.
Price, per 1000..........$40.00
Shells, " 22.00
Bullets, " 10.50

45-85—290.
Price, per 1000..........$37.00
Shells, " 22.00
Bullets, " 9.50

U.S. GOVT 45.70

45-70—405.
Price, per 1000..........$35.00
Shells, " 22.00
Bullets, " 12.00

WINCHESTER 45-60

45-60—300.
Price, per 1000..........$33.00
Shells, " 20.00
Bullets, " 9.00

50 EXPRESS

50-95—300.
Price, per 1000..........$40.00
Shells, " 23.00
Bullets, " Express.... 11.00

50-115—300.
Price, per 1000..........$90.00
Shells, " 50.00
Bullets, " Solid.......... 9.00

Rim-Fire Cartridges for Single-Shot Rifles

$5.00 per 1000.

$6.00 per 1000.

22 Ex LONG
$9.00 per 1000.

32 SHORT
$10.00 per 1000.

32 LONG
$11.50 per 1000.

32 Ex LONG
$16.50 per 1000.

38 SHORT
$10.00 per 1000.

38 LONG
$18.00 per 1000.

38 Ex LONG
$25.00 per 1000.

Center-Fire Cartridges for Single-Shot Rifles.

32 SHORT COLT
$11.00 per 1000.
Shells, 6.00 "

32 LONG COLT
$12.00 per 1000.
Shells, 6.50 "

32 Ex LONG BALLARD
$18.00 per 1000.
Shells, 8.00 "

38 SHORT COLTS
$13.50 per 1000.
Shells, 7.00 "

38 LONG COLTS
$14.50 per 1000.
Shells, 7.50 "

38 Ex LONG
$23.00 per 1000.
Shells, 11.00 "

32 — 40 BALLARD & MARLIN
$27.00 per 1000.
Shells, 15.00 "

38 — 55 BALLARD & MARLIN
$33.00 per 1000.
Shells, 18.00 "

22 WINCHESTER
$13.00 per 1000.
Shells, 9.00 "

32 WINCHESTER SOLID HEAD
$16.00 per 1000.
Shells, 9.00 "

The Single-shot Rifles are also chambered for the same cartridges as Repeaters.

The following letter is very faint (in the original form), therefore it has been typeset to allow easier reading. The letter is dated January 27 or 29, 1887, and is addressed to a Mr. Jackson, Esq., Upper Canada College, Toronto, Canada:

> Dear sir,
>
> Yours of the 25th Rec'd. Enclosed find slip showing the manner of securing the barrel in the detachable model single shot rifle. The rifle shown is intended for a standard or straight stock, that is not pistol grip and with a lever as shown & is intended for a low price [?]. We shall make them pistol grip & put on the same lever as used in our repeaters when preferred and sell them at $25 with open sights. The plain [model, rifle?] shown in cut will be $20. We have thoroughly tested this plan of screwing the barrel to the sleeve or [?]. The frame is cut out so as to receive the lug on a dovetail. Frame and lug secured by a screw. We are now making these guns in calibers up to 38. It will require a wider frame in 40 cal. When we get the smaller cals well under [?] we shall make the larger ones 40-45 & 50. In two or three weeks we shall be ready to go on with 40, 45 & 50 cals. The detachable and interchangeable will be only single loaders. The weights [?] can be made from 7½ lbs to 10 lbs. We do not make gun cases but can get anything you desire & send catalog from [? & ?]. Should you find what you want in their list please mark and return & we will get what you wish. The 40 cal is one of the best shooting rifles we make. The ammunition 75.258 is very flat shooting, has a trajectory of 10 in 200 yds. & for hunting purposes is excellent. For target shooting 200 or 300 yds or over where the most accuracy is desired we prefer a lighter powder charge 55 or 60 grs and a heavier bullet 300 grs. A 28 (inch) barrel is sufficiently long for 600 yds. We should prefer a rifle to use 75.258 or 60 to 300 lead not to be less than 8 lbs. We have a 38.45.190 rifle that weighs 6 lbs. This we think too light as the tendency is for the muzzle to throw upward at each discharge. An 8 lb 40 cal 28 in bbl would be a good weight. We test all our rifles. We do not shoot them all for accuracy at 200 yds. but do 100 yds. We do not send out a rifle that we are not satisfied will make a good target 200 yd that is put 10 consecutive shots in an 8 inch ring. We put on Lyman sights & when desired make a target. The price of 40-75 single shot detachable barrel rifle with pistol grip stock, rifle or shot butt & open sights 28 in barrel weight from 8 to 10 lbs as desired $25. Lyman sight $5. You can with this rifle have a barrel to interchange of larger cal but not smaller. Should you [choose] a 38 cal rifle you can have interchangeable barrels 32 or 22 cal. The extra barrels with lug and tip stock complete will be $10. You would then have two rifles and one stock for $35. We shall be pleased to give you any information we can & to make you a rifle.
>
> Yours truly
> Bullard R A Co.

Mr. Jackson did indeed buy a rifle, a .40-75, fully cased (See Plate 6-28). This letter came with the rifle, including the catalog with Wm. M. Cooper's business address stamped on it, and the letter addressed to Mr. Cooper from the factory. (See next page)

The letter is most important to the collector for two reasons. Firstly, it tells us that the factory did not make cases for its rifles but they would facilitate the making of a case to a customer's specifications through an outside source. Secondly, that there were large- and small-frame versions of the detachable single-shot rifle models. The differences though unlike the repeater's were less noticeable and according to the letter involved width of the receiver only. The author would like to hear from someone who has both a large-caliber and small-caliber detachable- barrel model single-shot rifle in order to compare widths and sizes of frames. Possibly these differences also occur on the solid-frame single-shots, but photographs have failed to bring this new discovery to light. This means that, like the repeaters, the small-frames were available in calibers up to and including .38 and the large-frames were available from .40 to .50 caliber.

Nov. 10, 1887
Mr. Wm. M. Cooper — Toronto, Canada

Dear Sir,

The detachable 40 cal which we sent to Qummbys [?] in N.Y. for a case was sighted before sending it. Please tell Mr. Jackson that for 100 yds it takes 6½ points elevation on Lyman rear, for 200 yds 10 points, 300-14, 400-18 & 500 yds 23 points, that is as high as it will go for regular measurement. We put a leaf sight on the rear of barrel. When this sight is elevated for 200 yds it will shoot 100 yds and so on subtract one from the number on sight and it will give distance. Unless your customer is well used to those distances, it will require a little practice to decide on them and he ought to indulge in a little target practice before trusting entirely to judgement,

Yours Truly
Bullard Repeating Arms Co.,
Field.

This letter was provided through the generosity of Ed Curtis. Mr. Cooper was the dealer and Canadian headquarters of the Bullard Repeating Arms Company.

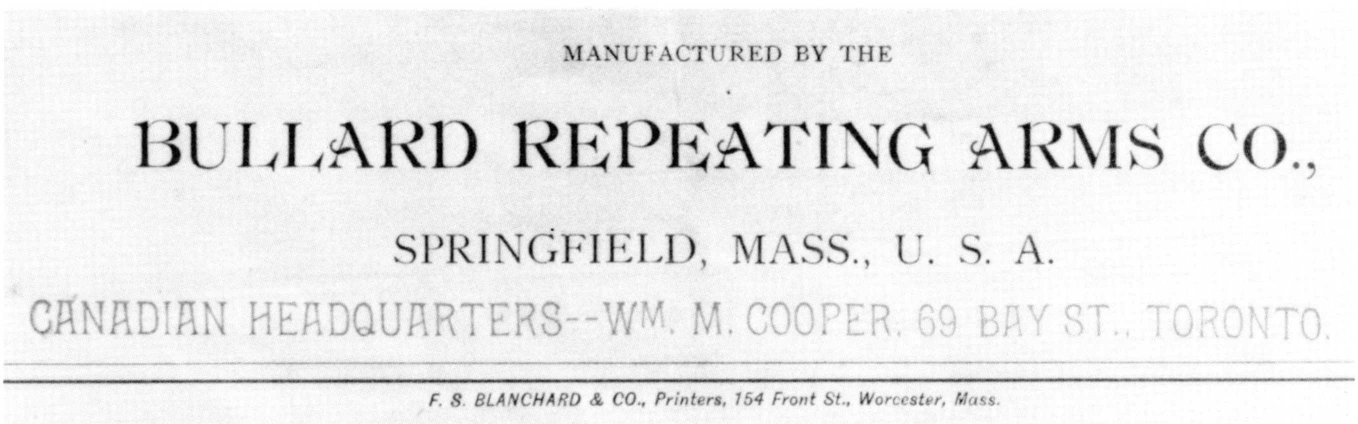

MANUFACTURED BY THE

BULLARD REPEATING ARMS CO.,

SPRINGFIELD, MASS., U. S. A.

CANADIAN HEADQUARTERS--WM. M. COOPER, 69 BAY ST., TORONTO.

F. S. BLANCHARD & CO., Printers, 154 Front St., Worcester, Mass.

Plate 9-10. This photograph of the title page is most interesting, for it contains the rubber stamping of the distributor of Bullard Arms in Canada. William M. Cooper was one of the two known dealers in Toronto (1879-89). Charles Stark of 52 Church Street was the other known dealer (1879-99). *Courtesy Ed Iwanchuk*

Many advertisements and articles about the Bullard Company and its products were printed in the various enthusiasts' publications of the day.

The author has not personally seen all the literature listed below. Some of the information has been compiled over the years from many sources, and due to the fragility of the material, it has not always been possible to see it first-hand. In some instances only the publication and date of the article are known but nothing further. The author is indebted to the National Rifle Association for most of *The Rifle* quotes.

The list has been compiled chronologically, that is with the date first, name of publication second and content last. This is not a complete list by any means and further study will shed more light on this subject.

October 20, 1883, *American Field*: Repeating rifle Advertisement (Plate 9-11)
December 29, 1883, *American Field*: Repeating rifle advertisement
March 20, 1884, *Forest & Stream*: Article on single-shot action
December 13, 1884, *American Field*: Farrow, Assistant Manager of Bullard. (Plate 9-12)
March 14, 1885, *American Field*: Article about W. Milton Farrow and .32-caliber cartridge from Bullard
May 16, 1885, *American Field*: A.G. Spalding advertisement (Plate 9-14)
June 1885, *The Rifle: Article (page 9) entitled "Ammunition for Indoor Shooting."* The cartridge was grooved to prevent the bullet from being seated past a certain point. The .45 cal mentioned contained 7 grains of wood powder. (Illustrated) (See Chapter 8 for full details.)
Sept. & Oct. 1885, *Forest & Stream*: Tests of Bullard rifles at Creedmoor.
January 30, 1886, *American Field*: Letter (page 113) about Freund sights and their Bullard application.
February 27, 1886, *American Field*: See Plate 9-13, A.C. White letter.
March 1886, *The Rifle*: Contains advertisement (page 13) shown in Plate 9-15. Page 11 mentions the shooter A.C. White and first clean score at Walnut Hill range on Jan. 30, 1886, using .35 cal Bullard.
October 16, 1886, *American Field*, See Plate 9-17, letter.
February 1887, *The Rifle*: Bullard rifles with detachable barrels (Page 11) — .22, .32, .38 now available; .40 & .50 will be available later.
April 1887, *The Rifle*: Page 5 mentions second Bullard shooting match. Editor complains scores not genuine.
May 1887, *The Rifle*: Contains letter to editor (page 5) and articles on Matches 3 & 4. Match 4 had 200 entries. Both S.K. Hindley and E.S. Field competed in Match 4.
May 1887, *The Rifle*: Contains advertisement (page 8) for detachable-interchangeable barrel model with detachable triggers! This rifle was also supplied with three barrels and two trigger sets. The article said that a sample rifle had been manufactured. It included the following specs: *1st barrel* was 30" long 14 pounds and .32 cal with set triggers; *2nd barrel* was 10 pounds & .32 cal with a regulation 3-pound pull trigger; *3rd barrel* was a .50 cal smoothbore .50-95 with ½ ounce of shot. (Plate 9-18)
June 23, 1887, *The Rifle*: Mentions third Bullard match in detail with complete scores of this match including the shooters. Some notable shooters mentioned were E.F. Richardson, F.J. Rabbeth, G.H. Wentworth, and O.M. Jewel. Also contained letter to editor.
August 1887, *The Rifle*: (Page 10) The Bullard military rifle. (Plate 9-20)

BULLARD REPEATING ARMS, SPRINGFIELD, MASS., U. S. A.

45-70 SPORTING RIFLE.

	Round Barrel.	*Half Octagon Barrel.*	*Octagon Barrel.*
Price, Plain..................................	$33.00	$35.00	$35.00
Length of Barrel..............................	28 inches.	28 inches.	28 inches.
Number of Shots...............................	11	11	11
Weight about..................................	9½ lbs.	9¾ lbs.	9¾ lbs.

45—75 } 45—60 } 40—60 } Same length Barrel, & Finish, same price.
Fancy Walnut Butt and Tip, "C with Globe and Peep sights, $5

We Challenge the World to produce a Repeater that can compare with ours in Ease of Operation, Safety, Quality of Workmanship, Style, Balance, Finish, and Rapidity of Firing.

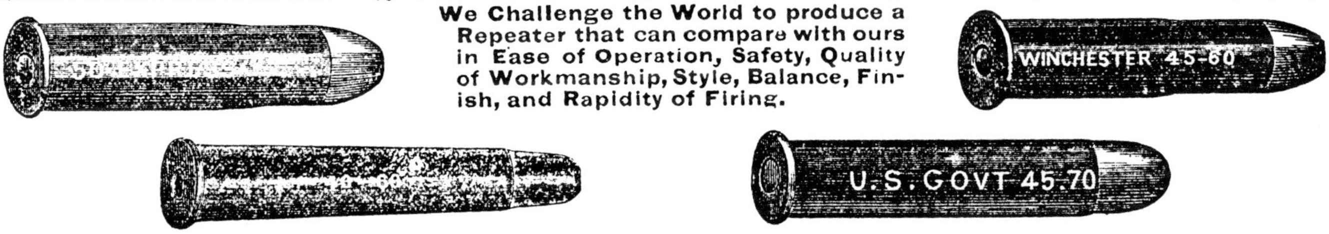

To an intending purchaser we will send one of our guns with the privilege of a three days' trial and examination on re five dollars as guarantee for express charges. **Please send for our Illustrated Catalogue.**

Plate 9-11. *American Field* advertisement, Oct. 20, 1883. This is one of the earliest known advertisements for Bullards yet found. *Courtesy George S. McCluney*

"There has been a long-felt want in the single breech-loading rifle, not any yet in the market being free from some most serious objections. The Ballard rifle cannot be opened if the hammer is first placed at full cock, and the most nicely adjusted pull is spoiled at once if it is attempted. Another objection shared by others is the great weight in the frame which should be in the barrel. For target shooting all rifles which have the barrel tip up between the sights, as in the Maynard, Stevens, Wesson, Franters and others, are defective in the fact that particles of dirt or powder crust, if falling on the proper spot will cause the barrel to spring to one side or prevent its closing down to the same exact point in each operation, and many are the inventions that have been experimented with, but only to run into some more serious objections.

I was shown last week a finished rifle, invented and patented by the famous rifle shot, W. Milton Farrow of Springfield, Mass., now assistant manager of the Bullard Arms Co. It has a number of new and novel points, as may be seen by referring to the en-,...

Plate 9-12. Reproduced from the December 13, 1884, issue of *American Field.* It announces the Farrow rifle, but more importantly, it states W. Milton Farrow's position in the Bullard Company.

EDITOR AMERICAN FIELD – A short time ago I noticed several inquiries about the Bullard single-shot rifle, and, as I purchased one of the first the company got out, I shall give my opinion. It is a .35-caliber and I must say it does the best shooting of any rifle I ever owned, and I have owned and shot every kind of rifle made in this part of the country. My first trial was made at the Manchester Fall meeting, before which time I had shot only seven shots from the rifle. The totals given are for strings of ten shots each on the Massachusetts target, and are as follows: 83, 82 and 86. At Walnut Hill, on October 11, shooting at the same target, strings of seven shots each, I made totals of 60 and 63. Again, at Walnut Hill, on January 30, shooting at the Hinman target, strings of ten shots, I made totals of 88, 83, and 87. These scores will prove to your correspondent the accuracy of this new Bullard rifle.
A. C. White"

Plate 9-13. "Letter to the Editor," Feb. 27, 1886, *American Field.* When Mr. White refers to purchasing one of the first the company got out, he was no doubt referring to the solid-frame model which was being advertised by May 1885 and which Mr. White had already shot at a match prior to October 11, 1885. As the detachable-interchangeable barrel model did not make an appearance until very late 1886, it could not have been one of these he was using.

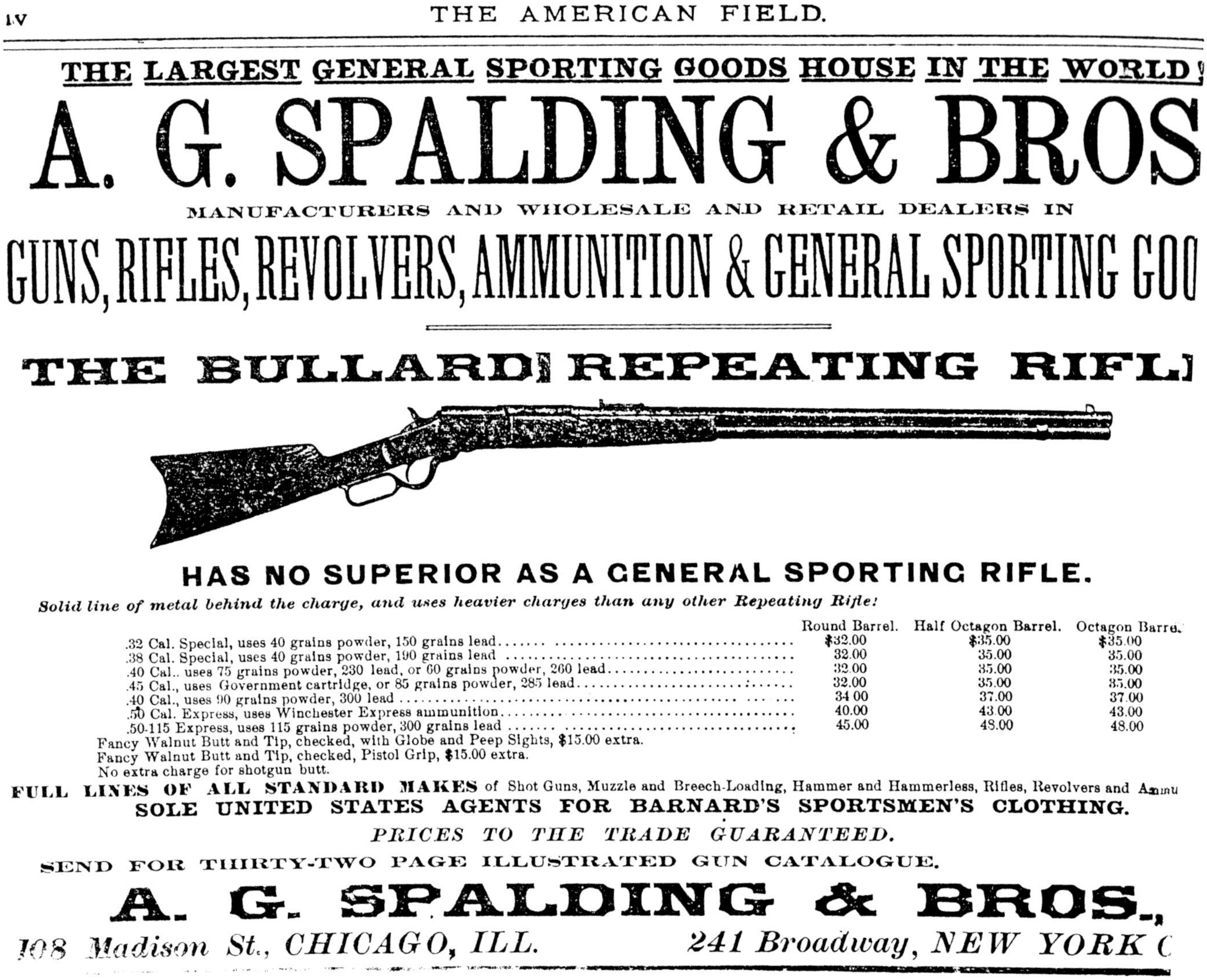

Plate 9-14. *American Field* advertisement, May 16, 1885. *Courtesy George S. McCluney*

Plate 9-15. This advertisement ran in the March 1886 to August 1887 issues of *The Rifle. Courtesy National Rifle Association and Joseph B. Roberts, Jr.*

CHAPTER XXVI.

THE SPORTSMEN'S OLD HAUNTS—Indian summer—on the mountain—a big horn chase—my largest ram.

In 1886 I was in Cooke City, where my dental work had called me. It was four years since I had indulged myself in a regular hunt. I had arranged for a hunting trip with some friends, and the appointed hour had passed and they had not appeared. My old Bullard stood in the corner. Cartridges, belt, knife and hunting suit, all were ready, and I felt the keenest disappointment.

Many changes had taken place in the sportsmen's old haunts. Where game had recently roamed the forests was now taken as ranges for domestic stock. The game had been driven far back into the high crags and mountains, such as the Index and Electric peaks, which stand eleven and twelve thousand feet above the level of the sea, the hunter being forced to climb those rugged cliffs to find satisfactory game.

The following morning even brought no news of my comrades. Anxiety and suspense overcame me. I grasped my faithful old friend, threw down the lever and surveyed the inside which gleamed like a mirror. It was my Bullard rifle, 45 calibre, eighty-five grains of powder, ten pounds weight and ten shots. I seized my cartridge belt, my knife, a cold lunch and started out.

The sun shone warmly and the smoke from the smelters gave an appearance of Indian summer. I soon gained a forest of pines whose boughs were loaded with nuts. The little

Plate 9-16. One of the few references to a Bullard rifle in contemporary literature: taken from *Adventures with Indians and Game or Twenty Years in the Rocky Mountains* by Dr. William A. Allen, published by A.W. Bowen & Co., 1903, Chicago. Allen was an avid Bullard enthusiast.

THE AUTHOR IN HIS "DEN."

EDITOR AMERICAN FIELD – "About three months ago I gave an order for a single shot target rifle, which, from its introduction to riflemen through the columns of the *American Field*, had, by its symmetrical proportions, struck me as being a rightful claimant for a great deal of consideration from the shooting public. And yet, after using one of the finest target guns made for a long time, it was with no little anxiety that I awaited the advent of the new "shooting-stick".

It finally came, the intermission between the time of giving the order and receipt of the rifle being very short even to one waiting with as much impatience as a boy for Christmas to come; and now after over two months thorough trial, using both patched and lubricated bullets, I am prepared to say that the rifle exceeds my expectations. It certainly is a beauty, both in finish and those qualities which make an elegant rifle. Weighing 9 3/4 pounds, with 28½ inch half octagon barrel, stock of finely selected walnut, with checkered pistol grip and forestock, the gun comes up to one's shoulder with such a fine balance that the pleasure of sighting along the polished barrel is enhanced by the perfect ease with which the gun is held to its place. I have no wonderful scores with which to startle your readers, but that is not the fault of the gun; it shoots with an accuracy that is perfection. Those acquainted with the Bullard repeater and its mechanism understand the working of the Bullard single shot rifle; the system is the same in both.

Plate 9-17. "Letter to the Editor," Oct. 16, 1886, *American Field*. It is sometimes difficult to tell if these were legitimate letters or unbridled attempts to scoop some free advertising in the media of the day. This letter could not have been more effective if it had been written by a modern-day marketing agency.

THE BULLARD RIFLE WITH INTERCHANGEABLE BARRELS.

THE Bullard Repeating Arms Co. has responded to the call for a heavy Scheutzen rifle, and has recently built one as a sample, weighing 14 lbs., having a 30-inch barrel, of .32 calibre, with set trigger and an interchangeable single trigger. The rifle has also an interchangeable barrel, .32-cal., which, when attached to the stock, reduces the weight of the arm to 10 lbs., which can be used with the set trigger, or the regulation three-pound pull. This arm is built for riflemen who desire to compete in Scheutzen festivals with the heavy rifle and set trigger, as well as in matches governed by the rules of the National Rifle Association, which can readily be done by changing the barrel from the heavy to the light one, and altering the trigger to the three-pound pull. The heavy barrel is also intended for rest-shooting, which is growing so popular in this country, the changing of the barrel transforming the heavy arm to a well-proportioned off-hand rifle.

In addition to the rifle barrels is one for shot, which is .50-calibre smooth bore, chambered for the .50–95-shell, which will hold two drachms of powder and half ounce of shot. The manufacturers of the Bullard rifle state they are ready to build rifles into the frame of which may be fitted Scheutzen, Creedmoor, Gallery rifles, and a shot barrel.

Plate 9-18. This column appeared in both the May and June 1887 issues of *The Rifle*.

Wind-gauge front sights with interchangeable discs were made in two basic types. The chief difference between the various models of these sights is in the manner of attaching and holding the sight discs in place. The interchangeable discs in these sights, as made by the Sharps Rifle Company, were held in place by a flat spring attached to the top of the shade tube, and the discs were positioned by square shoulders at each side of the upper half of each disc. In the sights made by E. Remington & Sons, the Marlin Firearms Company (makers of the Ballard rifle), Winchester Repeating Arms Company, the Massachusetts Arms Company (makers of Maynard rifles), the J. Stevens Arms Company, and the Bullard Arms Company, the interchangeable sight disc fitted a slot cut transversely across the center of the shade, and was held in place by a slide fitting across the rear and along the sides of the shade tube, in which the "arms" of the slide fitted small beveled slots cut in each side of the base of the sight disc. This arrangement held the sight discs more firmly in place than the method employed by the Sharps Rifle Company in their sights of this type and was, therefore, considered better by most match shooters.

Plate 9-19. Is this sight, made by the Bullard Co., the one mentioned in the advertisement of the August 1887 issue of *The Rifle*? Taken from the book *The Breech-Loading Single-Shot Match Rifle by Major Ned Roberts and Ken Waters.*

THE BULLARD MILITARY RIFLES.

The Bullard Repeating Arms Co. has perfected and are now prepared to supply a magazine and a single-shot military rifle. The accompanying illustrations show the repeating and single-shot rifle such as the company now have on exhibition in Europe. It is likely that the single-shot rifle will be welcomed by a number of individuals who have been anxious to secure a special military arm of smaller calibre than the regulation government .45. It is understood that the company are to supply these military rifles in .45, .40, and .38 calibre, in fact in all reasonable size of bore. The military arms of this company have been submitted to the severe government tests, which are as follows: —

The first test was the safety test, and consisted of firing the Standard cartridge; second, defective cartridges, which consist of firing regular cartridges, some with the head filed through to the powder crosswise on the head, and other cartridges filed through to the powder longitudinally. This is for the purpose of having the gas escape to the rear to test the system and to foul the chamber. Third, excessive charges: cartridges loaded with heavy powder charges, and sometimes one and sometimes two bullets. Dust test: the rifle is placed in a box made for the purpose and a blast of sand blown into the system. Two trials are made, one with system open and one closed. After these tests the guns were fired without cleaning. The endurance test consists of firing 500 rounds without cleaning. After every fifty rounds the guns were allowed to cool. The magazine gun used the 70.405 cartridge, and the single 70.500 cartridge. It took about four minutes to fire the fifty rounds, and six to eight minutes to cool. The rust test consists of dipping the stock and receiver up to the chamber in a solution of sal-ammoniac for ten minutes, then it is taken out and placed in a warm room to rust forty-eight hours, and then fired without cleaning. The guns were badly rusted, yet, after a little rapping, to break the rust on the bolt in the magazine gun, both guns were successfully fired. The final tests were excessive charges, and were five shots, 120 grs. powder and a slug of lead smaller than the inside of the barrel and weighing 1,200 grs. lead, fired without cleaning. The heads of the shell blew off, and great escape of gas to the rear, yet the guns came out all right. The rifles are now in the company's possession, and are in a serviceable condition. The trial demonstrated in the first place a few weak points, which were remedied, and in the final trial the guns passed successfully through the severe tests.

Plate 9-20. August 1887 issue of *The Rifle* extolls the virtues of the Bullard military offerings. The author has been unable to uncover the official government results for the single-shot tests.

Plate 9-21. This quote shows that Bullard target rifles were quite popular in their day and used by many marksmen. *Quote taken from The Breech-Loading Single-Shot Match Rifle by Major Ned Roberts and Ken Waters.*

" As an illustration of the popularity of the Maynard rifle, I may state that at the Fall Meeting of the Massachusetts Rifle Association at the Walnut Hill Range on October 6, 7, and 8, 1887, I made notes of the makes of rifles used by the 43 contestants, which showed that 16 men used the Maynard, 10 shot Ballards, 8 used the Bullard, 4 shot the Remington-Hepburn, 3 used Farrow rifles, 1 a Sharps-Borchardt, and 1 Warner muzzle-loading caplock rifle was used by William V. Lowe. It appears strange that not one shooter used the Winchester single-shot rifle in that match, although it had been on the market in various models and calibers for two years.

It is also interesting to note that, according to my records made at that time, the calibers of the rifles used by these 43 contestants were: 17–32–40, 9–32–35, 6–35–40, 6–38–55, 1–38–45 (Bullard), 1–38–50 (Maynard), 1–38–40 (Remington), 1–40–70 (Sharps), and 1–38 caliber muzzle-loader. All of the 32–35 and 35–40 rifles, and the one 38–50 caliber were Maynards. Had a checkup of the calibers of rifles used at a similar match been made in 1877, it would have shown that a large majority of the riflemen were using 40 or 45 caliber rifles, a small number using the 44 caliber, and probably none using smaller than 35 or 38 caliber. "

QUERIES AND REPLIES.

Subscriber, Chicago:—
Please let me know which rifle would be best for good, strong shooting, accuracy, durability of action and rapidity of fire: The 32-40 Marlin Repeater, 28-inch, or the 32-40 Bullard Repeater, 28-inch. I desire to use it on game and target, at all distances up to 500 yards. Please give me a definite answer.

The shooting qualities of the two rifles should be equal. A choice of the mechanism might make one arm preferred by some, while others' choice would be contrary to the first. Differences in ammunition might make a difference in the shooting of the two rifles, but it is probable that one is as good as the other.

Plate 9-22. Letter in *Shooting & Fishing*, July 5, 1888.

The following scores were made at 100 yards Saturday, June 23, by the Young Men's Club of this city, Creedmoor target:

S. Beals,	4	4	4	4	5=21
H. Landen,	4	4	5	4	4=21
J. Russell,	4	5	4	4	3=20
G. Carroll,	4	4	4	5	3=20
M. Wilder,	3	4	4	4	4=19
F. Robbins,	3	3	3	2	3=14

This club is composed of young men under 21 years of age. They have a very convenient 100-yard range close to the Bullard Repeating Arms factory, where they shoot every Saturday.

Plate 9-23. Article in *Shooting & Fishing*, July 5, 1888. Whether their club was sponsored by the Bullard factory is unknown, but due to the proximity of the range to the factory it is likely some form of patronage was given to the club.

SPRINGFIELD, MASS., Feb., 1888.

To the Editor of The Rifle:—

To meet the wants of the military authorities in this country and abroad for a magazine rifle having the magazine detachable, light, and readily attached, and dispensing with the magazine under the barrel, or in the butt-stock, the Bullard Repeating Arms Co. has succeeded in adapting such a system to the Bullard single-shot rifle. The receiver is cut open on the side and through the carrier-block to receive the cartridges. As a single-loader, one cartridge can be put in the rifle-chamber and another remain in the carrier-block; after firing, another can be put in the carrier, and always have two cartridges in the rifle. The important feature in this plan is always two shots in a single-loader. The opening on the side of the receiver is a most convenient place for loading, as it requires no special attention to place them in properly in the carrier-block. When the rifle is not in use a slide covers the loading part. The detachable magazine is a small box of the length of the cartridges. The interior of the box conforms to the shape of the cartridges, and is corrugated, so that when the cartridges feed down they keep the same relative position, which prevents jamming. Gravity carries them downward. The box has an opening corresponding to the one in the receiver and carrier-block, and is attached to the receiver by sliding the box in grooves on the receiver. The cartridge-box may be made of light material,—Russia iron, tin, paper, or any light material that can be moulded. The cartridges can be put up in these boxes, and after being used can be thrown away, or the box can be fixed on the rifle and loaded very conveniently. The cartridge-box can contain four to eight cartridges, as desired. The rifle has a stock similar to all single-loading military rifles, and will weigh about 9½ lbs., and is capable of being fired forty rounds in a minute.

S. H. HINDLEY.

Plate 9-24. March 1888 issue of *The Rifle*. Note the proofreader's error. It should be S.K. Hindley not S.H. In his letter, even Hindley seems somewhat confused. He says Bullard has succeeded in adapting such a system to the Bullard single-shot rifle but there is no carrier block in the single-shot. He really meant the repeater was adapted to the receiver-mounted magazine. it is easy to see where the rifle could be called a two-shot repeater, since even Hindley mentions that it can carry two rounds; one in the breech and one in the carrier when the magazine is not used, thereby always having two cartridges in the rifle.

August 1887, *The Rifle*: (Page 5) Bullard Match 3; again editorial complaint about false scores. Team that won should have been seventh, for they did not use the military rifles issued by their state. On July 26, 1887, Match 4 was held for the honor of winning the Bullard Champion Offhand Marksman Badge, a gold medal to be kept by a shooter after a third match win. Article states that matches were to be held twice each year from this point on, in May and October.

August 1887, *The Rifle*: (Page 12) "The Bullard Arms Company lately commenced manufacture of a rifle sight to speedily convert a target rifle to open sighted arm for hunting. The usual base into which the windgauge is screwed is affixed in the slot near the muzzle. The windgauge is removed and a neat and strong open sight is fitted into the windgauge base and with a few turns of a small screw is made fast which makes the rifle a perfectly open sighted rifle which can be used with the usual vernier peep sight or the vernier can be removed and a Lyman rear sight affixed or if a strictly front and rear open sight is preferred a rear step sight can be affixed on the barrel and cut down sufficiently to make for a clear sight when the rear vernier peep sight is used for a long shot."

September 1887, *The Rifle*: Article entitled "American Riflemen Abroad," written in Geneva, Switzerland. Both T.T. Cartwright of the Bullard Company and J.A. Huggins of Cincinnati won shooting prizes with their Bullard rifles. The Bullard rifle was much admired.

October 1887, *The Rifle*: (Page 4) Bullard Repeating Arms Company Match 4 second series held. The Gold Medal was again held up for grabs. These matches were all 200-yard offhand using the Hinman target.

March 1888, *The Rifle*: (Page 12) Letter from S.K. Hindley about his box magazine rifle. (Plate 9-24)

May & April 1888, *The Rifle*: Advertisement shown appeared again for these two months. May 1888 was last issue of *The Rifle*. (Plate 9-15)

June 5, 1888, *Shooting & Fishing*: First known issue of this magazine, and no more advertisements have been discovered at this time. A small article or letter in "Queries and Replies" does appear on page 11 and another very small Bullard mention appears on page 13. (Plates 9-22, 9-23)

Many fine target shooters used Bullard rifles in their careers. Amongst them was one man already mentioned, A.C. White. The October 1886 issue of *The Rifle* had this to say about the noted offhand shooter:

The subject of this sketch, Mr. A.C. White, has long been known to be a very fine rifle shot for in most of the contests in which he participated he was credited with the highest aggregate, and always a high and excellent score. Mr. White is 37 years old, is 5 feet 6½ inches in height, has blue eyes and brown hair and weighs 160 pounds. He has very practical ideas about rifle shooting; he believes that to secure good results under all circumstances and under various weather conditions, a man should first take a good shooting position. The one he adopts with the match rifle and the military arm is one which is sometimes seen in pictures but rarely on the rifle range. The position adopted by Mr. White is worthy of imitation. We have witnessed this gentleman record perfect scores of ten consecutive bullseyes with a match rifle and many other excellent scores, a number of which were made in high winds, which indicates the firmness of his position and his ability to hold steadily in this manner, which is easy, graceful and comfortable, and may be properly called a model position.

Mr. White is not wedded to any particular style or make of rifle. He acknowledges the excellence of most makes of American rifles, and believes with most all of them he could obtain good results, and he has well illustrated this by using a number of different makes with creditable results. He first used a Sharps 40-caliber rifle with a 26-inch barrel, taking a shell holding

Plate 9-25. Teddy Roosevelt (holding his .50-115 Bullard Repeater) flanked by Wilmot Dow (on the left) and William Sewall (on the right) sometime in March or April 1886, after capturing three thieves. *Courtesy Theodore Roosevelt Collection, Harvard College Library*

50 grains of powder, and using a bullet weighing 330 grains. Among the excellent scores made with this arm was one full score of ten consecutive bullseyes, and two of seven shots each, which will be recognized as fine work, but at the time they were made were considered unusually brilliant. Eight different makes of rifles have been used by Mr. White, and nearly all of them have been won in matches. The Bullard rifle used at present is 35-caliber; the shell holds 40 grains of powder, and the bullet used weights 250 grains and is made and patched by himself. He formerly swaged his bullets, but during the past years he has ceased to do this believing, as most of the Eastern riflemen do who make their own bullets, that he obtains fully as good results with unswaged bullets. The sights used on his match rifle are the usual rear peep sight, and for a front sight an aperture with quite a small hole and heavy rim. He formerly used an aperture sight with a very thin rim, but trying one with a thick rim made by himself, he found he could improve his scores nearly two points, and he states that he was prompted to devise this sight in order to improve his shooting two points, which was the amount he was handicapped in an important match.

His manner of loading is to use one shell, loading from a powder flask, filling the shell full, tapping it a few times to settle the powder, a wad placed over the powder seated slightly below the top edge, the bullet dropped into the rifle and seated with a plug seater. After firing, the gun is cleaned and wiped with considerable care to remove all fouling and moisture; the shell is scraped inside by a scraper to remove the fouling; it is then re-primed and the operation of reloading repeated. Mr. White has been a member of the Brattleboro Rifle Club, the Bridgeport Rifle Club, the Worcester Rifle Club, and the Massachusetts Rifle Association. He now resides in Worcester, Mass., and most of his shooting for the past year has been done on the Worcester range and at Walnut Hill.

Major Ned Roberts and Kenneth Waters, in *The Breech-Loading Single-Shot Match Rifle*, spoke of another fine offhand marksman, Elton F. Richardson:

Elton F. Richardson, who was one of the noted offhand marksmen during the period from 1880 to 1900, selected the Bullard target rifle in 32-40 Bullard caliber, loading a 165-grain bullet, as the best for his purpose after having thoroughly tested all of the other single-shot target rifles then on the market. Records show that he made many very high scores offhand with it during those years. He was also very partial to his 35-caliber Maynard rifle, with which he had made his record score in July, 1885.

Roberts and Waters also cited Adolf O. Niedner for his contribution:

Niedner became a member of the Massachusetts Rifle Association soon after coming to Malden and shot in a great many matches at the Walnut Hill Range, in which he won many prizes and made numerous high scores in 200-yard offhand matches on the various targets. On May 11, 1901, he broke the Walnut Hill Range record of 96 points for 10 shots on the Standard American target at 200 yards offhand, which had stood for fifteen years, by scoring a total of 97, for which feat the association presented him with a special silver cup.

During his rifle-shooting career, Niedner used a 43 caliber muzzle-loading caplock rifle by John Meunier until 1890, therefore testing the Remington-Hepburn, Maynard, Bullard, Winchester, and Ballard Schuetzen rifles in various calibers. He finally decided that the Winchester 38-55 and Ballard 32-40 best suited him for 200 yard offhand shooting. However, his favorite rifle, with which he won so many matches and made his highest scores, was the Ballard Schuetzen 32-40 caliber with special offhand stock.

Niedner used but one shell, reloading it and breech seating the bullet for each shot when shooting in matches. After Du Pont No. 1 Rifle Smokeless powder came on the market he loaded the 32-40 shell with 3 grains weight of that powder as a priming charge, filling the rest of the shell with FG black powder, with a card wad over the powder and a 200-grain bullet breech seated in the throat of the rifle, thereby saving the labor of cleaning the bore after each shot, as was necessary when using straight black powder loads in the Schuetzen rifle. He used a palm rest and shot from the hip-rest position, using a 5- or 6-power telescope in matches where this was allowed, or a vernier rear peep and shaded wind-gauge aperture front sight in matches where the telescope was barred.

These are only a very few of the many fine marksmen who used Bullard target rifles over the years.

GUIDE TO SHORT-FORM TERMINOLOGY USED IN TABLES

S.N. — serial number
Cal. — caliber
Loc. — location of caliber markings: A on barrel; B on breech at first reinforce; C on breech at second reinforce just ahead of dustcover; D on top facet of breech (single-shot only).
bbl. len — barrel length
bbl. type — barrel type: rd = round; pt rd = part round & part octagonal; oct = octagonal.
butt type — s = shotgun; c = crescent; sh = Schuetzen (Swiss); m = military style
pist grip — Y = yes; N = no
check — checkering: Y = yes; N = no
C/H — case hardening: 1 = lever; 2 = hammer; 3 = receiver; 4 = breechblock; 5 = trigger; 6 = buttplate; 7 = fore-end tip
add. type — address type:
1 = BULLARD REPEATING ARMS ASSOCIATION Springfield Mass. U.S.A. Pat. Aug. 16, 1881
2 = BULLARD REPEATING ARMS COMPANY Springfield Mass. U.S.A. Pat. Aug. 16, 1881
3 = BULLARD REPEATING ARMS COMPANY Springfield, Mass. Pat. Aug. 16, 1881
4 = BULLARD REPEATING ARMS COMPANY SPRINGFIELD MASS, U.S.A. PAT. JULY 6, 1886
See photographs for exact spacing and script styles of addresses shown. It's the deletions & additions of words & letters within these addresses that is important.
ext. hole — extractor removal hole: Y or N
Mod 1886 — Y or N: indicated whether it is marked Model 1886 or not.
S.F.S.S. — solid-frame single-shot: Y or N
det-int — detachable-interchangeable barrel model single-shot: Y or N
motif — if shotgun butt: does buttplate have T (turkey) or E (elk) motif
nick p. — nickel plating: 1 = receiver; 2 = buttplate
fin ser. — finger serrations on dustcover: A = 4 serrations; B = 3 serrations
grve — groove in bottom underside of carrier: Y or N
tang st — tang sight: Y or N
rec scre — receiver screws: meaning is it an 8- or 7-screw receiver; only applicable for large-frame repeaters
mag len. — magazine tube length: F = full-length; H = ½-length; B = button mag; T = ¾ mag
n/a — whenever this is used it means that it not applicabnle to that rifle
un — unknown: due to a lack of information, eg. early questionnaires were less than comprehensive and some info was taken from advertisements and therefore is sketchy on many details. The author felt that any verifiable information no matter how trivial was better than none.

10

PRODUCTION TABLES BY SERIAL NUMBER

The following table has some deficiencies, in that some of the sources of information were less than complete. Where information could not be verified, you will see *unkn* (unknown). In this way no guesses were made and only facts were included, even if it meant that most spaces were unknown for a particular serial number.

It is hoped that many people who have Bullards and who have not added theirs to the list will do so in the future by contacting the author.

LARGE-FRAME REPEATERS

SERIAL NO.	CALIBER	LOCATION	BBL. LEN.	BBL. TYPE	BUTT TYPE	PIST. GRIP	CHECK	C/H	ADD. TYPE	EXT. HOLE	MOD. 1886	S.F. / S.S	DET - INT	MOTIF.	NICK. P.	FIN. SER.	GRVE.	TANG ST	REC. SCRE.	MAG. LEN.	MISC.
23	.45	no	28"	rd	c	N	N	N	1	N	N	n/a	n/a	n/a	none	A	N	N	8	F	1
36	.40	no	28"	oct	c	N	N	125	1	N	N	n/a	n/a	n/a	none	B	N	N	7	F	
39	.45	A	28"	rd	c	N	N	N	1	N	N	n/a	n/a	n/a	none	A	N	N	8	F	
78	.40	un	un	un	un	un	un	un	1	un	un	n/a	n/a	un	un	un	un	un	un	F	
80	.40	A	28"	pt rd	c	N	N	12	1	N	N	n/a	n/a	n/a	none	A	Y	N	8	F	
85	.40	A	28"	oct	un	un	N	none	1	un	N	n/a	n/a	un	none	un	N	N	8	F	
109	.50-95	C	28"	oct	s	N	N	none	un	N	N	n/a	n/a	E	none	n/a	Y	N	8	H	
118	.45	A	28"	oct	c	N	N	N	un	n/a	N	n/a	n/a	n/a	none	A	N	N	8	F	
123	.40	n/a	26"	pt rd	s	N	N	n/a	un	un	un	n/a	n/a	un	none	un	un	N	un	F	2
132	.45	no	26"	pt rd	s	Y	Y	none	un	N	N	n/a	n/a	E	none	un	N	N	8	F	
133	.40	n/a	26"	pt rd	s	N	N	n/a	un	un	un	n/a	n/a	un	none	un	un	N	un	F	
153	.40	un	26"	pt rd	c	N	N	12	un	Y	un	n/a	n/a	n/a	n/a	un	un	N	8	F	
184	.50-115	C	26"	oct	s	un	N	N	un	N	N	n/a	n/a	E	none	un	Y	N	8	B	
201	.45	un	22"	un	c	N	N	12	un	un	un	n/a	n/a	n/a	1	un	un	N	un	F	3 4
203	.45-75	A	26.5"	rd	c	N	N	N	2	N	N	n/a	n/a	n/a	none	A	Y	N	8	H	
226	.40-90	un	28"	oct	s	Y	Y	3	un	Y	un	n/a	n/a	E	none	A	un	N	8	un	
231	.40	A	26"	pt rd	c	N	N	none	2	Y	N	n/a	n/a	n/a	none	A	un	N	8	F	5
250	.40-90	un	28"	pt rd	un	N	un	126	un	un	un	n/a	n/a	n/a	un	un	un	un	un	un	
260	.45	un	un	un	un	un	un	un	un	un	un	un	un	un	un	un	un	un	un	un	
269	.40	un	22"	pt rd	c	N	N	12	un	Y	N	n/a	n/a	n/a	none	un	un	un	8	F	6
270	.40-90	un	28"	pt rd	c	un	un	un	2	N	N	n/a	n/a	n/a	none	un	N	Y	8	un	
300	un	C	28"	pt rd	s	Y	Y	3	un	un	N	n/a	n/a	E	none	A	N	Y	8	H	

SERIAL NO.	CALIBER	LOCATION	BBL. LEN.	BBL. TYPE	BUTT TYPE	PIST. GRIP	CHECK	C/H	ADD. TYPE	EXT. HOLE	MOD. 1886	S.F. / S.S.	DET – INT	MOTIF.	NICK. P.	FIN. SER.	GRVE.	TANG ST.	REC. SCRE.	MAG. LEN.	MISC.
331	.45	A	24"	pt rd	c	N	N	1 2	un	N	N	n/a	n/a	n/a	none	none	Y	N	8	B	7
338	.40-90	un	26"	rd	s	N	un	1 2 5	un	N	un	n/a	n/a	un	none	un	Y	N	8	H	8 9
346	.45	un	24"	rd	s	Y	N	3	un	Y	un	n/a	n/a	un	none	un	Y	N	8	F	10 11
382	.40-90	B	26"	rd	s	Y	Y	1 2 5	2	N	N	n/a	n/a	E	none	A	Y	Y	8	un	
443	.45	C	26"	rd	c	N	N	none	un	N	un	n/a	n/a	n/a	none	un	N	N	8	F	
448	.45	C	27"	oct	c	N	N	1 2 6	2	N	N	n/a	n/a	n/a	none	A	N	Y	8	un	
461	.40-90	C	27"	oct	s	Y	Y	none	2	N	N	n/a	n/a	E	none	un	Y	N	8	F	12
491	.40	C	28"	oct	s	Y	Y	3	un	N	N	n/a	n/a	E	none	un	N	N	8	F	
521	.40	C	26"	rd	c	Y	N	2	2	Y	Y	n/a	n/a	n/a	none	B	N	Y	7	F	
539	.40-90	un	28"	oct	un	un	un	1 2	un	un	un	n/a	n/a	un	un	un	un	un	un	F	
546	.40	un	26"	pt rd	s	N	Y	3	un	Y	Y	n/a	n/a	un	none	un	Y	N	7	F	
555	.45	un	26"	oct	s	Y	Y	12357	un	N	N	n/a	n/a	E	none	un	N	N	7	F	13
557	.40	C	32"	oct	c	Y	Y	3	2	Y	N	n/a	n/a	n/a	N	A	N	N	7	H	
571	.45	un	28"	pt rd	s	Y	N	un	un	un	un	n/a	n/a	un	un	un	un	N	7	H	
582	.40	un	26"	pt rd	c	Y	N	1 2 5	un	N	N	n/a	n/a	n/a	none	un	N	N	7	F	14
585	.40	C	28"	oct	s	Y	Y	un	2	N	N	n/a	n/a	E	none	A	N	N	7	F	
602	.45	C	20"	oct	c	Y	N	12456	2	N	N	n/a	n/a	n/a	none	B	Y	N	7	F	
619	.45	C				N			2	N	N								7		
630	.45	C	26"	oct	c	Y	un	none	2	Y	N	n/a	n/a	n/a	none	A	N	Y	7	F	
654	.45	un	un	oct	c	N	Y	un	un	un	un	n/a	n/a	n/a	un	B	un	N	7	F	
656	.40-90	C	28"	oct	s	Y	N	none	un	N	un	n/a	n/a	E	none	un	Y	N	7	F	
658	.40	C	28"	oct	c	un	N	N	2	N	N	n/a	n/a	n/a	none	A	N	N	7	F	
734	.45	un	22"	oct	c	N	N	1 2	un	un	N	n/a	n/a	n/a	none	un	un	N	un	F	
770	.45	C	26"	oct	c	N	N	2 4	2	N	N	n/a	n/a	n/a	none	B	N	Y	7	F	
8XX	.45	un	28"	pt rd	s	un	un	un	un	un	un	n/a	n/a	un	un	un	un	Y	un	F	15
808	.50-115	C	28"	oct	s	N	Y	3	un	N	N	n/a	n/a	E	un	un	Y	N	7	F	16
811	.40-90	C	28"	oct	s	N	N	1 2	2	N	N	n/a	n/a	E	none	un	Y	N	7	F	17
812	.45	C	28"	rd	c	N	N	N	un	Y	un	n/a	n/a	n/a	none	un	N	N	7	F	
813	.40	C	24"	oct	c	N	N	1 2 3	2	N	N	n/a	n/a	n/a	N	un	N	N	7	F	
825	.45	C	28"	pt rd	s	Y	N	un	un	N	N	n/a	n/a	E	none		N	N	7	F	18 19
841	.45	un	un	pt rd	c	Y	Y	un	un	un	un	n/a	n/a	n/a	un	un	un	un	un	un	
844	.50-115	un	28"	oct	s	un	un	1 2	un	N	un	n/a	n/a	un	un	un	un	N	7	un	
874	.40	A C	26"	oct	s	Y	N	3	2	Y	Y	n/a	n/a	un	un	B	N	N	7	F	20
904	.45	C	24"	pd rd	c	Y	Y	1 2 4	2	N	N	n/a	n/a	n/a	none	B	N	Y	7	H	
932	.45	un	un	pt rd	c	N	N	un	un	un	un	n/a	n/a	n/a	un	un	un	N	7	F	
971	.45	un	24"	rd	s	N	N	1 3	un	un	un	n/a	n/a	un	none	un	un	N	un	T	21
974	.40	un	26"	oct	c	Y	N	1 2 3	2	Y	N	n/a	n/a	n/a	none	B	Y	N	7	F	22
977	.45	C	28"	pt rd	s	Y	Y	none	un	Y	un	n/a	n/a	n/a	none	un	N	Y	7	T	23
989	.45	B	28"	rd	s	N	N	1 2	2	N	N	n/a	n/a	E	none	A	N	Y	7	F	
1033	.50-115	un	27"	oct	s	Y	Y	un	un	Y	Y	n/a	n/a	T	un	un	un	Y	7	T	24
1054	.40	C	26"	pt rd	s	Y	Y	3	un	Y	un	n/a	n/a	T	un	un	N	Y	7	T	25
1070	.40	un	26"	oct	c	Y	Y	un	un	un	N	n/a	n/a	n/a	un	B	un	N	7	F	
1072	.50-115	C	28"	pt rd	c	N	N	1	un	Y	Y	n/a	n/a	E	N	A	Y	Y	7	H	
1073	.50-115	un	28"	oct	n/a	n/a	n/a	3	un	Y	Y	n/a	n/a	n/a	none	A	un	n/a	7	un	26
1100	.45	un	28"	oct	un	un	un	un	un	Y	un	n/a	n/a	un	un	B	un	un	7	F	
1105	.50-115	un	27"	rd	s	N	Y	1257	un	N	N	n/a	n/a	E	none	un	Y	N	7	H	
1128	.40	un	28"	oct	s	N	N	1 2	un	un	un	n/a	n/a	un	none	un	un	Y	un	F	
1134	.50-115	un	28"	pt rd	s	Y	N	1 2 5	2	un	N	n/a	n/a	E	none	un	un	N	un	un	27

SERIAL NO.	CALIBER	LOCATION	BBL. LEN.	BBL. TYPE	BUTT TYPE	PIST. GRIP	CHECK	C/H	ADD. TYPE	EXT. HOLE	MOD. 1886	S.F. / S.S.	DET - INT	MOTIF.	NICK. P.	FIN. SER.	GRVE.	TANG ST.	REC. SCRE.	MAG. LEN.	MISC.
1138	.45	un	28"	rd	c	Y	Y	un	2	un	un	n/a	n/a	n/a	un	un	Y	N	un	F	
1142	.40	B	28"	oct	s	Y	Y	1 2	2	N	N	n/a	n/a	E	none	A	N	N	7	H	
1145	.40	C	22"	rd	s	Y	Y	none	un	N	un	n/a	n/a	E	none	un	N	N	7	T	
1147	.45	un	28"	oct	c	un	un	un	un	un	un	un	un	un	un	un	un	un	un	un	
1181	.40-90	un	28"	rd	s	N	Y	12345	2	un	N	n/a	n/a	E	none	un	Y	N	un	H	
1233	.40	un	28"	oct	c	N	N	none	2	un	Y	n/a	n/a	n/a	none	un	Y	N	un	F	
1259	.40	C	28"	oct	s	N	N	un	2	N	N	n/a	n/a	E	none	B	N	Y	7	H	28
1272	.45	un	28"	rd	c	N	N	un	2	un	Y	n/a	n/a	n/a	none	un	un	N	un	un	29
1299	.40	C	24"	oct	c	N	N	3	2	N	N	n/a	n/a	n/a	none	B	N	N	7	F	
1305	.40	C	28"	oct	c	N	N	1 2	un	N	N	n/a	n/a	un	none	un	Y	N	7	F	
1349	.40	un	28"	oct	s	N	N	un	un	Y	Y	n/a	n/a	E	none	un	N	N	7	F	
1359	.45	C	28"	oct	c	N	N	12456	2	Y	Y	n/a	n/a	n/a	none	B	Y	N	7	F	
1367	.45	C	28"	oct	s	N	N	none	2	N	N	n/a	n/a	un	none	A	un	N	7	F	
1372	.50-115	un	20"	oct	s	N	N	none	2	Y	N	n/a	n/a	un	none	un	Y	N	7	F	30
1374	.45	C	26"	rd	s	N	N	un	2	un	N	n/a	n/a	E	un	un	un	N	7	H	
1413	.40	C	28"	rd	s	un	un	none	2	N	N	n/a	n/a	E	none	A	N	Y	7	un	
1460	.45	C	28"	pt rd	s	Y	Y	1237	2	Y	Y	n/a	n/a	T	N	un	N	Y	7	F	
1467	.45	un	28"	oct	un	N	un	un	un	un	un	n/a	n/a	un	un	un	un	un	un	F	
1480	.40	un	28"	oct	s	N	N	3	2	un	N	n/a	n/a	un	un	un	un	N	8	F	

SMALL-FRAME REPEATERS

SERIAL NO.	CALIBER	LOCATION	BBL. LEN.	BBL. TYPE	BUTT TYPE	PIST. GRIP	CHECK	C/H	ADD. TYPE	EXT. HOLE	MOD. 1886	S.F. / S.S.	DET - INT	MOTIF.	NICK. P.	FIN. SER.	GRVE.	TANG ST.	REC. SCRE.	MAG. LEN.	MISC.
1532	.32	un	26"	oct	sh	Y	Y	1 2 3	un	Y	Y	n/a	n/a	n/a	none	un	N	Y	n/a	F	
1552	.38	C	21"	rd	s	un	N	none	2	N	N	n/a	n/a	T	none	A	N	Y	n/a	*	31
1553	.38	un	un	un	un	un	un	un	un	un	un	un	un	un	un	un	un	un	un	un	
1559	dustcover only!																				
1563	.38	un	25"	pt rd	S	Y	Y	1 2 3	2	N	N	n/a	n/a	T	none	A	N	N	n/a	F	32
1567	.38	C	26"	oct	s	Y	N	none	un	N	un	n/a	n/a	T	none	un	N	N	n/a	F	
1604	.38	un	26"	pt rd	s	N	un	un	un	N	N	n/a	n/a	T	none	un	un	Y	n/a	F	
1625	.38	un	26"	oct	s	N	N	1 2 5	un	N	N	n/a	n/a	T	none	un	N	N	n/a	un	
1649	.32	un	24"	pt rd	s	N	N	1 2 5	un	N	N	n/a	n/a	T	none	un	N	Y	n/a	F	
1736	.38	C	26"	pt rd	c	Y	Y	un	2	un	N	n/a	n/a	n/a	none	un	un	Y	n/a	H	
1742	.38	un	28"	oct	un	N	un	un	un	un	Y	n/a	n/a	un	none	un	un	un	n/a	B	33
1751	.32	un	26"	oct	s	N	N	1 7	2	un	N	n/a	n/a	T	none	un	N	N	n/a	F	
1769	.38	C	un	rd	c	N	un	1 2 4	2	Y	Y	n/a	n/a	n/a	none	A	un	N	n/a	un	
1791	.38	un	26"	oct	s	Y	N	1 2	un	un	N	n/a	n/a	un	none	un	un	N	n/a	F	
1804	.38	un	24"	oct	sh	Y	Y	1 2	un	Y	Y	n/a	n/a	n/a	none	un	N	N	n/a	H	
1814	.32	C	26"	oct	c	Y	N	un	un	Y	Y	n/a	n/a	n/a	none	A	N	N	n/a	F	33A
1820	.32	C	26"	oct	c	N	Y	12367	2	un	Y	n/a	n/a	n/a	un	un	un	Y	n/a	F	
1823	.38	un	28"	pt rd	s	Y	Y	3	un	un	Y	n/a	n/a	un	un	un	un	Y	n/a	T	
1829	.32 spec	C	24"	rd	s	N	N	un	un	Y	Y	n/a	n/a	T	none	A	N	N	n/a	H	34
1853	.38	C	26"	rd	s	N	N	1 2	2	Y	Y	n/a	n/a	T	none	A	N	N	n/a	F	
1878	.38	C	26"	rd	s	un	N	none	4	Y	Y	n/a	n/a	T	none	un	N	Y	n/a	F	35
1971	.38	C	un	un	s	Y	Y	3	un	N	un	n/a	n/a	T	none	un	N	N	n/a	F	

SERIAL NO.	CALIBER	LOCATION	BBL. LEN.	BBL. TYPE	BUTT TYPE	PIST. GRIP	CHECK	C/H	ADD. TYPE	EXT. HOLE	MOD. 1886	S.F. / S.S.	DET - INT	MOTIF.	NICK. P.	FIN. SER.	GRVE.	TANG ST.	REC. SCRE.	MAG. LEN.	MISC.

LARGE FRAME REPEATERS

SERIAL NO.	CALIBER	LOCATION	BBL. LEN.	BBL. TYPE	BUTT TYPE	PIST. GRIP	CHECK	C/H	ADD. TYPE	EXT. HOLE	MOD. 1886	S.F. / S.S.	DET - INT	MOTIF.	NICK. P.	FIN. SER.	GRVE.	TANG ST.	REC. SCRE.	MAG. LEN.	MISC.
2331	.50-95	C	23"	rd	s	N	N	none	2	Y	Y	n/a	n/a	T	none	A	Y	N	7	H	36
2521	.45	C	22½"	rd	m	N	N	none	n/a	Y	N	n/a	n/a	n/a	none	A	N	N	n/a	F	37
2524	.45	un	28"	rd	m	N	N	none	un	un	un	n/a	n/a	n/a	none	un	un	N	un	box	38
2529	.45	C	28"	rd	m	N	N	none	un	Y	N	n/a	n/a	n/a	none	B	N	N	un	box	39
2538	.45-70	C	26"	rd	?	N	N	none	2	Y	un	n/a	n/a	n/a	none	A	N	N	7	H	40
2548	.50-95	C	28"	rd	c	N	N	none	2	Y	Y	n/a	n/a	n/a	none	A	Y	N	7	F	
2564	.45-70	C	un	rd	s	N	un	un	un	Y	Y	n/a	n/a	T	un	un	un	N	7	F	
2590	.45	un	26"	rd	s	N	N	3 7	3	Y	un	n/a	n/a	un	none	A	un	N	7	F	41
2597	.50-115	C	28"	oct	s	N	N	1 2 5	2	Y	Y	n/a	n/a	T	none	A	Y	N	7	F	
2622	.50-95	C	24"	rd	m	N	N	none	un	Y	Y	n/a	n/a	n/a	none	A	Y	N	7	H	42
2670	.50-115	C	24"	rd	s	N	N	1237	2	Y	Y	n/a	n/a	T	none	A	Y	N	7	H	43
2673	.50-115	un	23"	rd	s	N	N	un	2	Y	un	n/a	n/a	n/a	none	A	un	N	7	H	44
2684	.45	un	39"	un	s	Y	Y	un	un	un	Y	n/a	n/a	un	un	un	un	Y	7	F	45

SINGLE-SHOTS

SERIAL NO.	CALIBER	LOCATION	BBL. LEN.	BBL. TYPE	BUTT TYPE	PIST. GRIP	CHECK	C/H	ADD. TYPE	EXT. HOLE	MOD. 1886	S.F. / S.S.	DET - INT	MOTIF.	NICK. P.	FIN. SER.	GRVE.	TANG ST.	REC. SCRE.	MAG. LEN.	MISC.
3532	.40-90S	D	30"	pt rd	sh	Y	N	none	2	n/a	n/a	Y	n/a	n/a	2	n/a	n/a	un	n/a	n/a	46
3548	.38-45	D	28"	pt rd	sh	Y	Y	3?	2	n/a	n/a	Y	n/a	n/a	2	n/a	n/a	Y	m/a	n/a	
3559	.32-40	D	25"	pt rd	sh	Y	N	un	un	n/a	n/a	Y	n/a	n/a	un	n/a	n/a	N	n/a	n/a	47
3560	.22	D	un	un	c	Y	Y	3	2	n/a	n/a	Y	n/a	n/a	N	n/a	n/a	N	n/a	n/a	
3637	.22-7	D	26"	pt rd	sh	Y	un	none	un	n/a	n/a	n/a	Y	n/a	un	n/a	n/a	Y	n/a	n/a	
3642	.38-90	D	26"	oct	s	N	N	un	2	n/a	n/a	Y	n/a	un	un	n/a	n/a	N	n/a	n/a	
3657	.38-55	D	30"	pt rd	sh	Y	Y	none	4	n/a	n/a	Y	n/a	n/a	1	n/a	n/a	Y	n/a	n/a	
3660	.38-55	D	32"	rd	sh	Y	N	un	un	n/a	n/a	Y	n/a	n/a	un	n/a	n/a	N	n/a	n/a	48
3736	.32-40	D	un	un	sh	Y	Y	1 3	un	n/a	n/a	un	un	n/a	2	n/a	n/a	N	n/a	n/a	
3737	.32-40	D	28"	pt rd	sh	Y	Y	12345	4	n/a	n/a	n/a	Y	n/a	un	n/a	n/a	Y	n/a	n/a	
3739	.32-40	D	28"	pt rd	sh	Y	Y	3	un	n/a	n/a	n/a	Y	n/a	un	n/a	n/a	Y	n/a	n/a	
3757	.32-40	D	32"	pt rd	sh	Y	N	3	4	n/a	n/a	Y	n/a	n/a	2	n/a	n/a	Y	n/a	n/a	49
3761	.32-40	D	30"	pt rd	s	Y	Y	1 2 3	4	n/a	n/a	n/a	Y	T	N	n/a	n/a	Y	n/a	n/a	
3763	.32-40	D	un	un	s	un	un	un	un	n/a	n/a	un	un	T	un	n/a	n/a	un	n/a	n/a	
3780	.32-40	D	30"	pt rd	sh	Y	N	1 3	un	n/a	n/a	n/a	Y	n/a	2	n/a	n/a	Y	n/a	n/a	50
3788	.32-40	un	un	un	un	un	un	un	un	un	un	un	un	un	un	un	un	un	un	un	
3800	.32-40	D	26"	oct	s	N	N	1	4	n/a	n/a	n/a	Y	T	none	n/a	n/a	un	n/a	n/a	
3813	.38-45	D	26"	oct	s	N	N	?	4	n/a	n/a	n/a	Y	T	none	n/a	n/a	Y	n/a	n/a	
3831	.32-40	D	30"	pt rd	sh	Y	N	1 2 4	4	n/a	n/a	n/a	Y	n/a	N	n/a	n/a	Y	n/a	n/a	
3832	.32-40	D	30"	pt rd	sh	Y	N	none	4	n/a	n/a	n/a	Y	n/a	un	n/a	n/a	Y	n/a	n/a	
3835	.32-40	D	28"	pt rd	sh	Y	Y	1 2	4	n/a	n/a	n/a	Y	n/a	un	n/a	n/a	Y	n/a	n/a	
3856	.32-40	D	22½"	oct	s	N	N	un	4	n/a	n/a	n/a	Y	n/a	un	n/a	n/a	N	n/a	n/a	
3867	.22-15	D	un	pt rd	s	N	N	1234	4	n/a	n/a	n/a	Y	un	N	n/a	n/a	N	n/a	n/a	
386X	.22-15-60	D	26"	pt rd	un	un	un	un	un	n/a	n/a	un	un	un	un	n/a	n/a	un	n/a	n/a	
3878	.32-35	D	32"	oct	sh	Y	un	un	un	n/a	n/a	Y	n/a	n/a	2	n/a	n/a	Y	n/a	n/a	
3882	.50-95	D	28"	rd	s	N	N	none	un	n/a	n/a	n/a	Y	T	none	n/a	n/a	N	n/a	n/a	
3889	.32-40	D	28"	un	un	un	un	un	un	n/a	n/a	n/a	Y	un	un	n/a	n/a	Y	n/a	n/a	51
3897	.22-3	D	28"	pt rd	sh	Y	Y	1 2	4	n/a	n/a	n/a	Y	n/a	un	n/a	n/a	Y	n/a	n/a	52

SERIAL NO.	CALIBER	LOCATION	BBL. LEN.	BBL. TYPE	BUTT TYPE	PIST. GRIP	CHECK	C/H	ADD. TYPE	EXT. HOLE	MOD. 1886	S.F. / S.S.	DET - INT	MOTIF.	NICK. P.	FIN. SER.	GRVE.	TANG ST.	REC. SCRE.	MAG. LEN.	MISC.
3905	.22-5	D	un	oct	sh	N	Y	1	4	n/a	n/a	Y	n/a	n/a	2	n/a	n/a	Y	n/a	n/a	53
3908	.22-5	D	26"	oct	s	N	N	124	2	n/a	n/a	n/a	Y	T	N	n/a	n/a	Y	n/a	n/a	
3911	.32-40	D	un	un	un	un	un	un	un	n/a	n/a	n/a	Y	un	un	n/a	n/a	un	n/a	n/a	
3913	.25-20	D	28"	pt rd	sh	Y	un	none	un	n/a	n/a	Y	n/a	n/a	un	n/a	n/a	Y	n/a	n/a	
3915	.32-40	D	28½"	pt rd	sh	Y	Y	none	2	n/a	n/a	n/a	Y	n/a	1	n/a	n/a	Y	n/a	n/a	54
3941	.22-5	D	30"	oct	sh	Y	Y	1 2	4	n/a	n/a	n/a	Y	n/a	2	n/a	n/a	Y	n/a	n/a	
3953	.22-10	?	30"	pt rd	sh	Y	Y	1	un	n/a	n/a	n/a	Y	n/a	2	n/a	n/a	N	n/a	n/a	55
3964	.32-40	D	28"	pt rd	s	Y	Y	3	un	n/a	n/a	n/a	Y	un	un	n/a	n/a	N	n/a	n/a	56
3964	.32-40	D	26"	pt rd	s	Y	Y	3	un	n/a	n/a	n/a	Y	un	un	n/a	n/a	N	n/a	n/a	56
3981	.32-40	D	30"	pt rd	sh	un	un	1234	4	n/a	n/a	n/a	Y	n/a	2	n/a	n/a	Y	n/a	n/a	
3986	.32-40	D	26½"	pt rd	un	Y	N	un	un	n/a	n/a	n/a	Y	n/a	un	n/a	n/a	un	n/a	n/a	
4038	.32-40	D	28"	pt rd	?	N	N	none	4	n/a	n/a	n/a	Y	n/a	none	n/a	n/a	Y	n/a	n/a	57
4043	.38-55	D	28"	oct	s	N	N	Y	4	n/a	n/a	n/a	Y	T	none	n/a	n/a	Y	n/a	n/a	
4060	.32-40	D	28"	pt rd	sh	Y	Y	3	4	n/a	n/a	un	un	n/a	un	n/a	n/a	un	n/a	n/a	
4061	.22-15	D	24"	oct	sh	Y	Y	3	un	n/a	n/a	Y	n/a	n/a	un	n/a	n/a	Y	n/a	n/a	
4064	.32-40	D	26"	pt rd	s	Y	Y	3	4	n/a	n/a	n/a	Y	T	none	n/a	n/a	Y	n/a	n/a	58
4076	.22RF	D	28"	pt rd	s	Y	Y	un	un	n/a	n/a	n/a	Y	un	un	n/a	n/a	Y	n/a	n/a	59

As an addendum to the foregoing tables, the author would like to mention that almost all Bullard barrel lengths were uneven numbers (eg. 28⅛ inches; 27¾ inches), and the author has rounded them off to the nearest inch for the sake of clarity. Thus the first example given became 28 and the second 28 inches. Half inches were left as ½.

Magazine lengths were also a matter for deliberation. Descriptions ranged all the way from button to ⅔ and ¾ length. The author has tried where possible to give as accurate a length as possible in the column provided. However, the reason for listing the various lengths of magazines was not to foster controversy but to show the great variety of styles made by the Bullard factory.

Again for clarity's sake, pistol grips have been called that rather than semi-pistol grip, which is probably what they should technically be called. In charting their occurrence it was again to show their frequency in production, nothing more.

It should also be noted that these numbers can be added to the serial numbers tabled in the charts:

REPEATERS (All LARGE-FRAME)
7 Cal .45 Ser # Range 2XX to 19XX
8 Cal .40 Ser # Range 3XX to 19XX
3 Cal .50 Ser # Range 1XX to 11XX

SINGLE-SHOT RIFLES
5 Cal .32 Ser # Range 2XXX to 3XXX
2 Cal .22 Ser # Range 2XXX to 3XXX
2 Cal .38 Ser # Range 2XXX to 4XXX

It must be stressed that the above information is accurate; however the source of this information wishes to remain anonymous and I must respect his wishes.

It can be seen from the charts that no single-shots were logged with a serial number lower than 3532, which seems strange when dealing with the laws of probability and averages. It is odd that no single-shots turned up in the survey in the 2XXX range, except for those listed by the anonymous collector.

Mr. Lee R. Boddy, a man who has probably seen more Bullards "in the flesh" than any living person, states that he feels that a total of 1,541 single-shots were produced (serial numbers 2501 to 4041). While not fitting exactly into the confines of the foregoing tables, it is a statement that is certainly partially corroborated by the anonymous information. As well, it manages to stay within the limits of total Bullard production of all types (not much more than 4,100). It should also be noted that there are many gaps in the numbers from 1 through 4076. Future research may turn up the answers. The only "fly in the ointment" to all this would be if a military order were discovered to have been manufactured. That could add somewhat substantially to the total number of Bullards made. However, even at that the Bullard rifle in its many forms is rare; one of the rarest of the American breechloading cartridge longarms.

MISCELLANEOUS UNUSUAL FEATURE CODE

1 No caliber marked on gun but accepts .45-70 cartridge.
2 Two serial numbers are indicated here, for over the years the same rifle has appeared with two serial numbers. Have not been able to verify which is correct.
3 This rifle is a drummer's (salesman) cutaway model with chrome plating on the barrel, action, buttplate and forearm cap.
4 The lever, hammer, dustcover and carrier plate are case-hardened. Both sides of

the action are cut away completely with exception of ⅛ inch web between screws. This shows complete inner working of action. Action is operable.

5 Disregard

6 The extractor hole, drilled through from the right side of the receiver, may have been done by a gunsmith, as it differs from factory practice of drilling left side of receiver.

7 This rifle is marked with address that omits the A in USA. Since I do not have photographic proof, I have not included rifles with this example of address in the particular column on the chart, but rather they are listed as having an unknown address. A number of rifles appear to have this address style though and if it can be substantiated it will be included in the next edition.

8 Buttstock is a replacement

9 The rifle has sling swivels. The front one is brazed to the barrel.

10 Has factory swivels, also the extractor access hole is on right side of receiver.

11 Has factory swivels, also the extractor access hole is on right side of receiver.

12 Has magazine tube tenant only two inches from muzzle.

13 Has sling swivels.

14 Left side of receiver has two small screws filling holes tapped for sight.

15 Information taken from *Gun Report* advertisement of February 1982.

16 Has sling swivels.

17 Has magazine carrier groove two inches in length, and the magazine was factory lengthened from 15¾ inches to 27¾ inches and tube tenant moved with distance between new tenant mount and old tube tenant slot measured at 11¾ inches. The old slot is professionally filled in.

18 Has different dustcover and receiver has unslotted pins rather than slotted pins.

19 Screw retaining lower rack in receiver is larger in diameter than standard.

20 Caliber stamped twice on barrel (in front of dustcover and on barrel). The octagonal barrel is tapered.

21 Has British proof marks as well as several others.

22 Has 26 inch rapid-taper octagonal barrel.

23 Has cheekpiece on buttstock.

24 Forearm is about 19 inches long and ⅔ length magazine.

25 ⅔ length magazine.

26 British proof marks.

27 Has large exposed screw behind the trigger.

28 Forearm is about 14¾ inches long.

29 Has magazine cutoff button.

30 Barrel and magazine shortened, Freund sights added and was carried on horseback in Wyoming.

31 Magazine tip protrudes 3.3 inches from forearm cap.

32 Presentation rifle engraved by Gustave Young for D.B. Wesson from J.H. Bullard.

33 Magazine tip flush with fore-end cap.

33a Tapered octagonal barrel.

34 Barrel address with A omitted from USA.

35 Barrel address of style number 4 in Tables.

36 Has numerals 10138 stamped on wood and proof marks.

37 Company address stamped on upper tang not on left side of breech. This arm is a military-style carbine with magazine cutoff.

38 This arm is a military-style musket equipped for a side-mounted detachable magazine.
39 This arm is also a military-style musket equipped for a side-mounted detachable magazine. These two muskets vary on several details, including the sideplate that holds the magazine in place, the hammer and the lever one having the sharp V shape, the other the gentle curve discussed in Chapter 3.
40 Rifle has address stamped on upper tang, same as #37, and also has magazine cutoff device.
41 Has sling swivels and distinct ridge on receiver running the length from rear of cartridge carrier opening to back of receiver at breechblock.
42 Has sling swivels, has letters vp, and two crowns stamped by barrel address and letters CI stamped by the caliber.
43 Has British proof marks and a lightweight barrel.
44 This rifle may well have seen action in the Boer War.
45 The information on this rifle was taken from Jackson Arms catalog #20. The rifle shown has the highest serial number for a repeater of any yet seen by the author. Its overall length is a massive 54¼ inches!
46 This rifle is chambered for the .40-90 Sharp's Straight 3¼-inch cartridge. It is rifled for a greased bullet not a paper patched one. It is also the lowest serial number to be documented thus far for the single-shots.
47 Equipped with a Malcolm scope (Globe sight).
48 Equipped with John W. Sidle scope.
49 The owner has circled Solid-Frame Single-Shot on my spec sheet, but the patent date of 1886 indicates the det-int barrel model. Because the owner has indicated that it is a solid-frame, it has been left that way. A wrong date stamped on a rifle has already occurred once before.
50 Has a hole drilled into butt and has shot added to hole to balance rifle.
51 Late of the John T. Amber Collection. It had two barrels; one in .32-40 at 28 inches, the second in .22 cal at 26 inches.
52 Lead has been added inside butt.
53 Owner has this rifle marked as solid-frame, but with 1886 patent date. Same comments as in #49. There is a distinct possibility that both of these rifles may be take-down models.
54 Photographs of this rifle indicate that it is a det-int barrel model, however it has the 1881 patent date used on the solid-frame models. Just above the date on the address is a series of four marks beginning with a checkmark or V, then two Ls, and finally another V or U. As Plate 5-2 clearly shows, some part of the word Company must have been obliterated in order that this inscription could be placed directly above the date.
55 Originally bored to .22-10 then rebored to .25-21.
56 Information taken from ad in the April 1974 *Gun Report*. It gives the indication that the rifle is marked .32-40 Bul, but this cannot be substantiated.
57 Not original buttplate.
58 Trigger is checkered. Has right-hand cheekpiece and stands off slightly to the left when viewed from the butt end.
59 The highest serial-numbered Bullard rifle of any type yet discovered or seen by the author. It was originally chambered for .22 rimfire but was rebored and chambered by the Stevens A & T Co. to .28-30 caliber.

BULLARD PATENTS FROM 1877 TO 1889

This list only covers 12 years of James Bullard's life but clearly shows the manifest inventive abilities of the man.

187689	Revolving firearm; Bullard assigned to D.B. Wesson	Feb. 20, 1877
192676	Metallic cartridge	July 3, 1877
197823	Metallic cartridge	Dec. 4, 1877
198228	Revolving firearm	Dec. 18, 1877
210180	Machine for checking pistol handles; ass. to D.B. Wesson	Nov. 26, 1878
227481	Revolving firearm	May 11, 1880
237088	Sewing machine mechanisms	Feb. 1, 1881
241526	Sewing machine mechanisms	May 17, 1881
245700	Magazine firearm	Aug. 16, 1881
248701	Sewing machine mechanisms	Oct. 25, 1881
249155	Sewing machine mechanisms	Nov. 8, 1881
260655	Sewing machine mechanisms	July 4, 1882
287229	Magazine gun	Oct. 23, 1883
335740	Calipers	Feb. 9, 1886
339141	Tennis court marker	April 6, 1886
348151	Sewing machine mechanisms	Aug. 24, 1886
365788	Steam road vehicle	July 5, 1887
365789	Furnace	July 5, 1887
365790	Furnace	July 5, 1887
391735	Furnace	Oct. 23, 1888
411236	Furnace (re-issue)	(?) April 26, 18??
480068	Liquid receptacle & pump	(?) Aug. 2, 18??
483099	Furnace (re-issue)	(?) Sept. 20, 18??
483100	Furnace (re-issue)	(?) Sept. 20, 18??
483344	Pneumatic tire	(?) Sept. 27, 18??

The author's thanks to Mr. George S. McCluney for his suggestion that I list all the Bullard patents that have come to light during the research on this book.

ACKNOWLEDGEMENTS

The following list allows me to say thank you to the many individuals, companies and museums that assisted me in this book. I wholeheartedly urge readers to support the clubs, magazines and museums listed below. They freely gave of their time, expertise and editorial space to further our knowledge. They really care about preserving the history of firearms.

American Heritage, Richard F. Snow, Managing Editor; and James Long

American Single-Shot Rifle Association, 987 Jefferson Ave., Salem, Ohio 44460. George S. McCluney, President; Rudi Prusok, Editor; Larry B. Thompson, Secretary-Treasurer

Antique Arms Shows, Wallace Beinfeld, Show Manager

Antiques & Collecting Hobbies, Lightner Publishing Corp., 1006 S. Michigan Ave., Chicago, Illinois 60605. Frances L. Graham, Editor

Armémuseum, Royal Army Museum, Sweden. Johan Engström

The Armouries, H.M. Tower of London, England EC3N 4AB. Ruth R. Brown, Research Assistant; Peter-Tomas Gray, Photographic Librarian; M. Murray-Flutter, Museum Assistant; Sarah Barter-Bailey, Librarian

The Arms and Armour Society, Great Britain. A.R.E. North, Hon. Secretary

Arms Fairs Ltd., 121 Fawnbrake Ave., Herne Hill, London, England SE24 0BG. G.E. Bennett, Secretary

Department of the Army, United States Military Academy, West Point, New York, New York 10996-5000. Robert W. Fisch, Curator, West Point Museum

Department of the Army, U.S. Army Ordnance Center and School, Aberdeen Proving Ground, Maryland 21005. Daniel E. O'Brien, Curator, Ordnance Museum

Matthew A. Bailey

Baker Library, Harvard University, Boston, Massachusetts. Mark A. Mastromarino, Reference Assistant, Manuscripts & Archives; Florence Bartoshesky, Curator, Manuscripts & Archives

Emil Baronak

Felix A. Bedlan, Dealer, Box 244, 1318 E Street, Fairbury, Nebraska 68352

Jack H. Behn

Donald H. Bell

Dr. R.C. Bester

City of Birmingham (England), Public Libraries Department. Patrick Baird, Librarian, Local Studies Department

Donald M. Blyth

Lee R. Boddy

Boston Museum of Fine Arts, 465 Huntington Ave., Boston, Massachusetts 02115. Jonathan Fairbanks, Curator, American Decorative Arts & Sculpture

Boston Public Library, Boston, Massachusetts 02117. Betsey P. Anderson, Reference Librarian, Social Sciences

B.A. Boucher

Buffalo Bill Historical Center & Winchester Arms Museum, Cody, Wyoming. Herbert G. Houze, Curator, Winchester Arms Museum

Maurice L. Bullard

Robert L. Bullard

William T. Bundy, Jr.

Tom Burness

Ben R. Burnham

Butte-Silver Bow Public Archives, Butte, Montana. Mary Murphy, Director

Canadian Embassy, Ankara, Turkey. Stephen L. Bennett, Vice-Consul

Canadian Journal of Arms Collecting, published by Museum Restoration Service, P.O. Box 390, Bloomfield, Ontario K0K 1G0. S. James Gooding, Editor

Canadian War Museum, Ottawa, Ontario. V.J. Suthren, Associate Director

Douglas R. Carlson, Dealer, 1379 73rd St., Des Moines, Iowa 50311

Victor M. Carlstrom

Fred H. Cates

R.H. Chamberlain

Chicago Historical Society, Chicago, Illinois

Colt Industries, Firearms Division, P.O. Box 1868, Hartford, Connecticut 06101. Robert L. Fellows, Customer Service Representative

Allan C. Colville

Connecticut Historical Society, Hartford, Connecticut. Ruth M. Blair, Manuscript Cataloger

Connecticut Valley Historical Society, Springfield, Massachusetts. Juliette Tomlinson, Director

Connecticut State Library, Hartford, Connecticut. Nancy Peluso, General Reference; Marguerite A. Giguere-Davis, Assistant Archivist, Archives, History & Genealogy Unit

Leo H. Cook

E.R. "Ted" Cotton

Edward Curtis

J.M. Davis Gun Museum, P.O. Box 966, Claremore, Oklahoma 74017. Lee T. Good, Director of Exhibits & Programs
Department of External Affairs, Ottawa, Ontario K1A 0G2. Yves Brodeur, Political & Strategic Analysis Division
The Dixie Gun Works, Dealer, Gunpowder-Lane, Dept. 37, Union City, Tennessee 38261
Richard Donaldson
Eley Limited, Birmingham, England. C.A. Lewis
W.F. Evenson
Richard A. Failes
Joseph Fedor, Dealer, The Last Homestead, Route 3, Box 205, Reed City, Michigan 49677
O'Brien Forbes Ffrench
Field & Stream, 1515 Broadway, New York, New York 10036
Brian Flynn
Gary Frahm
Paul Fuog
James Goergen, Dealer, Goergen's Gun Shop, Route 2, Box 182 BB, Austin, Minnesota 55912
James Gould
James J. Grant
Great Lakes Gun, Knife & Militaria Collectors' Association, c/o John Cryderman, 136 Thames St., Chatham, Ontario N7L 2Y8
Robert Green
Wayne A. Grove
The Gun Report P.O. Box 111, Aledo, Illinois 61231. Kandy Liggett-Harrison, Publication Director; Charles R. Suydam, Staff Editor
Guns and Ammo, c/o Petersen Publishing Company, 8490 Sunset Blvd., Los Angeles, California 90069
Edward C. Haight
Harvard College Library, Cambridge, Massachusetts 02138. Wallace Finley Dailey, Curator of the Theodore Roosevelt Collection
William L. Harvey, Dealer, 251A Main Street, Falmouth, Massachusetts 02541
Heeresgeschichtliches Museum (Militarwissenschaftliches Institut). Dr. Franz Kaindl, Director
The Hermitage Museum, M Dvortsovaya Naberezhnaya, 34 Leningrad, U.S.S.R. Jouri A. Miller, Curator, Arms and Armour
Higgins Armory Museum, 100 Barber Ave., Worcester, Massachusetts 01606. Warren M. Little, Director
John Hintlian
Historical Firearms Society of South Africa, P.O. Box 145, 7725, Newlands, Republic of South Africa
Craig R. Hostetter
Robert Peter Howard
Kenneth Howell
Verle W. Hunt
Imperial War Museum, Lambeth Road, London, England SE1 6HZ. Victoria de Korda, Department of Exhibits and Firearms
Internal Revenue Service, Washington, D.C. Diane Lee; Earl Klema for Arnold B. Gordon, Acting Director, Disclosure and Security Division
Edward Iwanchuk
Roy G. Jinks, Shooting Sports Co-ordinator, Historian for Smith & Wesson
Ronald W. Johnson
Harold R. Jonah
Kawartha Arms, Dealer, Firearms Sales & Service, RR #4, Lindsay, Ontario K9V 4R4. R. Bradburn
Roy E. Keeler
Keenan Auction Company, Main Street, Kingfield, Maine 04947. Jim O'Regan
Gerald O. Kelver
Ted Knee
William R. Kurau, Jr.
Library of Congress, General Reading Rooms Division, Washington, D.C.
Joseph Logan
Verne W.L. MacRoberts
Maine Historical Society, 485 Congress St., Portland, Maine 04101
Maine Antique Digest, 71 Main St., Waldoboro, Maine 04572. "Meting Place" column.
Oak Grove Cemetery of Springfield Inc., 426 Bay St., Springfield, Mass. 01109.
John Maltese, Jr.
E. Lee Manning
The Marlin Firearms Company, North Haven, Connecticut, 06473-0903. William S. Brophy, Senior Technical Manager
Tommy M. Mason
The Commonwealth of Massachusetts, Office of the Secretary of State. Richard C. Kaplan, Reference Archivist
Massachusetts Institute of Technology, Cambridge, Massachusetts 02139. Merritt Roe Smith, Professor, History of Technology
Harold A. McCallum
Stuart W. McCallum
R. Bruce McDowell
McGregor Museum, 2 Egerton Road, P.O. Box 316, Kimberley, 8300, South Africa. G.C. Gates, Military Historian
Douglas B. Meier
Mexican Embassy, Ottawa, Ontario. Jose Luis Avila S., 3rd Secretary, Consular Section
Milwaukee Public Museum, 800 W. Wells St., Milwaukee, Wisconsin 53233. H. Michael Madaus, History Section
Ministry of Defence, Pattern Room, Royal Small Arms Factory, Enfield Lock, Middlesex, England EN36 JL. H.J. Woodend, Pattern Room Custodian for DGDQU
Donald H. Mitchell
Montana Historical Society Library, 225 N. Roberts St., Helena, Montana 59620. Rebecca Kohl, Photograph Archives Assistant; Dave Walter, Reference
Musée de l'Armée, Hôtel National des Invalides, 75007, Paris, France. Col. Neuville, Conservateur
Museo Historico Nacional, Casilla, 9764, Santiago, Chile. Johanna M. Theile, Encargada de Coleccion
National Archives and Records Service, U.S.A. Elaine C. Everly, Navy and Old Army Branch, Military Archives Division; Michael P. Musick, Navy and Old Army Branch, Military Archives Division; Barbara Burger, Still Pictures Branch, Special Archives Division; William J. Walsh, Legislative and Diplomatic Branch, Civil Archives Division; George C. Chalou, Assistant Chief, Military Reference Branch, Military Archives Division

National Museum of American History, Smithsonian Institution, Washington, D.C. 20560. Harry Hunter, Museum Specialist, Division of Armed Forces History
National Rifle Association of America, 1600 Rhode Island Ave. N.W., Washington, D.C. 20036. Mary E. Shelsby, Manager, Gun Collecting Department; William F. Parkerson III, Editor, *American Rifleman*; Joseph B. Roberts, Jr., Research Editor, *American Rifleman*; M.D. Waite, Technical Editor, *American Rifleman*
New York Public Library, The Research Libraries, Fifth Ave. at 42nd St., New York, New York 10018
The New Zealand Deerstalkers Association, P.O. Box 6514, Wellington, New Zealand. Mr. Shelby Grant
Alex J. Newberry
Ronald A. Ogan, Dealer, P.O. Box 153, Royal Palm Beach, Florida 33411
Oorlogsmuseum (War Museum) of the Boer Republics, P.O. Box 704, Bloemfontein 9300, South Africa. E.M. Wessels, on behalf of the Director
C.L. Parker
Paul Parsons
O.L. Peacock, Jr.
Al Perry
Charles Peterson
Steven Petris
K.E. Prince
J.P. Rehling
Man at Arms, E. Andrew Mowbray, Editor & Publisher, P.O. Box 460, Lincoln, R.I. 02865
Remington Arms Company Inc., Brandywine Building, 1007 Market St., Wilmington, Delaware 19898. J.D. Health, Supervisor, Consumer Information
Royal Ontario Museum, 100 Queen's Park, Toronto, Ontario M5S 2C6. Donald Blake Webster, Curator, Canadiana Department
Sagamore Hill National Historic Site, Cove Neck Road, Box 304, Oyster Bay, New York 11771. Gary G. Roth, Acting Museum Curator
Saint Paul's Rectory, 51 Fenimore Blvd., Springfield, Massachusetts 01108. Fr. Peter G. Loughran
Savage Arms Division, Emhart Corporation, Westfield, Massachusetts 01085. Earle G. Harrington, Sales Promotion Manager
Cliff Schisler, Cartridge & Gun Collector, Box 584, Orillia, Ontario
Scottish United Services Museum, Crown Square, The Castle, Edinburgh, Scotland EH1 2NG. William G.F. Boag, Assistant to the Keeper
S. Trent Seames
Robert N. Sears
Doyle N. Shaw
Paul Simkin
Allen Smith
Smith & Wesson, Springfield, Massachusetts. Roy G. Jinks, Shooting Sports Co-ordinator/Historian
Smithsonian Institution, Washington, D.C. 20560. Craddock R. Goins, Jr., Curator, Division of Military History
Sporting Shooters Association of Australia Inc., P.O. Box 210, Belgrave, Victoria 3160, Australia
Sports Afield, 250 West 55th St., New York, New York 10019. Anne Paugh, Reader Service
Springfield Armory National Historic Site, Springfield, Massachusetts 01105. William E. Meuse, Curator
Springfield City Library, 220 State St., Springfield, Massachusetts 01103. Dorothy Mozley, Genealogy & Local History Librarian; Edward Lonergan, Local History Department
The City of Springfield, Massachusetts. Rose A. Pollard, City Clerk
Cal Swazer
Swiss Institute of Arms and Armour, Granson, Switerland. Eugène Heer, Director
Syndicat National de l'Arquebuserie, du Commerce de l'Arme Historique, France
Joel M. Tastad
Allen Thompson
James C. Tillinghast, Dealer, Box 19-C, Hancock, New Hampshire 03449
Tojhusmuseet, 1220 Kobenhavn K, Frederiksholms Kanal 29, Denmark. Arne Orloff, Chief Curator
Edward A. Tolosky
Traders Den, Dealer, 397 Markham Road, Markham, Ontario. Doug Mitchell
Turkish Embassy, Ottawa, Ontario. Mengu Büyükdavras, Counsellor
Turkish Embassy, Office of the Air, Military and Naval Attache, 197 Wurtemburg St., Ottawa, Ontario K1N 8L9. Col. Ergun Cataloglu; Estelle Bernier, Secretary
Turkiye Askeri Muzesi (Turkish Military Museum), Istanbul, Turkey. Muhlis Ozmen, Infantry Colonel, Chief of the Military Museum
United States Department of Commerce, Patent and Trademark Office, Washington, D.C. Bernard H. Thomas, Manager, Patent Search Division
United Technologies, Hartford, Connecticut 06101. Harvey H. Lippincott, Corporate Historian
Victoria and Albert Museum, South Kensington, London, England SW7 2RL. A.R.E. North, Department of Metalwork
Wadsworth Anthenium, 600 Main Street, Hartford, Connecticut 06103. John W. Teahan
Francis E. Wall
Kenneth L. Waters
Richard Weaver
Michael E. Whiteley
Hampton Williams
R.L. Wilson
Winchester Gun Museum (now part of the Buffalo Bill Historical Center in Cody, Wyoming). T.E. Hall, Curator
John Witzel
Gary Woleslagle
Worcester Public Library, Salem Square, Worcester, Massachusetts 01608. Mary S. Gaudetts, Librarian, Worcester Collection
James Zupan

BIBLIOGRAPHY

Amber, John T. *The NRA Collector's Series*. Northfield, Illinois: Digest Books Inc., 1971.

Barnes, Frank C. *Cartridges of the World*. Northfield, Illinois: Digest Books Inc., 1972.

Behn, Jack. *".45-70" Rifles*. Highland Park, New Jersey: The Gun Room Press, 1972.

Bowman, Hank Wieand. *Famous Guns from the Winchester Collection*. Greenwich, Connecticut: Fawcett Publications Inc., 1958, 1964.

Burnside, Graham. "Marlin's .444 Starts Fire Under Cartridge Fans." *Shooting Times*, April 1965.

Butler, David F. *Winchester '73 '76: The First Repeating Centerfire Rifles*. New York: The Winchester Press, 1970.

Court Dockets of Civil Cases before the Police Court of Springfield, Case Numbers 6892, 6894, 6828.

Flayderman, E. Norman. *Flayderman's Guide to Antique American Firearms . . . and their values*. Northfield, Illinois: DBI Books Inc., 1977.

Gooding, James S. "The Bullard Line of Cartridges." *The Canadian Journal of Arms Collecting*, Vol. 20, No. 4, November 1982.

Gould, A.C. *Modern American Rifles*. Plantersville, South Carolina: Thomas G. Samworth, 1892 (republished 1946).

Hagel, Robert. "Head Designs for Centerfire Cases." *Handloader: The Journal of Ammunition Reloading*, No. 78, March-April 1979.

Hampden County Registry of Deeds, Books 326, 374, 399, 402, 408. Springfield, Massachusetts.

Howe, Gilman Bigelow. *Genealogy of the Bigelow Family*. Worchester, Massachusetts, 1890.

Inland Massachusetts Illustrated. Springfield, Massachusetts: The Elstner Publishing Co., 1890.

Kelver, Gerald O. *100 Years of Shooters and Gunmakers of Single Shot Rifles*. Ft. Collins, Colorado: Robinson Press Inc., 1975.

King, Moses. *King's Handbook of Springfield*. Springfield, Massachusetts: James D. Gill, 1884.

Leading Citizens of Hampden County. Boston: Biographical Review Publishing Co., 1895.

Leading Manufacturers and Merchants. Massachusetts Industries, 1887.

Manning, E. Lee. "Bullard Repeating Rifles" *Antiques & Collecting Hobbies*. Chicago: Lightner Publishing Corp., 1982.

National Archives. Record Group 156, File No. 1130 (May 2, 1888); Record Group 156, File No. 2845-3002 (1885); Record Group 156, File No. 2972 (1885). All files from the Ordnance Department, OCO-DOC File 1797-1894.

Nonte George. "New Brass for Old Charcoal Burners." *1973 Guns and Ammo Annual*.

Nutt, Charles. *History of Worcester and Its People*, Vol. 4. New York: N.Y. Lewis Historical Publishing Co., 1919.

Phinney, Wallace S. *James H. Bullard, Inventor (1842-1914)*. The Bulb Horn, 1967.

Progressive Springfield, June 1891, No. 5. Springfield, Massachusetts.

R.G. Dunn & Company Records, Baker Library, Harvard Business School, Boston.

Roberts, Major Ned H., and Kenneth L. Waters, *The Breech-Loading Single-Shot Match Rifle*. New York: Van Nostrand, 1967.

Roosevelt, Theodore. *Hunting Trips of a Ranchman*. New York: G.P. Putnam's Sons, 1885.

Sawyer, C.W. *Our Rifles*, Boston: The Cornhill Co., 1920.

Sears, Robert N. *N.R.A. Gun Collecting Newsletter*, No. 8, Spring 1983. Washington, D.C.: N.R.A. Gun Collecting Department.

Sharpe, Philip B. *The Rifle in America*. New York: Funk and Wagnalls Co., 1938.

Springfield (Massachusetts) *Daily News*, March 26, 1914.

Springfield and Environs: Commerce, Manufacturers and Resources of Springfield. National Publishing Co., June 1883.

Taylerson, A.W.F. *The Revolver 1889-1914*. New York: Crown Publishers Inc., 1971.

Tillinghast, James C. "Bullard Reloading Tools." *The Canadian Journal of Arms Collecting*. Vol. 21 No. 4, November 1983.

Tolosky, Edward A. "Exploded Views." *The American Rifleman*, August 1982. The words "Exploded Views" are copyrighted by the National Rifle Association and used with their permission.

Waters, Kenneth L. "Classic Rifles — the Bullard Single Shot." *The Rifle: Magazine for Shooters*, No. 35, September-October 1974.

West, William, *Marlin and Ballard — Firearms and History*. Whittier, California: Stockton Trade Press, 1968.

Wilson, R.L. *Theodore Roosevelt, Outdoorsman*. New York: The Winchester Press, 1971.

INDEX